AF505885

Optical and Electronic Properties of Fullerenes and Fullerene-Based Materials

Optical and Electronic Properties of Fullerenes and Fullerene-Based Materials

edited by

Joseph Shinar

Ames Laboratory, U.S. Department of Energy
Iowa State University
Ames, Iowa

Zeev Valy Vardeny

University of Utah
Salt Lake City, Utah

Zakya H. Kafafi

U.S. Naval Research Laboratory
Washington, D.C.

MARCEL DEKKER, INC.　　　　　NEW YORK · BASEL

ISBN: 0-8247-8257-7

This book is printed on acid-free paper.

Headquarters
Marcel Dekker, Inc.
270 Madison Avenue, New York, NY 10016
tel: 212-696-9000; fax: 212-685-4540

Eastern Hemisphere Distribution
Marcel Dekker AG
Hutgasse 4, Postfach 812, CH-4001 Basel, Switzerland
tel: 41-61-261-8482; fax: 41-61-261-8896

World Wide Web
http://www.dekker.com

The publisher offers discounts on this book when ordered in bulk quantities. For more information, write to Special Sales/Professional Marketing at the headquarters address above.

About fifteen years ago, Harry Kroto came to the laboratory of Rick Smalley at Rice University searching for a carbon chain molecule of potential importance in astronomy. Instead of finding this molecule, Kroto, Heath, O'Brien, Curl, and Smalley [1] witnessed the birth of a new form of carbon, C_{60}, in a molecular beam apparatus. They predicted its structure to be that of a hollow-soccer ball consisting of 12 five-membered rings separated by 20 hexagonal benzene-like rings. Because of the similarity between the geometry of C_{60} and the structure of the geodesic domes designed by Buckminster Fuller, C_{60} was named Buckminsterfullerene in honor of the late architect. It was not until a few years later, when Krätschmer, Lamb, Fostiropoulos, and Huffman [2] were able to synthesize C_{60} in bulk form, that scientists realized that this third allotrope of carbon is probably as old as graphite and diamond.

We dedicate this volume to the discoverers of C_{60} and C_{70}, and their extended family.

1. Harold W. Kroto, James R. Heath, Sean C. O'Brien, Robert F. Curl, and Richard E. Smalley, *Nature*, Volume 318, 162 (1985).
2. Wolfgang Krätschmer, L. D. Lamb, K. Fostiropoulos, and Donald R. Huffman, *Nature*, Volume 347, 354 (1990).

Preface

This volume provides a detailed description of the optical and electronic properties of fullerene-based materials. The explosion of activities and developments that this area has enjoyed since 1991 continues unabated, with intense attention now being focused on new fullerene-based materials and devices. The new materials include various buckytubes, compounds of fullerenes with other elements and structures, and polymerized fullerenes. Novel devices include fullerene/polymer composite photovoltaic cells and ordered arrays of buckytube field emitters. The optical and electronic properties of these materials are obviously the basis for understanding the behavior of the new materials and devices. It is therefore hoped that this volume will serve a dual purpose: that researchers established in these areas will find it to be a useful reference guide, as well as a tutorial on specific topics with which they are not familiar, and that scientists entering the field, particularly graduate students and postdoctoral fellows, will find it to be a useful introductory text.

Volumes on fullerenes are no longer scarce. A notable example is the recent monograph *Science of Fullerenes and Carbon Nanotubes*, by M. S. Dresselhaus, G. Dresselhaus, and P. C. Eklund (Academic Press, New York, 1995). That volume, however, provides a broad overview of fullerenes and nanotubes and incorporates major developments reported through 1994. The present volume focuses on providing a comprehensive survey of the optical and electronic properties of these materials that covers developments well into the late 1990s.

The volume begins with a review by Shuji Abe of theoretical studies on the properties of photoexcited states of fullerenes, including neutral states in the singlet and triplet manifolds, charged states such as polarons, electron–electron interactions, electron-vibration coupling, nonlinear optical (NLO) response, and the electronic structure of polymerized C_{60} and higher fullerenes.

Chapter 2, by Susan L. Dexheimer, reviews the ultrafast dynamics of electronic and vibrational excitations in fullerene solutions and films, which are characterized by a very high yield of triplet excitons. The chapter also addresses the issues of the induced absorption and exciton annihilation mechanisms, the dynamics of highly excited states in the singlet and triplet manifolds, and the ultrafast vibrational dynamics of C_{60}.

Chapter 3, by Ya-Ping Sun, Jason E. Riggs, Zhixin Guo, and Harry W. Rollins, reviews the recent photophysical studies of fullerenes and fullerene derivatives such as monofunctionalized C_{60} derivatives with an emphasis on the properties of excited singlet states. It also covers the inter- and intramolecular electron transfer properties associated with singlet-quenching processes.

Chapter 4, by R. Bruce Weisman, is devoted to optical studies of the triplet states in fullerenes. These states are important because their photogeneration yields are very high and their lifetimes are far longer than other electronic excited states. Through detailed solution-phase photophysics and photochemistry studies, the monomolecular and bimolecular processes that govern their dynamics are described and discussed.

Chapter 5, by F. P. Strohkendl and Zakya H. Kafafi, reviews the experimental work on the electronic third-order NLO properties of C_{60} and C_{70} films. A tutorial discusses the complementary information that can be obtained from third-harmonic generation, electroabsorption, two-photon absorption, and degenerate four-wave mixing. The advantages of tunable femtosecond degenerate four-wave-mixing for two-photon spectroscopy are emphasized. Results are discussed with respect to the underlying electronic states and applications to ultrafast all-optical switching.

Chapter 6, by R. Kohlman, V. Klimov, L. Smilowitz, and D. McBranch, describes studies of the closely related areas of optical limiting and excited-state absorption in fullerene solutions and doped glasses, which may result in significant technological applications. Specifically, the chapter reviews recent studies of both transient absorption spectroscopy and wavelength-dependent optical power limiting for neat and derivatized C_{60} in solution, films, and sol-gel glasses.

Chapter 7, by P. A. Lane, Zeev Valy Vardeny, and Joseph Shinar, reviews magnetic resonance studies of C_{60} and C_{70} films, C_{60} single crystals, and matrix-isolated molecules. It includes a brief tutorial on magnetic resonance spectroscopy and descriptions of studies on electron spin resonance (ESR), light-induced ESR (LESR), photoluminescence (PL)- and photoinduced absorption (PA)-detected magnetic resonance (PLDMR and PADMR, respectively). These studies have provided detailed insight into the structure and dynamics of the low-lying triplet states, and the effects of intermolecular coupling on these states.

Chapter 8, by M. S. Dresselhaus, G. Dresselhaus, P. C. Eklund, and R. Saito, provides an overview of electrons and phonons in fullerenes and carbon nanotubes, with reference to the underlying symmetry in these systems. The intermolecular coupling in fullerene-based solids is relatively weak and many of the properties are consequently related to the states of the isolated molecule. The periodic nature of the carbon nanotube can be exploited to derive many of its properties from the states of the flat two-dimensional graphene sheet by applying appropriate cylindrical boundary conditions and ignoring the corrections associated with the curvature of the nanotube wall.

Chapter 9, by Nobutsugu Minami and Said Kazaoui, describes the photoconductivity (σ_{ph}) of thin fullerene films and solids. It includes a tutorial on photocarrier generation and photocurrent action spectra, bimolecular recombination, absorption and PL in connection with σ_{ph}, PL quenching by electric fields, and the role of charge transfer excitons. It also describes carrier transport in fullerenes, including measurements of basic quantities such as carrier mobilities in films and MOSFET configurations. In addition, it also devotes a section to the effects of oxygen. These are known to be very considerable even at the low contamination levels that result if insufficient attention is devoted to avoid such contamination. Finally, the outlook for technological applications of fullerenes in xerographic photoreceptors and photovoltaic devices is discussed.

Chapter 10, by H. Kuzmany, B. Burger, and J. Kürti, is devoted to the optical and electronic properties of polymerized and dimerized products of fullerenes, as determined by X-ray and vibrational analysis, optical transmission and reflection studies, and electron spectroscopy. It covers the various forms of neutral and doped polymers, their preparation, and theoretical modeling of their properties.

Chapter 11, by N. Serdar Sariciftci, describes the electronic properties of fullerene/π-conjugated polymer composites. These systems exhibit ultrafast metastable photoinduced charge transfer from the polymer to the fullerene, with a quantum yield approaching unity. The chapter provides a detailed picture of the charge transfer and discusses a number of potentially interesting applications of this phenomenon, including photovotaic and nonlinear absorption (optical limiting) through excited-state absorption.

Joseph Shinar
Zeev Valy Vardeny
Zakya H. Kafafi

Contents

Contributors

Shuji Abe *Electrotechnical Laboratory, Tsukuba, Japan*

B. Burger *University of Vienna, Vienna, Austria*

Susan L. Dexheimer *Washington State University, Pullman, Washington*

G. Dresselhaus *Massachusetts Institute of Technology, Cambridge, Massachusetts*

M. S. Dresselhaus *Massachusetts Institute of Technology, Cambridge, Massachusetts*

P. C. Eklund *University of Kentucky, Lexington, Kentucky*

Zhixin Guo *Clemson University, Clemson, South Carolina*

Zakya H. Kafafi *U.S. Naval Research Laboratory, Washington, D.C.*

Said Kazaoui *National Institute of Materials and Chemical Research (AIST), Tsukuba, Japan*

V. Klimov *Los Alamos National Laboratory, Los Alamos, New Mexico*

R. Kohlman *Los Alamos National Laboratory, Los Alamos, New Mexico*

J. Kürti *University of Vienna, Vienna, Austria*

H. Kuzmany *University of Vienna, Vienna, Austria*

P. A. Lane *University of Sheffield, Sheffield, England*

D. McBranch *Los Alamos National Laboratory, Los Alamos, New Mexico*

Nobutsugu Minami *National Institute of Materials and Chemical Research (AIST), Tsukuba, Japan*

Jason E. Riggs *Clemson University, Clemson, South Carolina*

Harry W. Rollins *Clemson University, Clemson, South Carolina*

R. Saito *University of Electro-Communications, Tokyo, Japan*

N. Serdar Sariciftci *Johannes Kepler University of Linz, Linz, Austria*

Joseph Shinar *Ames Laboratory, U.S. Department of Energy, and Iowa State University, Ames, Iowa*

L. Smilowitz *Los Alamos National Laboratory, Los Alamos, New Mexico*

F. P. Strohkendl *University of Southern California, Los Angeles, California*

Ya-Ping Sun *Clemson University, Clemson, South Carolina*

Zeev Valy Vardeny *University of Utah, Salt Lake City, Utah*

R. Bruce Weisman *Rice University, Houston, Texas*

Optical and Electronic Properties of Fullerenes and Fullerene-Based Materials

1

Theoretical Studies of Photoexcitations in Fullerenes

Shuji Abe
Electrotechnical Laboratory
Tsukuba, Japan

1 INTRODUCTION

The C_{60} molecule had been a theoreticians' dream for a long time [1, 2] until its experimental discovery in the middle of the 1980s [3]. The rapid growth of experimental developments following the discovery, combined with the beauty of C_{60} with icosahedral symmetry, initiated many theoretical works on all aspects of fullerenes. By now the study of fullerenes has already grown to a large research field [4].

The purpose of the present chapter is to provide a brief, introductory survey on the optical properties of fullerenes from a theoretical point of view, so that the reader can grasp basic concepts and ideas to understand physics behind diverse experiments discussed in the rest of this book. For this purpose I will concentrate on the basic linear optical properties of C_{60} molecules and solids, leaving up-to-date topics to other chapters of this book. Therefore, the present chapter is by no means comprehensive and is in many respects complimentary to existing comprehensive reviews [4–8]. In particular, topics such as conduction properties, doped fullerenes, fullerene derivatives, and carbon nanotubes are not included in this chapter.

2 C_{60} MOLECULE

Fullerenes are fairly large molecules. The number of atomic orbitals and vibrational modes in a C_{60} molecule, for example, is quite large. This means that a full theoretical description of the electronic structure should be quite complicated, especially if we go beyond the one-electron approximation and take electron-electron and electron-vibration couplings into account. Therefore, we need to go step by step in order to grasp an overall physical picture of photoexcited states.

2.1 Hückel Calculations

To describe electrons in fullerenes, we may start with the simple Hückel model for organic molecules, considering only π electrons within the approximation of noninteracting electrons. Although σ and π orbitals are not completely decoupled in fullerenes because of curved molecular surfaces, it is still a good approximation to consider only π electrons for low-energy optical excitations.

Simple Hückel calculations for C_{60} were reported [2] long before the experimental discovery of C_{60}. The one-electron energy levels obtained in the Hückel model are shown in Figure 1 [9]: parts (b) and (c) are in the absence and presence of bond alternation, respectively. The bond alternation here implies a shorter bond ("double bond") between hexagonal faces and a longer bond ("single bond") between hexagonal and pentagonal faces of C_{60}. For comparison, the level scheme of free electrons on the sphere is shown in Figure 1a. The unit of energy is the standard resonance energy (transfer energy) of $\sim$2 eV in the simple Hückel model. The energy difference between the top t_g and the bottom a_g levels is about 10 eV. Sixty orbitals are organized into 16 multiply degenerate levels by the icosahedral symmetry of C_{60}. (There is an accidental degeneracy of the $2h_g$ and $1g_g$ levels in the case of no bond alternation.) The symmetry indices a, t_1, t_2, g, and h correspond to multiplicity 1, 3, 3, 4, and 5, respectively. Suffices g and u refer to gerade and ungerade states, respectively. For more details about icosahedral symmetry, see Refs. 10 and 11. In neutral C_{60} the 8 lower levels (30 orbitals) are just filled with 60 electrons, and the excitation gap lies between $1h_u$ and $2t_{1u}$ states.

With these multiplicities, the level structure becomes rather sparse with large gaps between them. If the 60 energy levels were distributed uniformly over the π bandwidth of about 10 eV, then the average level spacing would be about 0.2 eV. In this sense, the large optical gap of C_{60} is the result of its high symmetry. In fact, higher fullerenes with lower symmetry have smaller gaps [12–14].

An additional contribution to the optical gap comes from bond alternation. The energy gap between the highest occupied molecular orbital (HOMO) and the lowest unoccupied molecular orbital (LUMO) in Figure 1c is larger than that in Figure 1b. Hayden and Mele [15] used a tight-binding model with all $2s$ and

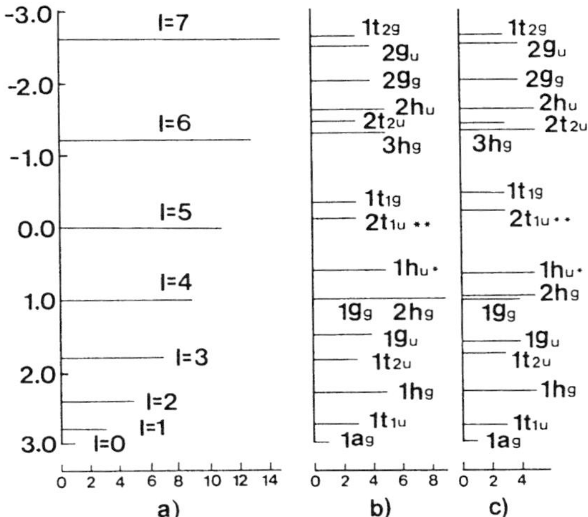

Figure 1   One-electron level schemes of C_{60}: (a) a free electron model; (b) a simple Hückel calculation without bond alternation; (c) a Coulson-Golebiewski self-consistent Hückel calculation with bond alternation. The abscissa represents the multiplicity of the level. The unit of energy is the standard transfer energy for graphite. I denotes the angular momentum. * and ** denote the highest occupied molecular orbitals (HOMO) and the lowest unoccupied molecular orbitals (LUMO), respectively. (*From Ref. 9.*)

$2p$ orbitals to describe bonding structure, while Harigaya [16] and Friedman [17] employed a π-electron model similar to the Su-Schrieffer-Heeger (SSH) model for conjugated polymers [18]. It should be noted that the bond alternation in C_{60} is governed by the presence of five-membered rings and it is not a self-breaking of symmetry. Yet the effect of bond alternation is important for quantitative argument of the electronic structure. The energy levels for arbitrary bond alternation were obtained in an analytical form [19]. The effect of electron-vibration coupling will be discussed later.

The most important feature of the level structure in Figure 1 is that both the HOMO and LUMO have ungerade symmetry, so that optical transitions between them are dipole-forbidden. For the many-body ground state of neutral C_{60} with A_g symmetry, dipole transitions are allowed for excitations with T_{1u} symmetry. (We use capital letters A, T, etc. to indicate the symmetry of a many-electron state.) The lowest dipole-allowed excitations originates from excitations between $1h_u$ and $1t_{1g}$ or between $2h_g$ and $2t_{1u}$ one-electron levels, since the product of h and t contains T_1 (actually, $h \otimes t_1 = T_1 + T_2 + G + H$). Thus the allowed optical gap is much larger than the HOMO-LUMO gap. This is another reason why C_{60} has a large optical gap.

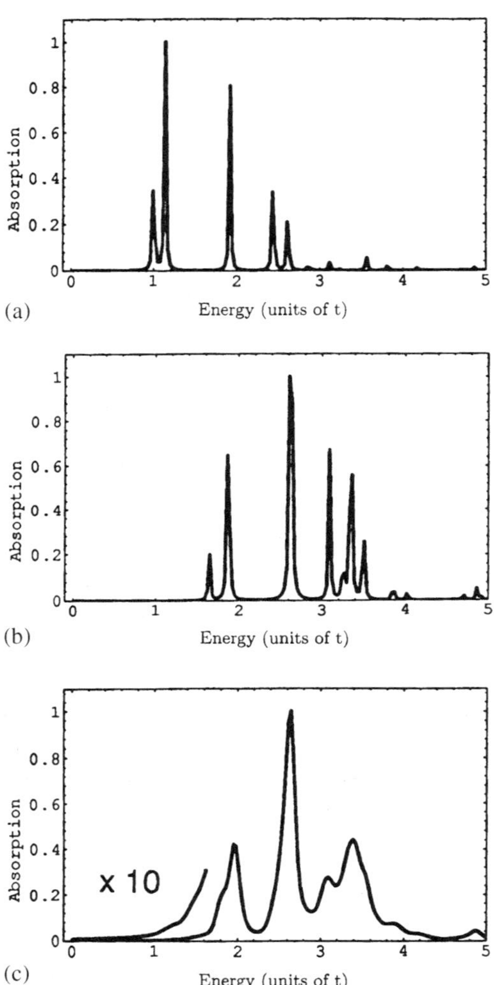

Figure 2 Theoretical absorption spectrum of C_{60} calculated (a) in the simple Hückel model, (b) in a PPP-type model, and (c) with bond disorder in the PPP-type model. The abscissa is scaled by the average transfer energy t. The Lorentzian broadening $\gamma = 0.01t$ is used. Bond alternation of $0.1t$ is assumed for all the cases. In (b) and (c), the Ohno potential with the on-site parameter $U = 4t$ and the off-site parameter $V = 2t$ are used. In (c), a Gaussian bond disorder with a standard deviation $0.09t$ is used. (*From Ref. 20.*)

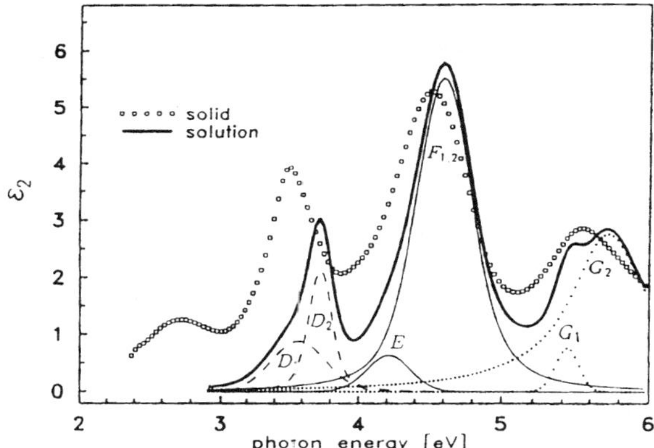

Figure 3 The imaginary part of the observed dielectric function of solid C_{60} (thick curve) and that of molecular C_{60}. The solution spectrum is decomposed into Gauss-Lorenz profiles. (*From Ref. 24.*)

The optical absorption spectrum of C_{60} in a simple Hückel model is shown in Figure 2a [20]. It should be compared with experimental absorption spectra [21–24], a typical example of which is shown in Figure 3 [24]. Apparently, the theory does not quite well reproduce the experimental spectra. This stems from the neglect of electron-electron interactions. This will become clear after we see calculations incorporating the electron-electron interactions in the next section.

2.2 Electron-Electron Interactions

Interactions among electrons are always an important ingredient in electronic structure calculations. Electron correlation is known to be relevant in one-dimensional conjugated polymers, while it does not seem to play a major role in two-dimensional graphite. In C_{60} molecule electron correlation turns out to be weak, but long-range Coulomb interactions affect its optical properties very much.

A quantum chemical calculation of the optical spectrum of C_{60} was given by Larsson et al. using the Complete Neglect of Differential Overlap for Spectroscopy (CNDO/S) method with configuration interaction (CI) [25]. CI was carried out between singly excited states of a singlet configuration, using 250 configurations with the lowest energies. Negri et al. [26, 27] also reported CNDO/S calculations with varying the number of configurations from 14×14 to 35×37 in the single-CI. Both calculations showed that there are many dipole-forbidden states in the range of 2–3.5 eV below the lowest dipole-allowed transition at 3.6 eV. The lowest allowed transition has quite small oscillator strength. Much stronger transitions are situated at higher energies. Although the energies and the oscillator strength distribution of the T_{1u} excited states in the energy region 4.0– 7.3 eV depended on the size of the CI matrix, all the calculations agree in that few T_{1u} states in the 4.0–6.5 eV region carry large oscillator strength. These calculations were within the single CI. Calculations with single and double excitations confirmed that the contributions of double excitations are negligible at least in the low energy region below ∼3.2 eV [28].

These quantum chemical calculations use the Ohno or Mataga-Nishimoto potential for intersite repulsion. By parametrizing the Ohno potential and reducing its overall strength within the Pariser-Parr-Pople (PPP) model [29], we obtain a theoretical absorption spectrum in reasonable agreement with experiment, as shown in Figure 2b [20]. One might ask whether the long-range interactions are really necessary or not in order to explain the absorption spectrum of C_{60}. A perturbational calculation of optical absorption in the Hubbard model [30] with only onsite Coulomb repulsion (U) indicated that, to reproduce experimental oscillator strengths at three major peaks up to 6 eV, an unreasonably large U of about 15 eV was required, although this is out of the range of the perturbational approach. The inadequacy of the Hubbard model was also demonstrated by a CI calculation [31]. The long-range part of interaction is essential in describing the excited states of C_{60}.

We note that the optical spectra calculated with CI are quite different from those of the simple Hückel model. The oscillator strength is transferred to high-energy excitations compared with the one-electron model. This is the result of large configurational mixing between low- and high-energy excitations. There is another way of looking at this effect. Due to the Coulomb interaction between electrons, the electromagnetic response of a molecule is substantially screened. A random-phase-approximation calculation [32] indicates blue shifts and oscillator strength reduction of low-energy excitations.

This is connected with the existence of a Mie-type plasmon in the C_{60} molecule [33, 34]. The 240 valence electrons of a C_{60} are many enough to be looked upon as a many-electron system. Barton and Eberlein [35] used a hydrodynamic model to discuss plasmons in C_{60} and C_{70}. They predicted π plasmons in the range between 6 and 8 eV and σ plasmons near and above 25 eV. These excitations with collective nature gains large oscillator strength in contrast to the low energy excitations. At low energies a dipole oscillation tends to be screened by many electrons, whereas at high energies the electrons tend to oscillate cooperatively, enhancing the dipole moment.

2.3 Electron-Vibration Coupling

Now we turn to the effect of electron-vibration coupling. Since the lowest optical transitions in C_{60} have gerade symmetry and is dipole-forbidden, optical transitions to these states must be assisted by appropriate ungerade modes of molecular vibrations (the Herzberg-Teller mechanism). Negri et al. [27] calculated couplings between various excited states and various molecular vibrations including Herzberg-Teller active modes and Jahn-Teller active modes. Vibronic interactions in C_{60} are in general smaller than the coupling calculated in aromatic compounds and polyenes. The lattice relaxation energy of the lowest excited state due to Jahn-Teller polaron distortion is on the order of 0.1 eV [16, 26]. The oscillator strengths gained by the Herzberg-Teller mechanism for the dipole forbidden states are four orders of magnitude smaller than the oscillator strength of allowed transitions [27]. Combined effects of electron-lattice coupling and electron correlation were shown to be important also by a perturbation calculation for the SSH-Hubbard model [36].

The weak absorption due to the Herzberg-Teller mechanism were detected by precise absorption measurements in solution [37] and by magnetic circular dichroism for C_{60} isolated in Ar matrices [38]. Structures in the 1.9–2.1 eV region were assigned to vibronic structures of dipole-forbidden singlet T_{1g} state with Herzberg-Teller couplings. Above this region there is a diffuse spectrum with many small peaks, which are ascribed to other forbidden singlets and triplets. Rather, it has been established that the threshold of the forbidden singlet transition is located at about 1.9 eV. The energy of the lowest triplet in C_{60} was estimated at about 1.6 eV by excitation quenching experiments [39, 40]. If we neglect the lattice relaxation energy, the singlet-triplet energy difference is about 0.3 eV. This is in agreement with calculations [26].

Another type of calculation concerning the effect of molecular vibrations on the optical spectrum was done by use of a static disorder model, in which the transfer energy of each bond is assumed to be a random variable with a Gaussian distribution [20]. The absorption spectrum is obtained by averaging over a large number of samples. This type of calculation simulates the effect of thermal fluctuations at high temperatures but may be used to discuss an overall feature of low-temperature spectra neglecting detailed vibronic structures. As shown in Figure 2c, the absorption spectrum thus obtained is in fairly good agreement with observed spectra over a wide energy range, the latter being shown by a thick curve in Figure 3. The reason why vibronic structures are not prominent experimentally in the region of main peaks is perhaps that a large number of vibration modes couple to the dipole-allowed T_{1u} excitations. Another reason may be short lifetimes of these excited states due to relaxation toward low-lying dipole-forbidden states.

The electron-vibration couplings are important also in understanding photoluminescence. Very weak emission from C_{60} in solution has been observed in the energy region extending from $\sim$1.9 eV down to $\sim$1.5 eV [41]. The fact that the lowest singlet excited state is dipole-forbidden implies a very long radiative lifetime of the luminescence. The quantum yield of the luminescence is fairly small, and this has been ascribed to nonradiative relaxation processes including intersystem crossing to the triplet manifold. The line shape of the luminescence spectrum is interpreted on the basis of Herzberg-Teller coupling [27]. Different suggestions based on strong polaron effects have also been proposed [42, 43]. However, a detailed comparison between the observed absorption and luminescence spectra [41] elucidates that the peak positions of the luminescence are in a mirror image of those of the forbidden absorption, strongly supporting the interpretation by Herzberg-Teller coupling. A puzzling feature is that the electronic origin, i.e., the position of the zero-phonon transition, of the luminescence must be red-shifted from that of the absorption by about 0.04 eV to achieve the mirror correspondence.

2.4 Nonlinear Optical Response

There has been a long-standing interest in π conjugated molecules as materials with large optical nonlinearity. Conjugated polymers, e.g., polyacetylene, is known to have cubic susceptibility $\chi^{(3)}$ of about 10^{-10} esu [44]. It is natural to expect similarly large $\chi^{(3)}$ or molecular hyperpolarizability γ in fullerenes. The observed values were on the order $\chi^{(3)} \sim 10^{-11}$ esu, or $\gamma \sim 10^{-33}$ esu per molecule in the off-resonance condition [45, 46].

There have been many calculations of hyperpolarizabilities of C_{60}. Hückel-type calculations gave overall γ in the off-resonant condition on the order of 10^{-33} esu [47, 48] in agreement with the experimental value. However, if we take electron-electron interactions into account, the theoretical value becomes smaller. Semiempirical quantum chemical calculations [49], the local density approximation (LDA) calculations [50, 51], and the time-dependent Hartree-Fock calculations [52] gave values on the order of 10^{-35} esu. The valence effective Hamiltonian calculations [53] gave values on the order of 10^{-34} esu.

This reduction of nonlinearity can be understood as a consequence of screening [54, 55], in accordance with the reduction of the oscillator strength in the linear absorption. The origin of the two-orders-of-magnitude discrepancy between theory and experiment is not yet known, although suggestions have been made such as local field corrections, dispersion effects or resonance enhancement [56, 54], etc. In case of C_{60} films, intermolecular interactions might also play a role.

Another interesting aspect of the nonlinear response of C_{60} is that a fairly large second harmonic generation has been observed in C_{60} films [57], in spite of the presence of inversion symmetry in the crystal structure of C_{60}. Although there are possible effects of surface, most of theoretical calculations indicate that the origin of large $\chi^{(2)}$ is the participation of magnetic dipole transitions [58–60]. Calculations taking both magnetic dipole and electric quadrupole contributions into account [61] confirmed that the dominant contribution comes from the magnetic dipole contribution.

3 C_{60} CRYSTAL

The C_{60} crystal [62] has a face-centered cubic (fcc) structure above a temperature ~ 250 K with rotating molecules. Below this temperature the orientations of the molecules become frozen, forming a simple cubic lattice with four molecules oriented in mutually different directions in a unit cell. C_{60} crystal is in a sense an ideal molecular crystal resembling rare gas solids.

The observed optical absorption spectrum of a C_{60} crystal is not very much different from that of a single C_{60} molecule, as shown in Figure 3 [24]. This indicates that the primary photoexcitations are molecular Frenkel excitons [63]. However, it is again useful to start from a one-electron picture of the C_{60} crystal.

3.1   Band Calculation

LDA calculations of the one-electron band structure of the fcc C_{60} crystal [64–67] indicate that the band dispersion is much smaller than the molecular level spacings. That is, the molecular character is well preserved in the solid state of C_{60}. The absorption spectrum was calculated by Ching et al. [65]. The calculated dielectric function and density of states (DOS) are shown in Figure 4a, b. In the DOS, the V1 and C1 bands mainly come from the $1h_u$ HOMO and $2t_{1u}$ LUMO, respectively. The V2 band originates from the $2h_g$ and $1g_g$ orbitals, while the C2 band corresponds to the $1t_{1g}$ orbitals (see Figure 1). The absorption band A in Figure 4a corresponds to transitions from V1 to C1, which are dipole-forbidden for an isolated C_{60} molecule. The forbiddenness is partially lifted in the crystal, because, for example, the band C1 mainly consists of t_{1u} molecular orbitals but also has an admixture of t_{1g} orbitals, etc. (In crystal, a Bloch state can be constructed by a linear combination of gerade and ungerade molecular orbitals except at special points in k space.) The absorption band B corresponds to transitions V1 $\rightarrow$ C2 and V2 $\rightarrow$ C1, both of which originate mainly from dipole-allowed molecular transitions.

The overall distribution of oscillator strength is similar to that for a single molecule in the simple Hückel model, Figure 2a; hence it is in poor agreement with experiment, as shown in the inset of Figure 4a. The band B seemed to coincide with the plateau in the experimental data, while the observed peak at 340 nm (~ 3.65 eV) was absent in the calculated curve. It was suggested at that time that the 340 nm peak could come from the presence of C_{70}, but later this possibility was ruled out. A reasonable assignment is that the band B corresponds to the 340 nm peak, and the band A corresponds to the plateau around 500 nm

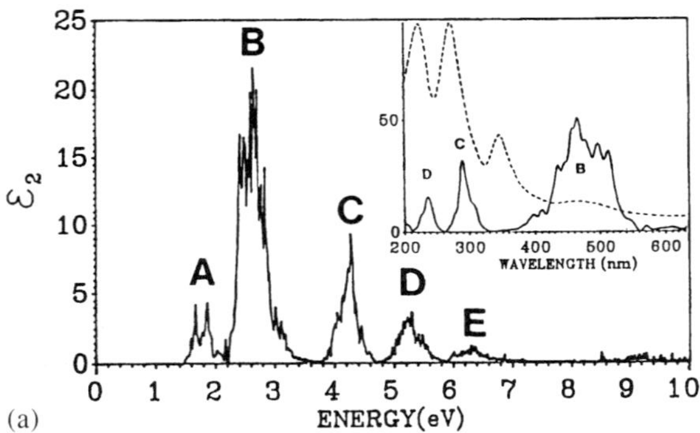

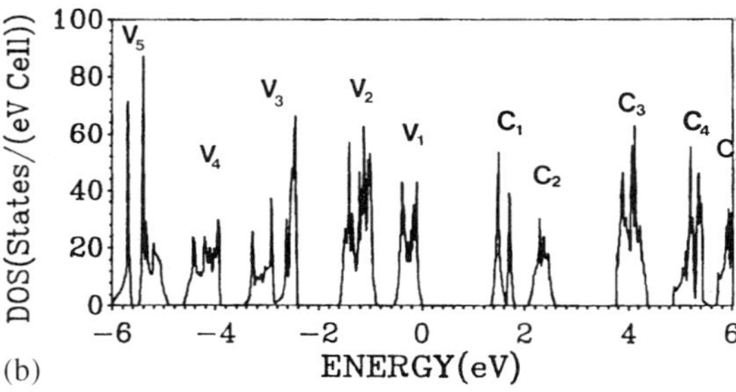

Figure 4 (a) The imaginary part of the dielectric function and (b) the density of states both obtained by a first-principles LDA calculation. (*From Ref. 65.*)

($\sim$2.5 eV). In fact, the plateau is missing in the observed spectrum of C_{60} molecule in solution (see Figure 3). The discrepancies in energies and oscillator strengths are due to the neglect of Coulomb interactions in excited states, as in the case of the isolated C_{60} molecule. However, we see that the appearance of the low-energy ''forbidden'' absorption is in accordance with the experiments in a qualitative sense.

3.2 Exciton Calculations

As mentioned before, the observed absorption spectrum of C_{60} crystal is very close to that of a C_{60} molecule, suggesting that the optical excited states of C_{60} are governed by Frenkel exciton states. This is also supported by CI calculations for a cluster of C_{60} molecules [68]. It was found that the intermolecular interactions do not change the major features of the absorption spectrum very much.

The Frenkel exciton picture is valid if the exciton binding energy, i.e., the energy required to take an electron from a molecule and to put it on a different molecule in the crystal, is much larger than the energy of exciton transfer between molecules. In C_{60} crystal this condition is well satisfied: The former is estimated as about 1 eV and the latter is ~ 0.1 eV. Experimentally, the binding energy of the lowest dipole-allowed exciton is estimated as 1 eV by comparing the position of the absorption peak with the energy difference between the corresponding peaks in the one-electron density of states obtained from photoemission and inverse photoemission measurements [69].

When molecules form a solid state in general, intermolecular charge transfer (CT) excitations become possible in addition to Frenkel excitons. A charge transfer excitation corresponds to a creation of an electron and a hole on different molecules. If the distance between the electron and the hole is very large, then the excitation is similar to the interband excitations described in the band calculation. If an electron and a hole are situated on nearest neighbor molecules and are bound to each other, then they are called a CT exciton. However, this does not mean that they are localized on two molecules. In the fcc lattice of C_{60}, the electron can be located on any one of the twelve equivalent nearest-neighbor molecules around the molecule occupied by the hole, and vice versa. The CT exciton here implies a Wannier-like exciton with a radius of about two molecules.

Tsubo and Nasu calculated the optical absorption spectrum of C_{60} crystal in an extended Hubbard model with weak short-ranged electron-electron interactions [70, 71]. Configuration space of intramolecular excitations and nearest-neighbor intermolecular excitations was taken into account. A calculated spectrum is shown in Figure 5. We see that the overall spectrum is fairly close to that of the band calculation shown in Figure 4a. The solid-state absorption in the low energy region was shown to have the character of CT-like exciton in the extended Hubbard model. This result is in harmony with the band calculation. The intermolecular electron transfer is the mechanism for the appearance of the low-energy absorption in both the calculations.

For realistic strong and long-ranged electron-electron interactions, this problem is more subtle. In molecular crystals in general, a CT absorption is usually very weak. It becomes sizable only when the CT excitation is strongly mixed with dipole-allowed Frenkel exciton states. A CI calculations for a linear cluster model [29] indicated that this mixing in C_{60} occurs due to accidental proximity in energy between the CT states of the forbidden $1h_u$-$2t_{1u}$ excitations and the Frenkel exciton states of the allowed $1h_u$-$1t_{1g}$ and $2h_g$-$2t_{1u}$ excitations. The absorption spectra calculated for this model with varying the intermolecular transfer energy is shown in Figure 6 [29]. Configurations for single-CI were limited to

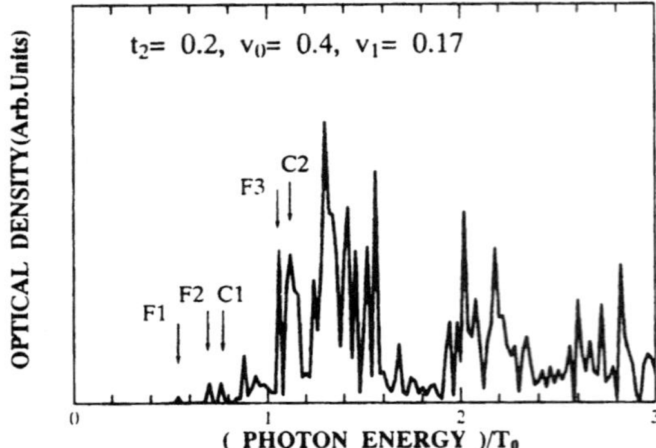

Figure 5   Absorption spectrum calculated in the extended Hubbard model for C_{60} crystal. T_0 is the intramolecular transfer energy ($\sim$2 eV) for a long bond, while t_2, v_0, v_1 are the intermolecular transfer energy, the on-site repulsion energy, the nearest-neighbor repulsion energy, respectively, in units of T_0. (*From Ref. 70.*)

excitations from V1 and V2 bands to C1 and C2 bands in Figure 4b. The two absorption peaks in Figure 6a correspond to the two lowest peaks in Figure 2b. Weak absorption peaks grow on the low- and high-energy sides of the two peaks with increasing the intermolecular transfer energy.

All these calculations remain at a qualitative level: the positions of the peaks in Figures 5 and 6 cannot be taken very seriously compared with the experimental spectra. Recently many calculations have been performed to estimate exciton binding energies more accurately. A calculation using a effective electron-hole interaction based on the one-electron band structure of the HOMO and LUMO bands gave a band gap at 2.5 eV and a singlet exciton at 1.6–2 eV depending on Coulomb parameters, with a singlet-triplet difference of about 0.1 eV [72]. Another calculation located the lowest forbidden excitons in the range between 1.6 eV and 2.3 eV, CT excitons at about 2.4 eV, the band gap at 2.5 eV [73]. The triplet excitons are located by about 0.3 eV below the corresponding singlet excitons. These excitation energies may be somewhat lower than the actual values, since the LDA usually underestimates the band gap.

A different approach to this problem was given by a calculation of polarization energies for charge transfer and band gap states [74]. The calculated single-charge polarization energy is about $P \sim 0.9$ eV. Combining with the observed ionization energy $I = 7.6$ eV and electron affinity $A = 2.6$ eV of the C_{60} molecule, the band gap was estimated as $E_g = I - A - 2P \sim 3.0$ eV. The energy of the nearest-neighbor charge transfer state is estimated as 2.7 eV.

There have been experimental reports on electroabsorption (EA) spectra of

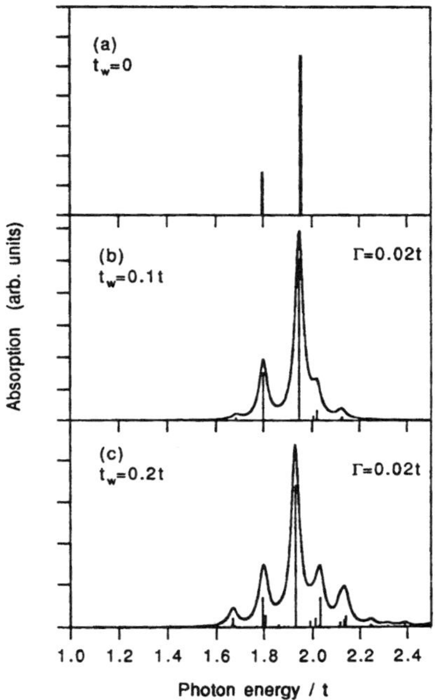

Figure 6 Low-energy absorption spectra calculated with the single-CI for a PPP-type model on a linear cluster of four C_{60} molecules with various intermolecular transfer t_w: (a) $t_w = 0$, (b) $t_w = 0.1t$, and (c) $t_w = 0.2t$, where t is the average intramolecular transfer energy ($t \approx 2$ eV). The same Ohno potential parameters as in Figure 2b are used. The smooth curves are obtained by assuming a damping constant Γ. (*From Ref. 29.*)

C_{60} solids [75–78]. Strong EA signals have been observed in the region of 2.2–2.8 eV, corresponding to the region of the absorption shoulder in the solid state. Possible assignments were given by Hess et al. [77] by using low-lying (mostly dipole-forbidden) molecular levels and relaxing the molecular selection rule. On the other hand, it was argued that the observed EA spectrum could not be explained without invoking CT components [79]. A calculation of the EA spectrum for a model cluster indicates that the low-lying states having CT components contribute to the EA spectrum significantly in spite of the small contribution to the linear absorption spectrum [80]. The actual assignment of EA peaks are still not easy, because it requires accurate calculations of the energy levels in the solid state including vibrational couplings. Possible assignments are discussed in Ref. 74. Electroabsorption spectra of C_{70} films [77, 81] have also been reported and discussed along similar lines.

The dispersion of exciton bands has been calculated for low-lying dipole-forbidden excitons with the mechanism of electron exchange [82]. It was applied to the calculation of the nonlinear optical susceptibility for second harmonic generation from a C_{60} film [82].

3.3 Relaxed Excitons

In molecular crystals, there is a question of exciton self-trapping. That is, a stabilized exciton in crystals can be either in a free (extended) state or in a self-trapped (small polaron) state, depending on the exciton band width, the exciton-vibration coupling constant, and the vibration frequencies. For detailed discussions on self-trapping, see Ref. 83.

For the low-lying dipole-forbidden exciton states in C_{60}, the exciton hopping due to ordinary dipole-dipole coupling is absent. Direct electron exchange leads to exciton bandwidths of ~0.03 eV [82], which is much smaller than the lattice relaxation energy of ~0.1 eV [16]. Therefore, the exciton in C_{60} is likely to be self-trapped, although the situation is in the weak coupling regime, where the relaxation energy is comparable to the vibration quantum, so that the free and self-trapped states are not separated distinctly. This is a typical situation in molecular crystals.

Weak photoluminescence in C_{60} films has been observed [84–86]. It has a similar spectra as that from C_{60} in solution, suggesting that the relaxed exciton is very similar to the relaxed excited state of a C_{60} molecule. A complication in the observed luminescence is that it is rather sensitive to morphology of the C_{60} sample, indicating strong influence by disorder and defects [86].

There has been an experimental report that the luminescence of C_{60} consists of fast and slow components with respective time constants of 0.5 and 1 ns [87]. It has been suggested that the fast component is the same origin as C_{60} in solution but the slow component originates from a self-trapped charge-transfer exciton. Similar experimental results were also reported in C_{70} solids [88]. An excimer model for the photoluminescence of C_{60} crystal has been proposed [89]. The vibronic structure of luminescence was interpreted by considering hopping of an exciton-polaron among two molecules. However, what is meant by the model is that the exciton-polaron is not confined in a single molecule but is mobile among molecules. In the fcc lattice of C_{60}, whether an excimer with symmetry-breaking lattice relaxation is responsible for the luminescence or not is still an open question.

3.4 Polymerized C_{60}

An especially unique feature among various properties of C_{60} solids may be the formation of covalent bonding between C_{60} molecules. The so called polymerized

C_{60} was produced by irradiation of light [90] or application of high pressure [91] on C_{60} crystal. A similar structure was found in the ground state of K_1C_{60} and Rb_1C_{60} [92]. See Ref. 8 for details. The formation of a C_{60} dimer was theoretically studied by various methods [93, 94].

Motivated by the possibility of π-electron delocalization over the C_{60} chain, Harigaya discussed the electronic structure of polymerized C_{60} by extending the SSH-type model [95, 96]. However, a tight-binding band calculation indicate that the bridging C—C bonds are almost pure σ-type bonding, so that the π conjugation is broken between C_{60} molecules [97, 98]. Fagerström and Stafström performed a semiempirical quantum chemical calculation to study the formation of a dimer [99]. The inspection of molecular orbitals elucidated that the LUMO of the dimer have the nonbonding character at the intermolecular bonding sites. They calculated also the absorption spectrum of the C_{60} dimer within the single-CI method. The effect of the dimer formation turned out to be similar to the effect of defects or disorder in C_{60} molecule. Sharp absorption peaks in the C_{60} molecule are broken up into many densely distributed transitions such that the overall effect resembles broadening of peaks. This is in accordance with the observed absorption spectra of photopolymerized C_{60}, which have main peak structures similar to pristine C_{60} but with much larger widths [100]. Also, the luminescence spectrum of polymerized C_{60} is very similar to that of pristine solid C_{60} [101], indicating the preservation of molecular character in photoexcited states.

4 HIGHER FULLERENES

The absorption spectra of C_{70} molecule were calculated in a tight binding model [102] and also by using the single-CI method with similar parameters as in C_{60} [20]. The major difference is that the spectrum consists of more peaks than that of C_{60} and the absorption edge is red-shifted to at about 1.8 eV, due to the lower symmetry of C_{70}. Another feature is the polarization dependence of the optical transitions. The dominant absorptions in C_{70} are located in the range 4–6 eV as in the case of C_{60}. The absorption peaks in the range 3–4 eV are also prominent in the theoretical spectra, but they are much weaker in the observed spectra [22].

In general, higher fullerenes have isomers [103]. For example, C_{78} have three possible structures that satisfy the isolated-pentagon rule, which demands that pentagonal faces are separated from each other. An interesting question is whether a structural difference can be identified by optical spectra or not. Geometric effects on optical spectra of higher fullerenes have been studied within the single-CI [104]. The absorption spectrum in the low energy region below about 4 eV are determined mainly by wave function amplitudes on pentagonal sites. However, no simple relationship between the arrangement of pentagonal faces and the optical spectrum was found. This implies that the spectrum depends on specific structure of each molecule.

Nonlinear optical susceptibilities of higher fullerenes have been calculated by Fanti et al. [105] and Harigaya [106]. Those calculations agree in general in that the off-resonant $\chi^{(3)}$ increases with going to larger carbon number, in harmony with decrease in the optical gap. This is in analogy with a similar tendency in one-dimensional conjugated polymers, where $\chi^{(3)}$ grows with increasing the conjugation length. However, again the story is not so simple in the sense that calculations for five isomers of C_{78} indicate strong effects of geometrical structures [107].

5   CONCLUSION

A variety of theoretical approaches have been taken to describe electronic properties and photoexcitations in fullerenes. Noninteracting electron models give simple overall pictures, while electron-electron interactions are essential in many aspects. Semiempirical quantum chemical calculations, density functional calculations, exciton model calculations among others have been used to clarify these effects. Although basic characteristics are reasonably well understood, still there are many unsolved theoretical questions, such as dynamical properties of photoexcitations, the effect of quantum lattice fluctuations, the influence of molecular rotations in the solid state, among others. Many attempts have already been in progress toward such directions, though not covered in this chapter.

ACKNOWLEDGMENT

The author thanks his colleagues at ETL, especially Kikuo Harigaya and Yoshihiro Asai, for many useful discussions.

REFERENCES

1.   E. Osawa, *Kagaku (Kyoto)* **25**, 854 (1970) [in Japanese].
2.   D. A. Bochvar and E. G. Gal'pern, *Dokl. Akad. Nauk SSSR* **209**, 610 (1973) [English translation, *Proc. Acad. Sci. USSR* **209**, 239 (1973)].
3.   H. W. Kroto, J. R. Heath, S. C. O'Brien, R. F. Curl, and R. E. Smalley, *Nature* **318**, 162 (1985).
4.   M. S. Dresselhaus, G. Dresselhaus, and P. C. Eklund, *Science of Fullerenes and Carbon Nanotubes*, Academic Press, San Diego, 1996.

5. H. J. Byrne, in *Physics and Chemistry of Fullerenes and Derivatives*, (H. Kuzmany, J. Fink, M. Mehring, and S. Roth, eds.), World Scientific, Singapore, 1995, p. 183.

6. J. H. Weaver and D. M. Poirier, in *Solid State Physics* (H. Ehrenreich and F. Spaepen, eds.), vol. 48, Academic Press, New York, 1994, p. 1.

7. W. E. Pickett, in *Solid State Physics* (H. Ehrenreich and F. Spaepen, eds.), vol. 48, Academic Press, New York, 1994, p. 349.

8. Y. Iwasa, in *Optical Properties of Low-Dimensional Materials* (T. Ogawa and Y. Kanemitsu, eds.), World Scientific, Singapore, 1995, chap. 7.

9. M. Ozaki and A. Takahashi, *Chem. Phys. Lett.* **127**, 242 (1986).

10. W. G. Harter and D. E. Weeks, *J. Chem. Phys.* **90**, 4727 (1989).

11. Chap. 4 of Ref. 4.

12. S. Saito and A. Oshiyama, *Phys. Rev. B* **44**, 11532 (1991).

13. S. J. Woo, E. Kim, Y. H. Lee, *Phys. Rev. B* **47**, 6721 (1993).

14. K. Nakao, N. Kurita, and M. Fujita, *Phys. Rev. B* **49**, 11415 (1994).

15. G. W. Hayden and E. J. Mele, *Phys. Rev. B* **36**, 5010 (1987).

16. K. Harigaya, *J. Phys. Soc. Jpn.* **60**, 4001 (1991).

17. B. Friedman, *Phys. Rev. B* **45**, 1454 (1992).

18. W. P. Su, J. R. Schrieffer, and A. J. Heeger, *Phys. Rev. Lett.* **42**, 1698 (1979).

19. C.-L. Wang, W.-Z. Wang, and Z.-B. Su, *J. Condens. Matter* **5**, 5851 (1993).

20. K. Harigaya and S. Abe, *Phys. Rev. B* **49**, 16746 (1994).

21. Y. Achiba, T. Nakagawa, Y. Matsui, S. Suzuki, H. Shiromaru, K. Yamauchi, K. Nishihara, M. Kainosho, H. Hoshi, Y. Maruyama, and T. Mitani, *Chem. Lett.* **1991**, 1233 (1991).

22. J. P. Hare, H. W. Kroto, and R. Taylor, *Chem. Phys. Lett.* **177**, 394 (1991).

23. S. L. Ren, Y. Wang, A. M. Rao, E. McRae, J. M. Holden, T. Hager, K. A. Wang, W. T. Lee, H. F. Ni, J. Selegue and P. C. Eklund, *Appl. Phys. Lett.* **59**, 2678 (1991).

24. J. Hora, P. Panek, K. Navratil, B. Handlirova, J. Humlicek, H. Sitter, and D. Stifter, *Phys. Rev. B* **54**, 5106 (1996).

25. S. Larsson, A. Volosov, and A. Rosen, *Chem. Phys. Lett.* **137**, 501 (1987).

26. F. Negri, G. Orlandi, and F. Zerbetto, *Chem. Phys. Lett.* **144**, 31 (1988).

27. F. Negri, G. Orlandi, and F. Zerbetto, *J. Chem. Phys.* **97**, 6496 (1992).

28. J. Fagerström and S. Stafström, *Phys. Rev. B* **48**, 11367 (1993).

29. S. Abe and K. Harigaya, SPIE Proceedings, vol. 2530, *Fullerenes and Photonics II*, 1995, p. 109.

30. B. Friedman and J. Kim, *Phys. Rev. B* **46**, 8638 (1992).

31. Y. Asai, Proceedings of Symposium on Recent Advances in the Chemistry and Physics of Fullerenes and Related Materials (K. M. Kadish and R. S. Ruoff, eds.), Electrochemical Society, 1994, p. 560.

32. G. F. Bertsch, A. Bulgac, D. Tomanek, and Y. Wang, *Phys. Rev. Lett.* **67**, 2690 (1991).

33. A. Bulgac and N. Ju, *Phys. Rev. B* **46**, 4297 (1992).

34. N. Ju and A. Bulgac, *Phys. Rev. B* **48**, 9071 (1993).

35. G. Barton and C. Eberlein, *J. Chem. Phys.* **95**, 1512 (1991).

36. M. I. Salkola, *Phys. Rev. B* **49**, 4407 (1994).

37. S. Leach, M. Vervloet, A. Despres, E. Breheret, J. P. Hare, T. J. Dennis, H. W. Kroto, R. Taylor, and D. R. M. Walton, *Chem. Phys.* **160**, 451 (1992).

38. Z. Gasyna, P. N. Schatz, J. P. Hare, T. J. Dennis, H. W. Kroto, R. Taylor, and D. R. M. Walton, *Chem. Phys. Lett.* **183**, 283 (1991).

39. J. W. Arbogast, A. P. Darmanyan, C. S. Foote, Y. Rubin, F. N. Diederich, M. M. Alvarez, S. J. Anz, and R. L. Whetten, *J. Chem. Phys.* **191**, 11 (1991).

40. R. R. Hung and J. J. Grabowski, *J. Phys. Chem.* **1991**, 6073 (1991).

41. Y. Wang, J. M. Holden, A. M. Rao, P. C. Eklund, U. Venkateswaran, E. Eastwood, R. L. Lidberg, G. Dresselhaus, and M. S. Dresselhaus, *Phys. Rev. B* **51**, 4547 (1995).

42. B. Friedman and K. Harigaya, *Phys. Rev. B* **47**, 3975 (1993).

43. W.-Z. Wang, C.-L. Wang, Z.-B. Su, and L. Yu, Phys. *Rev. Lett.* **72**, 3550 (1994).

44. F. Kajzar, S. Etemad, G. L. Baker, and J. Messier, *Solid State Commun.* **63**, 1113 (1987).

45. J. S. Meth, H. Vanherzeel, and Y. Wang, *Chem. Phys. Lett.* **197**, 26 (1992).

46. Z. H. Kafafi, J. R. Lindle, R.G.S. Pong, F. J. Bartoli, L. J. Lingg, and J. Milliken, *Chem. Phys. Lett.* **188**, 492 (1992).

47. K. Harigaya and S. Abe, *Jpn. J. Appl. Phys.* **31**, L887 (1992).

48. K. Harigaya and S. Abe, *J. Lumin.* **60, 61**, 380 (1994).

49. N. Matsuzawa and D. A. Dixon, *J. Phys. Chem.* **96**, 6241 (1992).

50. N. Matsuzawa and D. A. Dixon, *J. Phys. Chem.* **96**, 6872 (1992).

51. A. A. Quong and M. K. Pederson, *Phys. Rev. B* **46**, 12906 (1992).

52. A. Takahashi, H. X. Wang, and S. Mukamel, *Chem. Phys. Lett.* **216**, 394 (1993).

53. Z. Shuai and J. L. Bredas, *Phys. Rev. B* **46**, 16135 (1992).

54. E. Westin and A. Rosen, *Int. J. Modern Phys. B* **6**, 3893 (1992).

55. Y. Wang, G. F. Bertsch, and D. Tomanek, *Z. Phys. D* **25**, 181 (1993).

56. G. E. Whyman and M. M. Mestechkin, *Opt. Commun.* **109**, 410 (1994).

57. X. K. Wang, T. G. Zhang, W. P. Lin, S. Z. Liu, G. K. Wong, M. M. Kappes, R. P. H. Chang, and J. B. Ketterson, *Appl. Phys. Lett.* **60**, 810 (1992).

58. B. Koopmans, A. Anema, H. T. Jonkman, G. A. Sawatzky, and F. van der Woude, *Phys. Rev. B* **48**, 2759 (1993).

59. B. Koopmans, A.-M. Janner, H. T. Jonkman, G. A. Sawatzky, and F. van der Woude, *Phys. Rev. Lett.* **71**, 3569 (1993).

60. S. Qin, W.-M. You, Z.-B. Su, *Phys. Rev. B* **48**, 17562 (1993).

61. Z. Shuai and J. L. Bredas, *Mol. Cryst. Liq. Cryst.* **256**, 801 (1994).

62. W. I. F. David, R. M. Ibberson, J. C. Matthewman, K. Prassides, T. J. S. Dennis, J. P. Hare, H. W. Kroto, R. Taylor, and D. R. M. Walton, *Nature* **353**, 147 (1991).

63. A. S. Davydov, *Theory of Molecular Excitons*, Plenum, New York, 1971.

64. S. Saito and A. Oshiyama, *Phys. Rev. Lett.* **66**, 2637 (1991).

65. W. Y. Ching, M.-Z. Huang, Y.-N. Xu, W. G. Harter, and F. T. Chan, *Phys. Rev. Lett.* **67**, 2045 (1991).

66. N. Troullier and J. L. Martins, *Phys. Rev. B* **46**, 1754 (1992).

67. B.-L. Gu, Y. Maruyama, J.-Z. Yu, K. Ohno, and Y. Kawazoe, *Phys. Rev. B* **49**, 16202 (1994).

68. K. Harigaya and S. Abe, *Mol. Cryst. Liq. Cryst.* **256**, 825 (1994).

69. R. W. Lof, M. A. van Veenendaal, B. Koopmans, H. T. Jonkman, and G. A. Sawatzky, *Phys. Rev. Lett.* **68**, 3924 (1992).

70. T. Tsubo and K. Nasu, *Solid State Commun.* **91**, 907 (1994).

71. T. Tsubo and K. Nasu, *J. Phys. Soc. Jpn*, **63**, 2401 (1994).
72. X. Jiang and Z. Gan, *Phys. Rev. B* **54**, 4504 (1996).
73. E. L. Shirley, L. X. Benedict, and S. G. Louie, *Phys. Rev. B* **54**, 10970 (1996).
74. A. Eilmes, R. W. Munn, B. Pac, and P. Petelenz, *Chem. Phys.* **214**, 341 (1997).
75. K. Pichler, S. Graham, O. M. Gelsen, R. H. Friend, W. J. Romanow, J. P. McCauley Jr., N. Coustel, J. E. Fischer, and A. B. Smith III., *J. Phys.: Condens. Matter* **3**, 9259 (1991).
76. S. Jeglinski, Z. V. Vardeny, D. Moses, V. I. Srdanov, and F. Wudl, *Synth. Metals* **49–50**, 557 (1992).
77. B. C. Hess, D. V. Bowersox, S. H. Mardirosian, L. D. Unterberger, *Chem. Phys. Lett.* **248**, 141 (1996).
78. S. Kazaoui, R. Ross, and N. Minami, SPIE Proceedings, vol. 2530, *Fullerenes and Photonics II*, 1995, p. 41.
79. P. Petelenz, M. Slawik, and B. Pac, *Synth. Metals* **64**, 335 (1994).
80. S. Abe, SPIE Proceedings, vol. 2854, *Fullerenes and Photonics III*, 1996, p. 88.
81. N. Minami, S. Kazaoui, C. Wen, H. J. Byrne, SPIE Proceedings, vol. 2854, *Fullerenes and Photonics III*, 1996, p. 76.
82. R. Eder, A.-M. Janner, and G. A. Sawatzky, *Phys. Rev. B* **53**, 12786 (1996).
83. M. Ueta, H. Kanzaki, K. Kobayashi, Y. Toyozawa, and E. Hanamura, *Excitonic Processes in Solids*, Springer Verlag, Berlin, 1985, Chap. 4.
84. C. Reber, L. Yee, J. McKiernan, J. I. Zink, R. S. Williams, W. M. Tong, D. A. A. Ohlberg, R. L. Whetten, and F. Diederich, *J. Chem. Phys.* **95**, 2127 (1991).
85. M. Matus, H. Kuzmany, and E. Sohmen, *Phys. Rev. Lett.* **68**, 2822 (1992).
86. W. Guss, J. Feldmann, E. O. Göbel, and C. Taliani, *Phys. Rev. Lett.* **72**, 2644 (1994).
87. M. Ichida, M. Sakai, T. Yajima, and A. Nakamura, *Prog. Cryst. Growth Charct.* **33**, 125 (1996).
88. M. Ichida, M. Sakai, T. Yajima, A. Nakamura, and H. Shinohara, *Chem. Phys. Lett.* **271**, 27 (1997).
89. P. M. Pippenger, R. D. Averitt, V. O. Papanyan, P. Nordlander, and N. J. Halas, *J. Phys. Chem.* **100**, 2854 (1996).
90. A. M. Rao, P. Zhou, K.-A. Wang, G. T. Hager, J. M. Holden, Y. Wang, W. T. Lee, X.-X. Bi, P. C. Edlund, D. S. Cornett, M. A. Dincan, and I. J. Amster, *Science* **259**, 955 (1993).
91. H. Yamawaki, M. Yoshida, Y. Kakudate, S. Usuba, H. Yokoi, S. Fujiwara, K. Aoki, R. Ruoff, R. Malhotra, and D. C. Lorents, *J. Phys. Chem.* **97**, 11161 (1993).
92. P. W. Stephens, G. Bortel, G. Fagel, M. Tegze, A. Janossy, S. Pekker, G. Oszanyi, and L. Forro, *Nature* **356**, 383 (1992).
93. D. L. Strout, R. L. Murry, C. Xu, W. C. Eckhoff, G. K. Odom, and G. E. Scuseria, *Chem. Phys. Lett.* **214**, 576 (1993).
94. M. Menon, K. R. Subbaswamy, and M. Sawtarie, *Phys. Rev. B* **49**, 13966 (1994).
95. K. Harigaya, *Phys. Rev. B* **52**, 7968 (1995).
96. K. Harigaya, *Phys. Rev. B* **54**, 12087 (1996).
97. K. Tanaka, Y. Matsuura, Y. Oshima, T. Yamabe, Y. Asai, and M. Tokumoto, *Solid State Commun.* **93**, 163 (1995).
98. K. Tanaka, Y. Matsuura, Y. Oshima, T. Yamabe, H. Kobayashi, and Y. Asai, *Chem. Phys. Lett.* **241**, 149 (1995).

99.   J. Fagerström and S. Stafström, *Phys. Rev. B* **53**, 13150 (1996).
100. Y. Wang, J. M. Holden, A. M. Rao, W.-T. Lee, X. X. Bi, S. L. Ren, G. W. Lehman, G. T. Hager, and P. C. Eklund, *Phys. Rev. B* **45**, 14396 (1992).
101. U. D. Venkateswaran, M. G. Schall, Y. Wang, P. Zhou, and P. C. Eklund, *Solid. State Commun.* **96**, 951 (1995).
102. J. Shumway and S. Satpathy, *Chem. Phys. Lett.* **211**, 595 (1993).
103. See, for example, chap. 3 of Ref. 4.
104. K. Harigaya and S. Abe, *J. Phys. Condens. Matter* **8**, 8057 (1996).
105. M. Fanti, G. Orlandi, and F. Zerbetto, *J. Am. Chem. Soc.* **117**, 6101 (1995).
106. K. Harigaya, *Jpn. J. Appl. Phys.* **36**, L485 (1997).
107. X. Wan, J. Dong, and D. Y. Xing, *J. Phys. B* **30**, 1323 (1997).

2
Ultrafast Dynamics of Electronic and Vibrational Excitations in Fullerenes

Susan L. Dexheimer
Washington State University
Pullman, Washington

1 INTRODUCTION

The ultrafast dynamics of the fullerenes have attracted interest from a number of perspectives. The promise of optical and electronic applications has motivated studies of the fundamental properties of this new class of molecular materials. Time-resolved spectroscopic measurements have been carried out on picosecond and femtosecond time scales to determine the speed of the optical response as well as to understand the physical mechanisms involved in the optical nonlinearities, and fast time-resolved measurements have also contributed to the understanding of the electronic transport properties of these materials.

A wide range of studies has been carried out to investigate ultrafast dynamics in the fullerenes. Most of this work has focused on C_{60}. The relaxation dynamics of photoexcited states have been investigated in the solution phase, as well as in thin solid films and in crystalline material. This chapter will review research performed on femtosecond and picosecond time scales to investigate the fast dynamics of photoexcitations. The review will focus primarily on the electronic dynamics of solid C_{60}, which exhibits an unusual nonexponential response. The dynamics of highly excited electronic states of C_{60} and the femtosecond vibrational response are also discussed.

2   TIME-RESOLVED MEASUREMENT TECHNIQUES

Much of the work on fast dynamics in fullerenes has been carried out using the pump-probe technique. In this well-established method for time-resolved optical studies, a short pump pulse excites the sample and a time-delayed probe pulse measures the resulting change in optical properties as a function of pump-probe delay [1]. The change in optical transmission is usually the detected optical property in the experiments, although time-resolved reflectivity and polarization-rotation measurements are also possible. In the simplest case, the transmitted (or reflected) probe pulse is detected without any wavelength resolution. This method is usually employed in experiments with pulses of relatively narrow spectral bandwidth. Broadband probe pulses, either having the inherent spectral width associated with a transform-limited ultrashort pulse or otherwise produced by nonlinear continuum generation, afford the possibility of measuring the optical response over the full bandwidth of the spectrum of the probe pulse. In these experiments, the probe pulse is resolved in wavelength after it interacts with the sample and yields a measurement of the transient absorption spectrum (that is, the change in the absorption spectrum resulting from excitation by the pump pulse) at a given pump-probe delay time. Another technique that has been employed for investigating fast dynamics in fullerenes is degenerate four-wave mixing (DFWM) [1]. Depending on the experimental arrangement, this technique can be used to investigate either the electronic dephasing (T_2) of the initial photoexcitations, or the population dynamics, in a manner analogous to a single-wavelength pump-probe measurement.

An interesting limiting case of time-resolved optical measurements results when the light pulses are short compared to the periods of the characteristic vibrations of the material. In this case, the vibrational modes coupled to the excited electronic transition are impulsively excited, creating vibrational wavepackets that consist of coherent superpositions of vibrational states [2–4]. These non-stationary vibrational wavepackets, which can be formed on both the ground and excited potential energy surfaces, oscillate at the characteristic vibrational frequencies of the material and are observable as a time-dependent modulation of the transient absorption signal. The time course of this modulation reflects the vibrational dynamics of the material.

3   OPTICAL PROPERTIES OF C$_{60}$

The optical properties of isolated C_{60} molecules are well understood in terms of the molecular orbitals expected for a system of icosahedral symmetry, and the optical absorption bands of C_{60} in solution have been analyzed in detail [5]. In the solid state, C_{60} forms a van der Waals–bonded crystal with weak interactions between the covalently bonded molecules. As a result, the electronic states largely maintain their molecular character. The optical absorption spectrum of solid C_{60} films shows transitions that correspond well to those of C_{60} in solution, although the thin film spectra are red-shifted and broadened as a result of solid-state interactions [6]. The molecular nature of the electronic states in the solid form of C_{60} is also reflected in the optical absorption spectrum of crystalline C_{60}, which is essentially identical to that of the thin solid films despite the much higher degree of long-range order [7].

Much of the ultrafast time-resolved work on C_{60} has investigated the dynamics following photoexcitation into the first excited electronic state. In isolated C_{60} molecules, this corresponds to the transition between the h_u highest occupied molecular orbital (HOMO) to the t_{1u} lowest unoccupied molecular orbital (LUMO). Because of the symmetries of these states, the HOMO-LUMO transition is formally electric dipole–forbidden; however, Herzberg-Teller coupling to infrared-active intramolecular vibrational modes results in a weak transition dipole moment. The lowest absorption band in solid C_{60} is also assigned to this vibronically assisted transition [6].

4 DYNAMICS OF THE PRIMARY PHOTOEXCITATIONS OF C_{60} IN SOLUTION

Time-resolved optical measurements of C_{60} in solution have been recently reviewed [8]. Because of the limited solubility of C_{60}, the molecules are well separated in solution, and the observed ultrafast dynamics reflect intramolecular decay processes rather than the intermolecular processes that dominate the femtosecond and picosecond response in solid C_{60}. For C_{60} in solution, the relaxation following excitation into the HOMO-LUMO optical band is dominated by intersystem crossing from the initially excited singlet state to the lowest lying triplet state. This process proceeds with near unity quantum yield, as a result of the small energy splitting between the S_1 and T_1 states, the low fluorescence rate from the S_1 state, and the large spin-orbit interaction in the molecule [9]. Measurements of the intersystem crossing time have given values ranging from 650 ps [10] to 1.2 ns [11]. The measured rate may depend on the oxygen content of the solution, since O_2 effectively quenches the triplet electronic state [9]. Additional faster components on time scales of ~ 100 ps or less have been attributed to intramolecular energy relaxation (see, for example, Ref. 10).

5 DYNAMICS OF THE PRIMARY PHOTOEXCITATIONS
OF SOLID C$_{60}$

5.1 Experimental Results

Much more rapid and complex dynamics occur in solid C$_{60}$. Pump-probe or time-resolved DFWM studies of thin films of C$_{60}$ on picosecond and femtosecond time scales were undertaken in a number of research groups [12–25], revealing nonexponential relaxation dynamics following photoexcitation into the first excited electronic state. The time-resolved optical response following excitation into the HOMO-LUMO optical band corresponds to an induced absorption, or decrease in sample transmission following excitation, with a nonexponential decay that includes subpicosecond temporal components as well as components that extend to much longer time scales.

The time-resolved response of solid C$_{60}$ was discovered to be strongly dependent on excitation density [14–16], as shown in Figure 1, which presents pump-probe measurements of the time-resolved induced absorption of a C$_{60}$ thin film. These measurements were made using pulses 60 fs in duration centered at 620 nm in the HOMO-LUMO optical band at a series of laser fluences that correspond to initial excitation densities ranging from ~0.5 to ~5% of the molecules

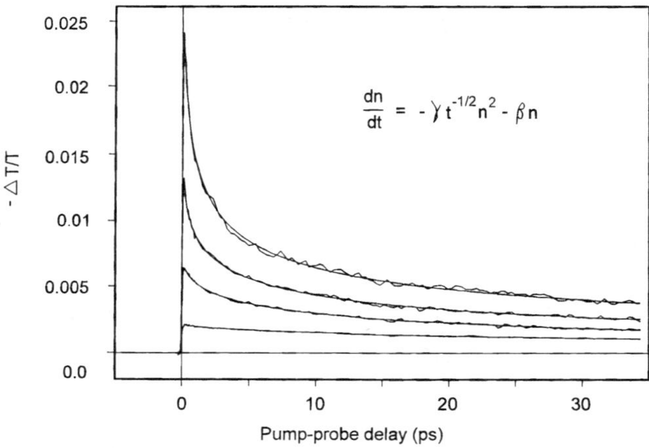

$$\frac{dn}{dt} = -\gamma\, t^{-1/2} n^2 - \beta n$$

Figure 1 Time-resolved negative differential transmittance at 620 nm of a C$_{60}$ thin film following photoexcitation into the HOMO-LUMO optical band with pulses 60 fs in duration. Laser fluences (from top) are 2.3, 1.3, 0.6, and 0.2 mJ/cm^2, corresponding to approximate excitation densities of 8.1×10^{19}, 4.4×10^{19}, 2.1×10^{19}, and 7.0×10^{18} cm^{-3}. The solid lines represent the results of fits to the time-dependent exciton annihilation model presented in Eqs. (6) and (7). (*From Ref. 24.*)

in the solid [24]. At the higher initial excitation densities, the initial decay is rapid, and the overall response is highly nonexponential. As the initial excitation density is decreased, the initial relaxation slows, and the response appears to approach a single exponential decay.

This unusual nonexponential response led to a great deal of speculation as to the physical origin of the relaxation processes, and a wide range of mechanisms was proposed as contributing to the dynamics, including electronic relaxation processes such as carrier trapping processes characteristic of disordered semiconductors, free carrier scattering, intramolecular relaxation processes, excitonic interactions, and lattice relaxation processes. Over the past few years, many studies of the optical and electronic properties of C_{60} and of its fast photoinduced response have contributed to the general consensus that the excitations and their dynamics are molecular in nature, and that the excitation-density-dependent nonexponential relaxation arises from exciton annihilation processes, discussed below, that involve the interaction of localized photoexcitations in the solid. The apparent variability of the experimental results obtained in different laboratories and under different experimental conditions has contributed to the difficulty of interpreting the fast dynamics in solid C_{60}. As is discussed in more detail below, factors that may be responsible for this variability include sample preparation methods, measurement artifacts resulting from high sample optical density and from high laser repetition rates, and photoinduced sample changes.

5.2   Induced Absorption Mechanism

The observation of an induced absorption immediately following photoexcitation into the lowest optical band can be understood in terms of the energy level structure of C_{60}, shown schematically in Figure 2. The lowest optical band corresponds to the weak symmetry-forbidden transition between the bands derived from the h_u highest occupied molecular orbital (HOMO), or S_0 state, and the t_{1u} lowest unoccupied molecular orbital (LUMO), or S_1 state. In a pump-probe experiment, the pump pulse will populate the t_{1u} LUMO state. This state has an allowed transition to a higher-lying excited state that is accessible by absorption of the $\sim$2 eV probe pulse. Since the transition between the two excited states has a higher absorption cross section than the HOMO-LUMO transition, the measured change in transmission following the pump pulse is negative, corresponding to an induced absorption. The pump pulse also removes population from the HOMO level, opening the final state for transitions from the lower-lying g_g and h_g bands, and a contribution to the absorption of the probe pulse from this transition may be possible [14]. Given the presence of an accessible excited state transition, two-photon transitions from the HOMO levels to the excited states at $\sim$4 eV may be expected to be excited by the pump pulse as the excitation intensity is increased. This effect has been observed as a saturation in the initial signal inten-

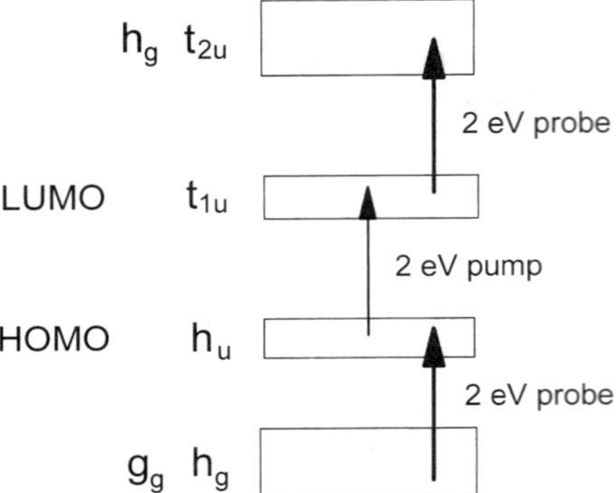

Figure 2 Schematic electronic energy levels in solid C_{60}. (*Adapted from Refs. 14 and 26.*)

sity at high excitation intensity [26, 27]. At moderate excitation densities, the dominant contribution to the time-resolved induced absorption signal is from excitation of the HOMO-LUMO transition, and the observed signal level corresponds largely to the population of the first excited singlet state.

5.3 Exciton Annihilation Mechanism

The observation that the relaxation of the excited state population depends strongly on excitation density suggests that the relaxation dynamics involve interactions between photoexcitations. An excitation-density-dependent process that is well established in molecular crystals is exciton annihilation [28], and a number of authors have proposed that exciton annihilation may be responsible for the nonexponential relaxation dynamics in solid C_{60} [15, 16, 19, 24, 25]. In this case, the term *exciton* refers to a Frenkel-type exciton, or localized molecular excitation, as is expected to occur in solid C_{60}. Exciton annihilation is a nonradiative process involving the interaction of two excitations that results in the net depopulation of the excited state. One possible pathway is

$$S_1 + S_1 \rightarrow S_0 + S_n \tag{1}$$

in which two molecules in the first excited singlet state interact, driving one molecule down to the ground electronic state, and the other to a higher-lying excited state. In general, one or both excitons may be de-excited, and the excess energy goes into electronic and/or vibrational degrees of freedom. Typically, excitons that are promoted to higher-lying excited states rapidly relax back to the S_1 state by nonradiative internal conversion processes. As long as appropriate conservation laws are met, singlet exciton annihilation may also produce triplet states or generate free carriers. Exciton annihilation may also take place between pairs of triplet-state excitons or between singlet and triplet excitons.

In general, the process of exciton annihilation involves both the transport of the excitons within the medium and the interaction that gives rise to annihilation. A rigorous theory for exciton annihilation must include both the spatial and temporal dependence of the excitation density and the spatial dependence of the annihilation interaction. Theoretical frameworks for treating the problem of exciton annihilation have been developed (for example, Refs. 29, 30), and various limiting cases have been identified that depend on the physical nature of the transport processes, the relative magnitudes of the rates for transport, annihilation, and first-order population decay, and the physical structure of the medium. For excitations in systems such as solid C_{60}, transport would be expected to occur by an incoherent hopping mechanism, so that it can be treated as a diffusion process. Limiting cases of exciton annihilation involving diffusive transport are outlined below.

A physical mechanism that is responsible for excitation transport in a wide range of molecular crystals and that can, in some cases, be responsible for the annihilation interaction is the Förster dipole-dipole interaction [28, 31, 32]. In Förster energy transfer, the electronic excitation is transferred from a donor to an acceptor with a rate:

$$k \propto \tau^{-1}\kappa^2 \left(\frac{R_0}{R}\right)^6 \int d\nu \ \nu^{-4} f_d(\nu) \ \varepsilon_a(\nu) \tag{2}$$

where τ is the donor lifetime in the absence of the transfer process, κ reflects the mutual orientation of the transition dipoles of the donor and acceptor, R is the distance between the donor and acceptor, and R_0 is a characteristic length scale for the process. The integral in the expression for the Förster energy transfer rate reflects the overlap between the fluorescence spectrum of the donor, $f_d(\nu)$, and the absorption spectrum of the acceptor, $\varepsilon_a(\nu)$. This nonradiative process results from the interaction between the induced dipoles that correspond to the transition moments of the donor and acceptor molecules. Förster energy transfer can result in spatial diffusion of excitons when the electronic excitation is transferred from a ''donor'' molecule in the S_1 state to an ''acceptor'' molecule in the S_0 state. Since the transfer rate drops rapidly with distance, the dominant contribution to exciton diffusion in a molecular crystal is from transfer between nearest neighbors.

The Förster process can also give rise to exciton annihilation when the dipole interaction takes place between two excitons. In this case, one molecule in the S_1 state is driven down to the ground S_0 state, and the other molecule is driven from the S_1 state up to a higher-lying electronic state, a process which was shown in the example in Eq. (1). The relevant spectra in the overlap integral in the expression for the rate are the $S_1 \to S_0$ emission spectrum and the $S_1 \to S_n$ excited-state absorption spectrum. Annihilation processes consistent with the Förster mechanism have been observed in a number of molecular systems [33, 34]. The rate of Förster annihilation follows the $1/R^6$ distance dependence characteristic of dipole-dipole interactions, which drops rapidly with increasing inter-exciton separation.

Regardless of the details of the physical interaction processes giving rise to the dynamics, the following limiting cases for exciton annihilation can be identified:

Exciton Annihilation in the Homogeneous Binary Collision Model

The most commonly made assumption in modeling exciton annihilation dynamics is that the density of excitations remains spatially homogeneous, so that the excited-state population density (in three dimensions) can be modeled by the rate equation

$$\frac{dn}{dt} = -\gamma n^2 - \beta n \tag{3}$$

where the annihilation rate is described by a constant γ multiplied by the square of an excitation density that is assumed to be a function only of time. A term linear in density has been added to account for a monomolecular depopulation process, such as radiative decay. This model requires that the excitations undergo sufficiently rapid diffusion to maintain a spatially homogeneous excitation density throughout the time course of the population decay, or, in other words, that the diffusion rate is much greater than the annihilation rate. Pairs of excitons are assumed to annihilate only when they diffuse into sufficiently close proximity, and the annihilation rate constant γ depends directly on the diffusion rate. Equation (3) has been successfully applied to measurements of exciton dynamics in numerous molecular systems; however, it is valid only in the limit of fast diffusion.

Exciton Annihilation in Systems of Reduced Dimensionality

The dimensionality of the system directly influences the exciton annihilation dynamics by virtue of its effect on diffusive transport. When the migration of the excitons is confined to a dimension d (where $1 \leq d \leq 2$ to include the possibility of fractal dimensionality [35]), and diffusion is assumed to be rapid, as in the above model, the exciton decay can be modeled as

$$\frac{dn}{dt} = -\gamma t^{-h} n^2 - \beta n \tag{4}$$

where the exponent h is related to the dimensionality d by

$$h = 1 - \frac{d}{2} \tag{5}$$

This treatment gives rise to an effective time-dependent annihilation rate γt^{-h}. Photoinduced dynamics in quasi-one-dimensional molecular solids, including conjugated polymers and linear chain complexes, have been successfully modeled using exciton annihilation rates proportional to $t^{-1/2}$ [36, 37].

Exciton Annihilation in Confined Domains

Another limiting case results when the diffusion of the excitons is restricted to domains of finite size, with rapid diffusion of excitons within a domain but with no interaction between excitons in different domains. This limiting case was initially developed to model exciton dynamics in photosynthetic antenna complexes [38, 39], and has also been applied to molecular aggregates [40, 41]. By using a master equation approach and by assuming Poisson statistics for the number of excitations initially present in a given domain, a functional form consistent with an effective time-dependent annihilation rate can be derived.

Exciton Annihilation in the Limit of Static Excitations

An effective annihilation rate with a characteristic time dependence also results from the limit that the exciton diffusion rate is sufficiently slow that the excitations effectively remain fixed in location throughout the time scale of the population decay. If the annihilation process proceeds by the distance-dependent Förster dipole-dipole interaction described above, the effective rate can be determined by integrating the Förster rates over a distribution of randomly distributed, but spatially fixed excitations. In a three-dimensional system, the resulting integral gives rise to an effective rate proportional to $t^{-1/2}$ [42], and the population decay can be modeled with the rate equation:

$$\frac{dn}{dt} = -\gamma t^{-1/2} n^2 - \beta n \tag{6}$$

which has the solution:

$$n(t) = n(0) \exp\left(-\beta t\right) \left\{ 1 + \frac{\pi^{1/2} \gamma n(0)}{\beta^{1/2}} \, \mathrm{erf}\left[(\beta t)^{1/2}\right] \right\}^{-1} \tag{7}$$

where erf is the Gaussian error function. In this limit of exciton annihilation, the distribution of rates will reflect the strong distance dependence of the Förster process, and nearby exciton pairs will annihilate rapidly, while pairs of excitons with progressively larger separations will give rise to progressively slower interaction rates. Since the excitons remain fixed throughout the population decay process, the spatial distribution of excitons will depend on the time following the initial excitation, and the process is inherently non-Markovian. This case for exciton annihilation dynamics has been used to describe the rapid photoinduced dynamics in phthalocyanine-based systems [43, 44].

5.4 Experimental Considerations for Quantitative Data Interpretation

Since the time-resolved induced absorption in solid C_{60} corresponds to the time course of the excited state population density, it should be possible to test the validity of the models for exciton annihilation described above by fitting the measured pump-probe signals to the appropriate rate equations. However, a number of experimental effects, which are outlined below, can distort the measured response and complicate the interpretation of the response.

Optical Density

When the relaxation dynamics depend strongly on excitation density, as is the case in exciton annihilation, the measured temporal response in a pump-probe or DFWM transmission experiment will be distorted unless the sample is optically thin. If the sample thickness is comparable to or greater than the absorption depth for the transition excited by the pump pulse, the excitation density generated by the pump pulse, and correspondingly, the resulting dynamics, will vary throughout the depth of the sample. Since the probe beam traverses the full depth of the sample, the measured response will be an average of the dynamics occurring over the range of excitation densities present in the sample. In general, this effect will give rise to a slower observed decay curve, as slow components from the less strongly excited back portion of the sample are averaged in with the more rapid components from the front. The observed averaged response will also, in general, deviate from the simple functional forms expected from the rate equations shown above. For the lowest optical band in solid C_{60}, the absorption depth is less than 1 μm, and the sample thickness must be substantially less than this to avoid a distorted measurement.

Repetition Rate

If the photoinduced dynamics of a material result in the formation of long-lived excitations, the measured response may be distorted if the repetition rate of the pairs of pump and probe pulses (in other words, the repetition rate of the laser system used for the measurements) is faster than or comparable to the decay rate of the long-lived states. If successive pump pulses arrive before the previously generated excitations return to the ground state, the pump-probe signal may contain contributions from multiply excited states. In addition, if the dynamics involve interactions between excitations, the presence of a background density of excitations remaining from previous pump pulses can influence the measured response.

In C_{60}, the initially excited singlet state can undergo intersystem crossing to a long-lived triplet electronic state, as seen in studies of the dynamics of C_{60} in solution. If the repetition rate of the laser system is faster than or comparable to the rate for the triplet state to return to the ground state, a substantial background population of triplet-state excitations will remain in the sample. The absorption spectrum of the C_{60} triplet state has been measured in longer time scale pump-probe experiments [23, 45], and has a small component at the wavelengths used in the HOMO-LUMO pump-probe measurements. As a result, a residual population of triplet-state excitations would be expected to produce a background signal in a time-resolved measurement that would appear as a dc signal level at negative pump-probe delay times, and may also result in additional fast components superimposed on the singlet-state response that would correspond to relaxation of doubly excited triplet states. Even more problematic for the interpretation of the nonexponential response in C_{60} is the potential for the interaction of newly excited singlet-state excitations with the residual population of triplet-state excitations. Singlet-triplet annihilation processes are well documented [28], and the similarity of the optical spectra of the singlet and triplet states in C_{60} would act to enhance the probability of this interaction.

The lifetime of the triplet state in C_{60} is most likely dependent on the oxygen content of the samples, given its strong quenching activity [9]. Measurements reported for the triplet-state lifetime for C_{60} in oxygen-free solutions are ~ 40 µs [9, 46]. The laser systems used for the HOMO-LUMO pump-probe measurements on solid C_{60} typically have repetition rates of ~ 100 MHz, corresponding to a 10 ns time between pulses, or ~ 8 kHz, corresponding to 125 µs between pulses. Since the triplet states are expected to be formed with high quantum yield (at least for excitons that do not undergo annihilation), as has been seen to be the case for C_{60} in solution, a substantial steady-state population of triplet-state excitations would be expected to be present in the high repetition rate experiments, where the microsecond time-scale triplet state lifetime is far longer than the ~ 10 ns time between excitation pulses. Since the optical absorption of the triplet state is considerably lower than that of the singlet state at the 600–620 nm wavelengths used in these experiments, even a small dc background signal would correspond to a large triplet-state population.

Pulse Duration

A final, and more obvious, technical consideration in a time-resolved optical measurement is the duration of the optical pulses relative to the time scale of the dynamics. Pulses longer than the fast components of the dynamics will average over the initial part of the response, and will make measurements involving excitation density–dependent dynamics difficult to interpret quantitatively, since the excitation density will change within the duration of the pump pulse. HOMO-LUMO pump-probe measurements on C_{60} thin films have revealed excitation-density-dependent relaxation components on subpicosecond time scales [24], indicating that pulses ~ 100 fs or less in duration are required to observe the full ultrafast electronic response.

Photopolymerization

Solid C_{60} has been observed to undergo a light-induced polymerization reaction, in which covalent bonds form between adjacent C_{60} molecules [47]. Measurements of the optical properties of photopolymerized C_{60} indicate that the electronic transitions still correspond closely to those of monomeric C_{60}, with some broadening of the peaks in the absorption spectra. The broadening has been attributed to the lifting of the degeneracies of the electronic energy levels due to the lower symmetry, as well as the presence of a distribution of oligomeric units of varying length and connectivity [6].

The extent of photopolymerization in solid C_{60} depends on temperature. The polymerization reaction is reversible upon heating, and for illumination at a given temperature, there will be some equilibrium between the forward and reverse processes [48]. In addition, the photopolymerization reaction is strongly inhibited at temperatures below ~ 260 K. This corresponds to the transition temperature for the order-disorder phase transition in solid C_{60}, in which the free rotation of the molecules begins to freeze out. The absence of photopolymerization below the phase transition temperature has been attributed to the lowered probability for adjacent molecules to achieve the appropriate orientations for bond formation [49]. Photopolymerization is also inhibited in C_{60} that has been exposed to oxygen. Atmospheric O_2 has been observed to diffuse into the interstitial spaces in solid C_{60}, and is expected to interfere with the rotational motion of the C_{60} molecules even at room temperature, leading a similar inhibition mechanism as that proposed for low temperatures [50].

The presence of photopolymerization can be most easily detected by Raman spectroscopy. The high-frequency A_g pentagonal pinch vibrational mode is particularly sensitive to polymerization. In monomeric C_{60}, this mode gives an intense peak at 1467 cm^{-1}. Upon photopolymerization, the peak broadens and shifts to ~ 1458 cm^{-1} while also losing some intensity [47].

The possibility that photopolymerization may affect the measured response in a time-resolved optical measurement has been raised in recent work. The time-resolved DFWM response of a C_{60} film that had been polymerized by previous intense light exposure was reported to differ from that of a "pristine" film [25], but, unfortunately, the comparison may have been complicated by optical density effects, given the different film thicknesses for the two samples. Fleischer et al. [51] reported changes in the time-resolved response of a C_{60} film dependent on the total illumination time during the course of a series of pump-probe measurements, and attributed the changes to rapid photopolymerization by the excitation pulses used in the measurements. Some authors have reported irreversible changes in the measured time-resolved response under very high-power excitation [14, 27], and these effects may be related to polymerization. However, other measurements argue against polymerization effects under less extreme conditions in time-resolved experiments on C_{60}. The nonexponential dynamics in C_{60} have been reported by a number of groups [12, 18, 24] to be identical from room temperature to temperatures well below the order-disorder phase transition. Since photopolymerization does not take place below the phase transition temperature, the observed nonexponential dynamics cannot be attributed to material in the polymerized phase. Vibrationally impulsive excitation measurements directly show the vibrational response along with the electronic relaxation processes, and the frequency of the oscillations due to the vibrational response provides a clear indication of the state of the sample. Pump-probe measurements in the vibrationally impulsive limit are discussed below, and these measurements clearly show a vibrational modulation frequency of 1467 cm^{-1}, consistent with the presence of monomeric rather than polymeric C_{60} [16].

5.5 Modeling of the Time-Resolved Dynamics

With the experimental considerations discussed above in mind, the time-resolved response of C_{60} can be compared to the models for exciton annihilation. The measurements displayed in Figure 1 were made using a thin film of low optical density ($\sim$0.14) to avoid a significant variation in the excitation density throughout the depth of the sample, and the sample was excited with pulses 60 fs in duration at an 8 kHz pulse repetition rate to allow relaxation of the excitations between laser shots. The signals are presented as the negative of the normalized differential transmittance ($-\Delta T/T$), where ΔT is the change in transmission due to excitation by the pump pulse, and T is the transmission in the absence of the pump pulse. In the small signal limit, this normalized signal is proportional to the excited-state population density $n(t)$, which is the quantity modeled in the rate equations.

A striking property of the set of excitation density dependent decay curves is their non-Markovian behavior. If, for example, the initial slope (dn/dt) of a low fluence decay curve is compared to the slope at a point corresponding to the same signal level, but farther out in time on a higher fluence decay curve, it is clear that the slopes of the curves are unequal. This demonstrates that the dynamics are not simply a function of the total excited state population density, but rather that they depend on the history following excitation. The appropriate model for the dynamics in solid C_{60} must reflect the non-Markovian nature of the relaxation. The first case discussed above for exciton annihilation, in which the excitons are assumed to undergo rapid diffusion, clearly cannot correctly model the data, because, in Eq. (3), dn/dt is a function only of the total excitation density n. This conclusion was verified by a poor fit of Eq. (3) to the set of data traces.

The data in Figure 1 were successfully fit using Eq. (7), the solution to the rate equation with an effective time-dependent annihilation rate $\gamma t^{-1/2}$ [24]. As discussed above, this functional form for the time-dependent rate may correspond to (1) exciton annihilation confined to one dimension, (2) exciton annihilation within noninteracting domains of limited size, or (3) the Förster annihilation mechanism in the limit of static excitations. The first possibility may be excluded based on the structure of the material. The issue of domains within the sample may be addressed by comparing measurements on thin-film and crystalline samples. Thin films prepared by thermal sublimation typically show poor long-range order, with a lattice coherence length on the order of only a few fcc unit cells [52], and may be expected to have a large number of grain boundaries within the excited volume in a time-resolved optical measurement. However, the time-resolved response in thin films was found to be identical to that observed in crystalline C_{60} samples having a substantially higher degree of long-range order [24], indicating that the presence of grain boundaries or their associated domains do not affect the ultrafast dynamics.

A Förster annihilation mechanism in the limit of static excitations is consistent with the understanding of the electronic transitions in C_{60}. Exciton diffusion in solid C_{60} would be expected to be slow: The rate for the Förster energy transfer process responsible for exciton diffusion depends on the intensity of the absorption and emission bands involving the S_0 and S_1 electronic states, and in C_{60}, this transition is only weakly allowed. In contrast, the annihilation rate depends on the overlap of the S_1 to S_0 emission band with the S_1 to S_n absorption band, which is strongly allowed. As a result, the annihilation rate, even for relatively large interexciton separations, would be expected to be significantly higher than the diffusion rate, and the nonexponential dynamics in solid C_{60} can be understood in terms of Förster exciton annihilation in the limit of slow diffusion.

The rate equation (6) also includes a rate β that reflects a monomolecular decay process. Fits to Eq. (7), as well as fits to data acquired over longer time ranges at low excitation density, give a value $\beta \approx 10^{10}$ s^{-1}. This linear decay time of approximately 100 ps is of the same order as the lifetime of the initially excited singlet state for C_{60} in solution, where the lifetime is determined by intersystem crossing to the triplet state. This process would also be expected to take place in solid C_{60}, and is consistent with the time scale of the observed linear rate.

It is important to note that, in order to draw a definitive conclusion about the validity of a particular rate equation model for an excitation-density-dependent process such as exciton annihilation, a single consistent set of fitting parameters must describe the decay curves for data corresponding to a range of initial excitation densities. In particular, an important test is whether the model successfully predicts both the shape of the decay curve and the initial signal intensity, or $n(0)$, for each trace. This requires that the data be appropriately normalized (for example, $\Delta T/T$) so that the magnitudes of the decay traces can be compared. The full set of data in Figure 1 can be fit with the parameters $\gamma = (7 \pm 1) \times 10^{-15}$ cm^3 s$^{-1/2}$ and $\beta = 10^{10}$ s^{-1}, demonstrating that the time-dependent exciton annihilation model accurately reflects the dynamics [24].

The interpretation of the nonexponential dynamics in solid C_{60} in terms of exciton annihilation is consistent with a number of additional experimental observations. The nonexponential relaxation dynamics do not show a strong dependence on the probe wavelength, indicating that the same relaxation process dominates throughout the optical band [24]. In contrast, energy relaxation within the excited electronic state would be expected to show a strong dependence on probe wavelength. The nonexponential relaxation dynamics were also observed to be largely independent of temperature [12, 24], arguing against a major role for thermally activated processes. The Förster mechanism typically shows only a weak dependence on temperature through the overlap of the absorption and emission spectra, which may shift slightly and decrease in width at low temperature. The assignment of the nonexponential ultrafast dynamics to exciton annihilation in C_{60} is also supported by observation of the same relaxation dynamics in thin film and crystalline samples [24], which rules out structural disorder as the origin of the nonexponential behavior.

The nonexponential dynamics in C_{60} have also been discussed in terms of the stretched exponential function $n(t) = \mathrm{n}(0)\,\exp\left[-(t/\tau)^{\beta}\right]$ that has been previously invoked to describe the distribution of relaxation rates found in disordered systems such as amorphous semiconductors [12, 18]. However, modeling of excitation-density-dependent dynamics in C_{60} does not result in a consistent set of fitting parameters, and as the initial excitation density increases, the resulting values of τ become unphysically small [7]. In disordered semiconductors, the stretched exponential functions reflect the relaxation of photoexcited carriers as they become trapped in localized sites that have a distribution of energies [53]. Low-energy sites may be present in C_{60} thin films, for example, at grain boundaries, and they have been proposed to contribute to observed photoluminescence properties [6]; however, as the comparison of thin film and crystalline C_{60} dynamics and the success of the exciton annihilation model indicate, this type of trapping process does not appear to contribute significantly to the nonexponential dynamics observed on femtosecond and picosecond time scales.

6 DYNAMICS OF HIGHLY EXCITED STATES OF C_{60}

Additional studies have probed the dynamics of C_{60} upon excitation to higher-lying electronic states [16]. Figure 3 shows the spectral evolution on a femtosecond time scale following excitation into the dipole-allowed transition above the HOMO-LUMO transition. Measurements of the time-resolved absorption spectrum of a C_{60} thin film were made at a series of pump-probe delay times using a 40 fs pump pulse centered at 500 nm and a broadband 10 fs probe pulse. The measurements, which are presented as a positive differential transmittance, show an induced transmittance in the blue part of the spectrum, in contrast to the induced absorption observed for excitation of the HOMO-LUMO transition. This induced transmittance rapidly decays on a time scale of ~ 100 fs, while at longer wavelengths, an induced absorption grows in and then decays on a longer time scale.

The observation of an induced transmittance is consistent simply with bleaching of the optical transition by transferring population from the ground state to the final state of the transition. Intramolecular energy relaxation, in which the highly excited state undergoes internal conversion to a lower-lying excited electronic state, would be expected to contribute to the fast decay observed at short wavelengths. The initial decay of the induced transmittance parallels the growth of the induced absorption at longer wavelengths. This induced absorption is observed in the same wavelength range as the induced absorption that results

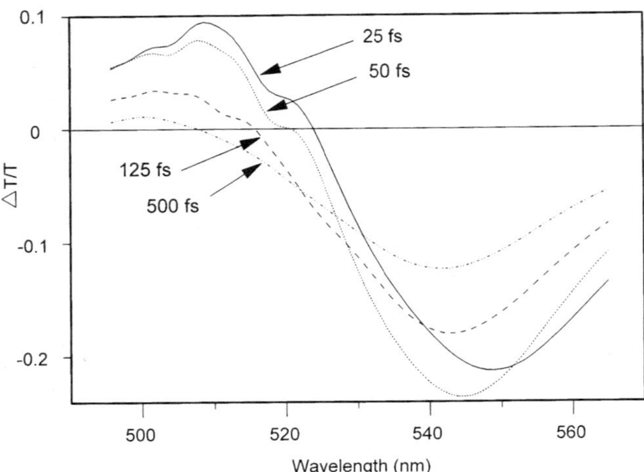

Figure 3 Time-resolved spectral measurements of the differential transmittance of a C_{60} thin film following excitation with pulses 40 fs in duration at 500 nm. (*From Ref. 16.*)

from population of the LUMO level, suggesting that the initially excited molecules relax to this state, which would then be expected to give rise to the dynamics described earlier for excitation directly into this level. The observed relaxation dynamics are highly nonexponential throughout the detected wavelength range, and depend in detail on the probe wavelength, suggesting that exciton annihilation occurs concomitantly with rapid intramolecular energy relaxation. The possibility that intermolecular charge-transfer excitons, in which charge is separated between molecules in adjacent sites, may occur in the energy range excited in this experiment has been discussed [54–56], and, if present, these would be reflected in the short-lived decay components.

7   ULTRAFAST VIBRATIONAL DYNAMICS IN C_{60}

Experimental studies in which C_{60} is excited with pulses short compared to its characteristic vibrational modes reveal its ultrafast coherent vibrational response. Figure 4 shows the time-resolved induced absorption of a thin film of C_{60} following excitation with pulses 12 fs in duration centered at 620 nm, in the HOMO-LUMO optical transition. As can be seen in the figure, the response is modulated by regular oscillations that persist on a picosecond time scale.

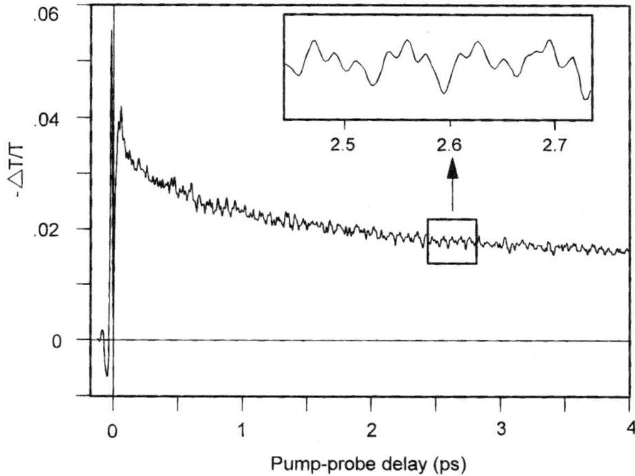

Figure 4   Time-resolved negative differential transmittance of a C_{60} thin film following impulsive excitation with pulses 12 fs in duration centered at 620 nm. The transmitted component at 580 nm was detected. (*From Ref. 16.*)

The frequency components of the oscillatory response can be resolved by Fourier analysis. Figure 5 shows the Fourier power spectrum of the oscillatory part of the time-resolved signal. Since the excitation pulses are resonant with an optical transition, the fully symmetric A_g Raman active modes should contribute to the impulsively excited vibrational response, and the peaks in the Fourier power spectrum at 495 cm^{-1} and 1467 cm^{-1} correspond to the fully symmetric intramolecular modes of C_{60} identified by Raman spectroscopy [57].

A number of mechanisms for the generation of coherent phonons in condensed media have been identified [2–4, 58]. In C_{60}, the excited vibrational modes are clearly intramolecular in nature, and the molecular wavepacket picture is appropriate for interpreting the response [3]. Resonant impulsive excitation of a molecule in general produces nonstationary vibrational states, or vibrational wavepackets, on the potential energy surfaces of both the excited and ground electronic states. The dependence of the oscillatory response on probe wavelength reflects the origin of the coherent vibrational response [59, 60], and in C_{60}, the enhanced amplitude of the high-frequency mode in the wings of the pulse spectrum and its near absence at the central probe wavelength suggests that the observed oscillatory response results from vibrational coherence in the ground electronic state created by a resonant impulsive Raman process. This assignment is consistent with the observed time dependence of the amplitude of the oscillations, which remain strong throughout the signal trace, rather than damping with the decay of the excited-state population. The absence of an oscillatory response

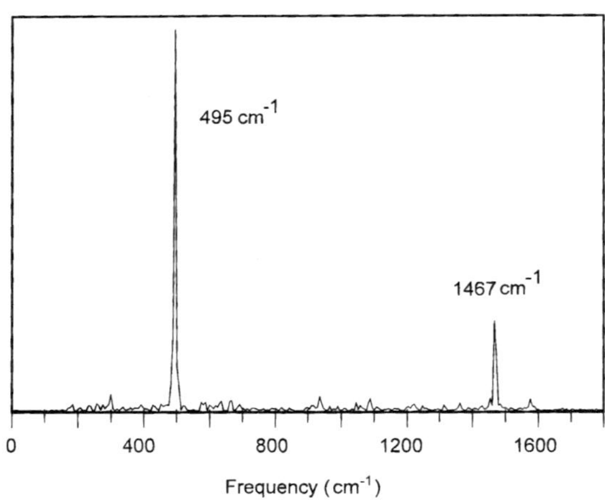

Figure 5   Fourier power spectrum of the oscillatory component of the time-resolved response presented in Figure 4, showing the A_g radial breathing mode at 495 cm^{-1} and the A_g pentagonal pinching mode at 1467 cm^{-1}. (*From Ref. 16.*)

from the excited-state potential energy surface suggests the presence of rapid vibrational dephasing processes in the excited electronic state.

It is particularly interesting to note that, in this measurement on C_{60}, the simultaneous detection of the excited-state population dynamics along with the coherent vibrational response allows the state of the sample to be monitored during the course of the time-resolved measurement. The vibrational frequencies of C_{60} change as a result of photopolymerization, a process that has been a serious concern in the application of optical techniques to study this material. The observation of modulation frequencies characteristic of monomeric, rather than polymerized, C_{60} clearly indicate the state of the sample.

8 CONCLUDING REMARKS

Time-resolved optical studies of C_{60} on femtosecond and picosecond time scales have revealed complex relaxation dynamics. The unusual nonexponential response of solid C_{60} has been successfully interpreted in terms of exciton annihilation and intramolecular relaxation processes, reflecting the molecular nature of this material. Issues that, as yet, remain unresolved include those that involve solid-state effects, such as the possible role of charge-transfer excitons, as well as the detailed nature of the effects of photoinduced changes in the material.

ACKNOWLEDGMENTS

The author would like to thank her previous collaborators on her experimental work on C_{60}: C. V. Shank, R. W. Schoenlein, and D. M. Mittleman on the femtosecond time-resolved measurements, and A. Zettl, W. A. Vareka, and X.-D. Xiang for sample preparation.

REFERENCES

1. J.-C. Diels and W. Rudolph, *Ultrashort Laser Pulse Phenomena: Fundamentals, Techniques, and Applications on a Femtosecond Time Scale.* Academic Press, San Diego, 1996.
2. Y.-X. Yan, E. B. Gamble, and K. A. Nelson, *J. Chem. Phys.* **83**, 5391–5399, 1985.
3. W. T. Pollard, S.-Y. Lee, and R. A. Mathies, *J. Chem. Phys.* **92**, 4012–4028, 1990.
4. H. J. Zeiger, J. Vidal, T. K. Cheng, E. P. Ippen, G. Dresselhaus, and M. S. Dresselhaus, *Phys. Rev. B* **45**, 768–778, 1992.
5. S. Leach, M. Vervloet, A. Despres, E. Breheret, J. P. Hare, T. J. Dennis, H. W. Kroto, R. Taylor, and D.R.M. Walton, *Chem. Phys.* **160**, 451–466, 1992.

6. Y. Wang, J. M. Holden, A. M. Rao, P. C. Eklund, U. D. Vekateswaran, D. Eastwood, R. L. Lidberg, G. Dresselhaus, and M. S. Dresselhaus, *Phys. Rev. B* **51**, 4547–4556, 1995.

7. S. L. Dexheimer, unpublished results.

8. M. S. Dresselhaus, G. Dresselhaus, and P. C. Eklund, *Science of Fullerenes and Carbon Nanotubes*, Academic Press, New York, 1996.

9. J. W. Arbogast, A. P. Darmanyan, C. S. Foote, Y. Rubin, F. N. Diederich, M. M. Alvarez, S. J. Anz, and R. L. Whetten, *J. Phys. Chem.* **95**, 11–12, 1991.

10. R. J. Sension, C. M. Phillips, A. Z. Szarka, W. J. Romanow, A. R. McGhie, J. P. McCauley Jr., A. B. Smith III, and R. M. Hochstrasser, *J. Phys. Chem.* **95**, 6075–6078, 1991.

11. T. W. Ebbesen, K. Tanigaki, and S. Kuroshima, *Chem. Phys. Lett.* **181**, 501–504, 1991.

12. R. A. Cheville and N. J. Halas, *Phys. Rev. B* **45**, 4548–4550, 1992.

13. M. J. Rosker, H. O. Marcy, T. Y. Chang, J. T. Khoury, K. Hansen, and R. L. Whetten, Chem. *Phys. Lett.* **196**, 427–432, 1992.

14. S. D. Brorson, M. K. Kelly, U. Wenschuh, R. Buhleier, and J. Kuhl, *Phys. Rev. B* **46**, 7329–7332, 1992.

15. S. R. Flom, R. G. S. Pong, F. J. Bartoli, and Z. H. Kafafi, *Phys. Rev. B* **46**, 15 598–15 601, 1992.

16. S. L. Dexheimer, D. M. Mittleman, R. W. Schoenlein, X.-D. Xiang. W. A. Vareka, A. Zettl, and C. V. Shank, Ultrafast Pulse Generation and Spectroscopy, *SPIE* v. **1861**, 328–332, 1993.

17. S. B. Fleischer, E. P. Ippen, G. Dresselhaus, M. S. Dresselhaus, A. M. Rao, P. Zhou, and P. C. Eklund, *Appl. Phys. Lett.* **62**, 3241–3243, 1993.

18. T. Juhasz, X. H. Hu, C. Suarez, W. E. Bron, E. Maiken, and P. Taborek, *Phys. Rev. B* **48**, 4929–4932, 1993.

19. T. N. Thomas, R. A. Taylor, J. F. Ryan, D. Mihailovic, and R. Zamboni, *Europhys. Lett.* **25**, 403–408, 1994.

20. T. W. Ebbesen, Y. Mochizuki, K. Tanigaki, and H. Hiura, *Europhys. Lett.* **25**, 503–508, 1994.

21. B. C. Hess, E. A. Forgy, S. Frolov, D. D. Dick, and Z. V. Vardeny, *Phys. Rev. B* **50**, 4871–4874, 1994.

22. V. M. Farztdinov, Y. E. Lozovik, Y. A. Matveets, A. G. Stepanov, and V. S. Letokhov, *J. Phys. Chem.* **98**, 3290–3294, 1994.

23. D. Dick, X. Wei, S. Jeglinski, R. E. Benner, Z. V. Vardeny, D. Moses, V. I. Srdanov, and F. Wudl, *Phys. Rev. Lett.* **73**, 2760–2763, 1994.

24. S. L. Dexheimer, W. A. Vareka, D. Mittleman, A. Zettl, and C. V. Shank, *Chem. Phys. Lett.* **235**, 552–557, 1995.

25. S. R. Flom, F. J. Bartoli, H. W. Sarkas, C. D. Merritt, and Z. H. Kafafi, *Phys. Rev. B* **51**, 11 376–11 381, 1995.

26. V. M. Farztdinov, Y. E. Lozovik, and V. S. Letokhov, *Chem. Phys. Lett.* **224**, 493–500, 1994.

27. I. V. Bezel, S. V. Chekalin, Y. A. Matveets, A. G. Stepanov, A. P. Yartsev, and V. S. Letokhov, *Chem. Phys. Lett.* **218**, 475–478, 1994.

28. J. B. Birks, *Photophysics of Aromatic Molecules*, Wiley-Interscience, New York, 1970.
29. A. Suna, *Phys. Rev. B* **1**, 1716–1738, 1970.
30. V. M. Kenkre, *Phys. Rev. B* **22**, 2089–2097, 1980.
31. T. Forster, Delocalized excitation and excitation transfer, in *Modern Quantum Chemistry, Part III: Action of Light and Organic Crystals* (O. Sinanoglu, ed.), Academic Press, New York, 1965, pp. 93–137.
32. R. C. Powell and Z. G. Soos, *J. Lumin.* **11**, 1–45, 1975.
33. C. E. Swenberg and N. E. Geacintov, Exciton interactions in organic solids, in *Organic Molecular Photophysics* (J. B. Birks, ed.), Wiley-Interscience: New York, 1973.
34. K. Kamioka and S. E. Webber, *Chem. Phys. Lett.* **133**, 353–357, 1987.
35. P. W. Klymko and R. Kopelman, *J. Phys. Chem.* **87**, 4565–4567, 1983.
36. R. L. Blakley, C. E. Martinez, M. F. Herman, and G. L. McPherson, *Chem. Phys.* **146**, 373–380, 1990.
37. R. Kroon, H. Fleurent, and R. Sprik, *Phys. Rev. E* **47**, 2462–2472, 1993.
38. G. Paillotin, C. E. Swenberg, J. Breton, and N. E. Geacintov, *Biophys. J.* **25**, 513–534, 1979.
39. D. Gulen, B. P. Wittmershaus, and R. S. Knox, *Biophys. J.* **49**, 469–477, 1986.
40. V. Sundstrom, T. Gillbro, R. A. Gadonas, and A. Piskarskas, *J. Chem. Phys.* **89**, 2754–2762, 1988.
41. P. J. Reid, D. A. Higgins, and P. F. Barbara, *J. Phys. Chem.* **100**, 3892–3899, 1996.
42. T. Forster, *Z. Naturforschung* **4a**, 321–327, 1949.
43. B. I. Greene and R. R. Millard, *Phys. Rev. Lett.* **55**, 1331–1334, 1985.
44. A. Terasaki, M. Hosoda, T. Wada, A. Yamada, H. Sasabe, A. F. Garito, and T. Kobayashi, *Nonlinear Optics* **3**, 161–168, 1992.
45. K. Pichler, S. Graham, O. M. Gelsen, R. H. Friend, W. J. Romanow, J. P. McCauley, N. Coustel, J. E. Fischer, and A. B. Smith III, *J. Phys.: Condens. Matter* **3**, 9259–9270, 1991.
46. Y. Kajii, T. Nakagawa, S. Suzki, Y. Achiba, K. Obi, and K. Shibuya, *Chem. Phys. Lett.* **181**, 100–104, 1991.
47. A. M. Rao, P. Zhou, K.-A. Wang, G. T. Hager, J. M. Holden, Y. Wang, W.-T. Lee, X.-X. Bi, P. C. Eklund, D. S. Cornett, M. A. Duncan, and I. J. Amster, *Science* **259**, 955–957, 1993.
48. Y. Wang, J. M. Holden, X.-X. Bi, and P. C. Eklund, *Chem. Phys. Lett.* **217**, 413–417, 1994.
49. P. Zhou, Z.-H. Dong, A. M. Rao, and P. C. Eklund, *Chem. Phys. Lett.* **211**, 337–340, 1993.
50. P. Zhou, A. M. Rao, K.-A. Wang, J. D. Robertson, C. Eloi, M. S. Meier, S. L. Rent, X.-X. Bi, P. C. Eklund, and M. S. Dresselhaus, *Appl. Phys. Lett.* **60**, 2871–2873, 1992.
51. S. B. Fleischer, B. Pevzner, D. J. Dougherty, E. P. Ippen, M. S. Dresselhaus, and A. F. Hebard, *Appl. Phys. Lett.* **69**, 296–298, 1996.
52. A. F. Hebard, R. C. Haddon, R. M. Fleming, and A. R. Kortan, *Appl. Phys. Lett.* **59**, 2109–2111, 1991.

53. J. Tauc and Z. Vardeny, *Solid State and Materials Sciences* **16**, 403–416, 1990.

54. S. Kazaoui, R. Ross, and N. Minami, *Phys. Rev. B* **52**, R11 665–R11 668, 1995.

55. G. P. Zhang, R. T. Fu, X. Sun, X. F. Zong, K. H. Lee, and T. Y. Park, *J. Phys. Chem.* **99**, 12301–12304, 1995.

56. P. Petelenz, *Chem. Phys. Lett.* **273**, 42–46, 1997.

57. D. S. Bethune, G. Meijer, W. C. Tang, and H. J. Rosen, *Chem. Phys. Lett.* **174**, 219–222, 1990.

58. W. Kutt, W. Albrecht, and H. Kurz, *IEEE J. Quant. Electr.* **28**, 2434–2444, 1992.

59. W. T. Pollard, S. L. Dexheimer, Q. Wang, L. A. Peteanu, C. V. Shank, and R. A. Mathies, *J. Phys. Chem.* **96**, 6147–6158, 1992.

60. S. L. Dexheimer, Q. Wang, L. A. Peteanu, W. T. Pollard, R. A. Mathies, and C. V. Shank, *Chem. Phys. Lett.* **188**, 61–66, 1992.

3

Photoexcited State and Electron Transfer Properties of Fullerenes and Related Materials

Ya-Ping Sun, Jason E. Riggs, Zhixin Guo, and Harry W. Rollins
Clemson University
Clemson, South Carolina

1 INTRODUCTION

Since the discovery and large-scale production of fullerenes, there have been extensive experimental and computational investigations of the photophysical and photochemical properties of fullerene molecules and related materials [1–7]. It is now well established that the high molecular symmetry and spheric π-electronic systems of fullerenes have profound effects on their electronic transitions and excited-state processes. In fact, fullerenes represent a new and unique class of molecules with conjugated electronic structures for testing the limits of the existing photophysical and photochemical principles.

The electronic transitions and photoexcited state processes of fullerenes are directly related to their other properties. For example, fullerenes, especially [60] fullerene, were found to be excellent optical limiters in room-temperature solution [8–17]. The strong optical limiting responses were attributed to nonlinear absorptions due to higher excited state absorption cross sections than ground state absorption cross sections [15]. However, a recent study suggested that in addition to the strongly absorptive excited singlet and triplet states, other transient species that are most likely associated with photoexcited state bimolecular processes may also be responsible for the observed strong optical limiting responses of fullerenes in solution [16, 17]. Thus, a thorough understanding of the electronic transitions and photoexcited state processes in fullerenes is required for the mechanistic details of the optical limiting and other properties.

Photoexcited fullerenes are excellent electron acceptors, exhibiting strong interactions with a variety of electron donors [1, 2, 4, 5, 7]. In fact, the study of photoinduced intermolecular and intramolecular electron transfers in fullerene systems is among the most popular topics in fullerene research. The unique properties of fullerenes with respect to photoinduced electron transfers include the low solvent reorganization energy [18] and the ability of accepting multiple electrons by a single fullerene cage [19]. In addition, photochemical reactions of fullerenes through an electron transfer mechanism often yield products that are different from those expected on the basis of known reactions of typical aromatic systems. For example, completely unexpected products were obtained from the photochemical reactions of [60] fullerene with tertiary amines as the electron donors [20]. Generally speaking, fullerenes often behave quite differently from polyaromatic compounds and other electron acceptors such as porphyrins in reactions of a photoinduced electron transfer mechanism. The recent development of a variety of cage functionalization methods [21–27] has made it possible to prepare redox active fullerene derivatives for fundamental studies of photoinduced electron transfers and charge separations and for potential technological applications such as solar energy conservation. Fullerene molecules and derivatives have also been used as charge generators in photoconductive polymer thin films [28–39].

In this chapter, we will review the recent development in photophysical studies of fullerenes and related materials, with emphases on the new results obtained since our previous review [7]. Because the excited triplet-state properties of fullerenes are discussed in detail in Chapter 4, our review covers primarily excited singlet-state properties of fullerenes, as well as intermolecular and intramolecular electron transfers that are associated with the quenching of excited singlet states of fullerenes and related materials.

2 ABSORPTION AND EMISSION PROPERTIES OF THE PARENT FULLERENES

With extensive experimental investigations [1–3, 6, 7, 40–52], the electronic absorption spectra and related parameters of carefully purified C_{60} and C_{70} in solution are now well known. The fluorescence properties of C_{60} and C_{70} are also well understood. As summarized in Table 1, both C_{60} and C_{70} are only weakly fluorescent, with C_{70} having a slightly higher fluorescence yield. In the absence of electron transfer interactions, fluorescence quantum yields of the fullerene molecules in solution under ambient conditions are essentially solvent independent. Fluorescence lifetimes of the fullerene molecules are also insensitive to solvent changes. In addition, intermolecular interactions between fullerene molecules in solution are insignificant to the extent that there are no solution concentration effects on fluorescence properties [52, 53]. The fluorescence decays of C_{60} in dilute toluene solution and in saturated o-dichlorobenzene and 1-chloronapthalene solutions (up to 8×10^{-3} M) at room temperature show excellent agreement, yielding the same fluorescence lifetimes upon deconvolution [53].

An interesting new finding in the photophysics of fullerenes is the thermally delayed fluorescence emissions of C_{60} and C_{70} [54]. The delayed emissions are due to the fullerene excited singlet states thermally repopulated from the excited triplet states. With measurements of the delayed fluorescence intensities at different temperatures, the singlet-triplet energy gaps of C_{60} and C_{70} were determined to be 35 ± 2 kJ/mol and 26 ± 2 kJ/mol, respectively [54], which are in good agreement with the results from other methods. The delayed fluorescence is more significant in C_{70} than in C_{60}, which is probably due to the fact that C_{70} has a much longer-lived excited triplet state and a significantly smaller singlet-triplet energy gap.

The excited singlet-state transient absorptions of C_{60} and C_{70} are now well characterized [1–3, 6, 7, 55–60]. Results from picosecond flash photolysis measurements show that the singlet-singlet absorption spectrum of C_{60} has peaks at 513, 759, and 885 nm with molar extinction coefficients of 4500, 3700, and 6300 M^{-1} cm^{-1}, respectively [1, 58].

There have been only a few photophysical studies of higher fullerenes including C_{76}, C_{78}, C_{82}, and C_{84} [61–69]. For C_{76}, Benssasson et al. investigated the photophysical properties of the compound in a racemic mixture [61]. The excited singlet-state lifetime of C_{76} was determined by Ito and coworkers in transient absorption measurements [63]. The lifetime of 1.7 ns from the decay of singlet-singlet absorption is significantly longer than the 1.25 ns value obtained by Guldi et al. [63, 64]. Guldi has also studied the photoexcited state and electron transfer properties of C_{76} and C_{78} using laser flash photolysis and pulse radiolysis methods [65]. However, there have been no reports of any fluorescence emissions from the higher fullerene molecules [69]. The absence of any fluorescence emissions remains an interesting mechanistic question in further photophysical investigations of higher fullerenes.

Unlike C_{60} and C_{70}, each of the higher fullerenes has different isomers. The separation of isomeric mixtures is a difficult task. Despite recent efforts, it re-

Table 1 Fluorescence Properties of C_{60} and C_{70}

Solvent	C_{60}			C_{70}		
	Φ_F (10^{-4})	τ_F (ns)	Ref.	Φ_F (10^{-4})	τ_F (ps)	Ref.
Hexane	3.3		52	5.9		52
Benzene		1.17,[a] 1.2[b]	40, 58		627,[c] 450[b]	56a,b
Toluene	3.2	1.2,[c] 1.17[c]	52, 55a	5.7		52, 55a, 57
o-Xylene	2.9		52	5.7		52
1,2,4-Trimethylbenzene	3.1		52	5.8		52
Decalin		1.14[a]	40			
Carbon Disulfide	2.6		52	4.2		52
Chlorobenzene	2.8		52	5.3		52
o-Dichlorobenzene	3.0		52	4.8		52
Dichloromethane	3.3		52	5.4		52
1-Methylnaphthalene	3.7		52	6.2		52
1-Chloronaphthalene	3.4		52			
Benzonitrile		1.16[a]	40			

[a] From the rising of triplet transient absorption.
[b] From the decay of singlet transient absorption.
[c] From time-correlated single photon counting.

mains a significant challenge to obtain pure individual isomers of the higher fullerenes. As a result, some of the photophysical properties reported in the literature are probably average properties of different isomers.

3 ABSORPTION AND EMISSION PROPERTIES OF FULLERENE DERIVATIVES

Since the author's last review [7], a large number of mono- and multiple-functionalized fullerenes have become available. Absorption and emission studies of a large number of C_{60} derivatives in different classes have significantly enriched the understanding of derivatization effects on the fullerene photophysical properties and related excited-state processes.

3.1 Monofunctionalized C_{60} Derivatives

Sun and coworkers carried out a systematic investigation of absorption and emission properties of different classes of monofunctionalized C_{60} derivatives [70–73]. For methano-C_{60} derivatives (**1–9**), absorption spectra and molar absorptivities of the derivatives with different substituents are rather similar (Figure 1; Table 2), indicating that transitions in the visible region are dictated by the elec-

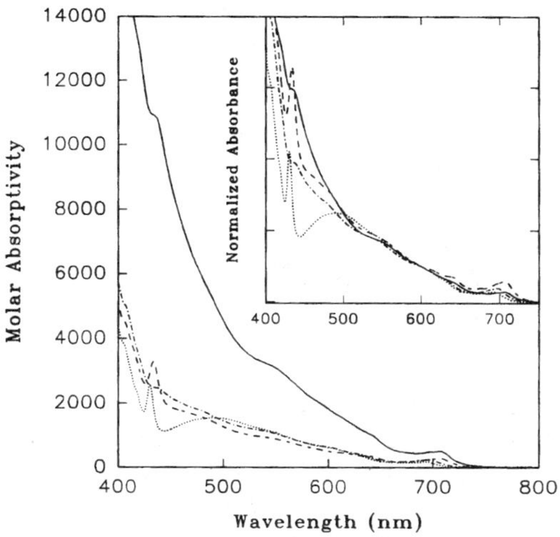

Figure 1 The absorption spectrum of the C_{60}-dimer **21** (—) is compared with those of the methano-C_{60} derivative **5** (•••), pyrollidino-C_{60} derivative **10** (---), and amino C_{60} derivative **11** (-•-) in room-temperature solutions. Shown in the inset is a comparison of the normalized absorption spectra. (*From Ref. 90.*)

Table 2 Photophysical Properties of Methano-C_{60} Derivatives

Compound	Solvent	$ABS_{0\text{-}0}$ (nm)	$\varepsilon_{0\text{-}0}$ ($M^{-1}cm^{-1}$)	$FLSC_{0\text{-}0}$ (nm)	Φ_F (10^{-3})	τ_F (ns)
C_{60}	Toluene				0.32	1.2
1	Toluene	687	210	711	1.07	1.51
2	Toluene	693	230	718	1.14	1.49
3	Toluene	695	240	720	1.19	1.47
4	Toluene	695	210	719	1.18	1.46
5	Toluene	692	190	713	1.0	1.49
6	THF	694	185	716	0.8	1.46
7	Chloroform	692	176	713	0.99	1.45
8	Chloroform	688	195	715	1.0	1.48
9	THF	690		715		

Source: From Refs. 70 and 72.

tronic structures of the functionalized fullerene moiety. A particularly noticeable common feature in the absorption spectra of different methano-C_{60} derivatives is a weak absorption peak at ~700 nm. It prompted a conclusion that the lowest electronic transitions in the C_{60} derivatives have similar energies and transition probabilities [70, 72]. The conclusion is also supported by the fluorescence results. The fluorescence spectra (Figure 2), quantum yields, and lifetimes of the different methano-C_{60} derivatives are all quite similar (Table 2).

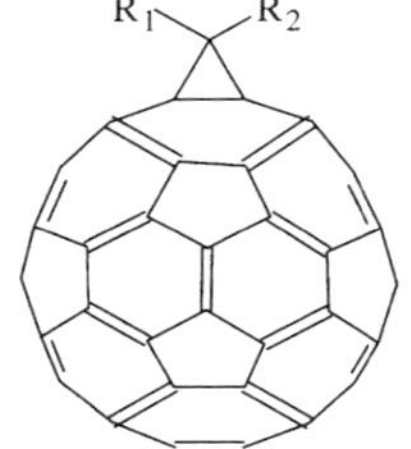

1: R_1=H, R_2=COOCN
2: R_1=H, R_2=COOEt
3: R_1=H, R_2=COPh
4: R_1=H, R_2=COC(CH$_3$)$_2$CH$_2$COOEt
5: R_1=H, R_2=COOC(CH$_3$)$_3$
6: R_1=H, R_2=COOH
7: R_1=H, R_2=CONEt$_2$
8: R_1=R_2=COOEt
9: R_1=R_2=COOH

The comparison of absorption and fluorescence properties was extended to include different classes of monofunctionalized fullerenes including pyrrolidino-C_{60}, amino-C_{60}, and other C_{60} derivatives (**10–14**) [70, 71, 73]. Despite their different molecular structures, with significantly different functional groups on the fullerene cage, the C_{60} derivatives have similar absorption and fluorescence emission parameters (Figures 1 and 2; Table 3). The results further support the conclusion that the electronic transition energies and probabilities in monofunctionalized C_{60} derivatives are dictated by the functionalized fullerene cage and are little affected by the functional groups [70–73].

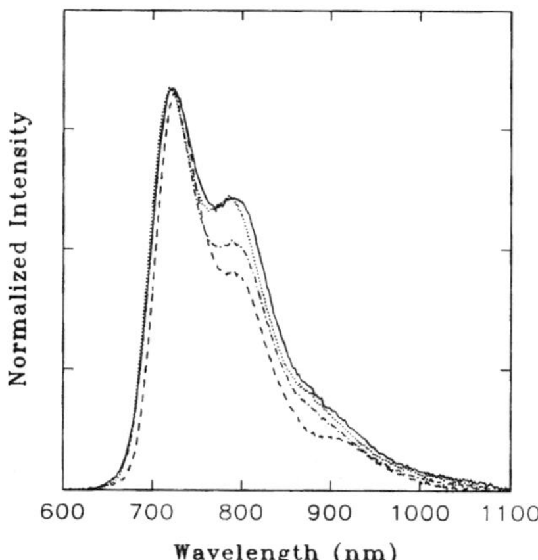

Figure 2   The fluorescence spectrum of the C_{60}-dimer **21** (—) is compared with those of the methano-C_{60} derivative **5** (•••), pyrollidino-C_{60} derivative **10** (---), and amino C_{60} derivative **11** (-•-) in room-temperature solutions. (*From Ref. 90.*)

Table 3   Photophysical Properties of Monofunctionalized C_{60} Derivatives

Compound	Solvent	ABS$_{0-0}$ (nm)	FLSC$_{0-0}$ (nm)	Φ_F (10^{-3})	τ_F (ns)	Ref.
C_{60}	Toluene			0.32	1.2	52
Pyrrolidino **10**	Toluene	707	726	1.05	1.5	71,72
Pyrrolidino **15**	MCH[a]	701[b]	702	0.60	1.25[c]	79
Pyrrolidino **16**	Benzene	705	711	1.06	1.3	83
Pyrrolidino **17**	Benzene	702	708	1.18	1.2	83
Pyrrolidino **18**	Benzene	701	708	1.10	1.3	83
Pyrrolidino **19**	Benzene	706	712	1.05	1.3	83
Pyrrolidino **20**	Benzene	707	714	1.0	1.2	83
Pyrrolidino **22**	MCH[a]	701[b]	702	0.44	1.05[d]	79
Other **13**	Chloroform	694	712	0.93	1.6	70
Other **14**	Hexane		730	1.2	1.25	73
Amino **11**	Toluene	694	713	0.91	1.3	71
Amino **12**	Hexane	695	699	1.00	1.3	71
Dimer **21**	Toluene	698	720	0.92	1.5	90

[a] Methylcyclohexane glass at 77 K.
[b] In room-temperature dichloromethane.
[c] From the transient decay in toluene at 880 nm.
[d] From the transient decay in toluene at 890 nm.

Monofunctionalized C_{60} derivatives have also been studied by other research groups [74–89]. Thomas et al. investigated the photophysical properties of a monofunctionalized pyrrolidino-C_{60} derivative **15** [79]. The fluorescence spectrum of **15** has a maximum at 702 nm, which is typical of a monofunctionalized C_{60} derivative (Table 3). The fluorescence quantum yield of **15** was estimated to be 6×10^{-4} in methylcyclohexane at 77 K. The observed yield is lower than those of other mono-functionalized C_{60} derivatives including the pyrrolidino-C_{60} derivative **10** (Table 3). According to results from picosecond flash photolysis measurements, an excited singlet state lifetime of 1.25 ns was obtained for **15** in toluene [79].

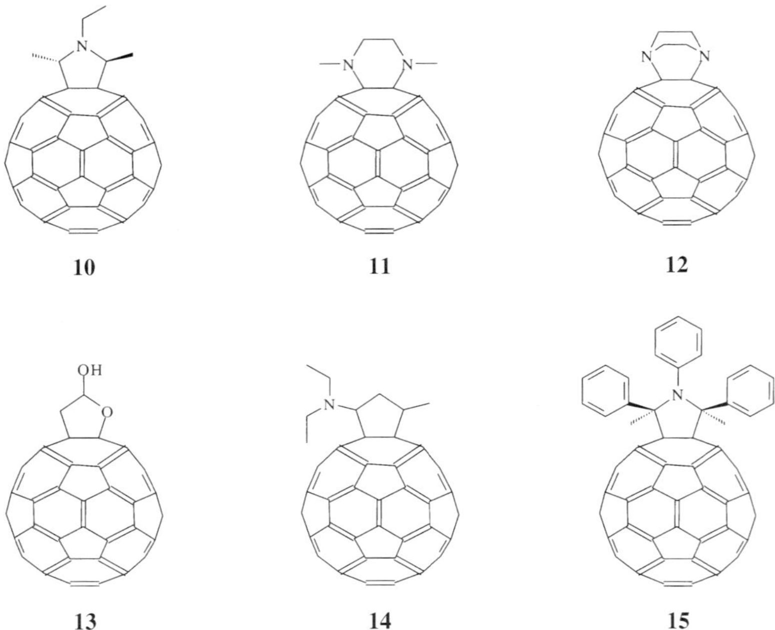

Luo et al. examined substituent and solvent effects on the photoexcited state properties of several N-methylpyrollidino-C_{60} derivatives **16–20** [83]. The fluorescence intensities of the derivatives in nonpolar solvents were found to be quite similar among themselves, but several times higher than that of the parent C_{60}. The fluorescence quantum yields of **16–20** in benzene are in the range of 1.05×10^{-3} to 1.18×10^{-3} [83], similar to those of other monofunctionalized C_{60} derivatives (Table 3). Results from time-resolved measurements show that the fluorescence lifetimes of **16–20** in benzene are in the range of 1.1 to 1.3 ns

(Table 3). Changes in the solvent polarity have only minor effects on the fluorescence properties of the derivatives, except for **20**. The solvent-dependent photophysical properties of **20** were explained in terms of effects associated with intramolecular electron transfer [83], though the same conclusion had been reached by Williams et al. [87].

16: R= H
17: R= p-C_6H_4CHO
18: R= p-$C_6H_4NO_2$
19: R= p-C_6H_4OMe
20: R= p-$C_6H_4NMe_2$

The fact that all of the derivatives have a weak absorption band at ~700 nm suggests that absorptions in the ~400–650 nm wavelength region are due primarily to contributions of electronic transitions other than the 0-0 transition. Under the assumption that the absorption due to the lowest electronic transition may be estimated by the mirror image of the observed fluorescence spectrum, there is a qualitative agreement between the calculated fluorescence radiative rate constants $k_{F,c}$ on the basis of transition probabilities and the experimental rate constants $k_{F,e}$ obtained from fluorescence quantum yield and lifetime results in the C_{60} derivatives [70, 72]. Thus, a qualitative conclusion is that for the C_{60} derivatives in room-temperature solution the lowest-energy absorption and the fluorescence emission are associated with the same excited singlet state [70, 72]. In fact, the mirror-image relationship between the 0-0 absorption and fluorescence bands is more evident in the derivatives than in the parent C_{60}.

Interestingly, the absorption spectra of the parent C_{60} and C_{60} derivatives show a characteristic solvent dependence in the solvent series of methyl substituted benzenes [52, 72]. The intense second absorption band undergoes significant solvatochromic shifts, while the absorption in the visible region (480–750 nm) is essentially unchanged. A convincing explanation for the observed characteristic solvent effects in a specific solvent series remains to be found.

The fluorescence properties of monofunctionalized C_{60} derivatives are apparently more inert to the molecular structural differences in the derivatives (Table 3). The fluorescence spectra of the different classes of C_{60} derivatives are rather similar, with only minor changes in the relative intensities of the vibrational bands (Figure 2). The fluorescence quantum yields and lifetimes of the different C_{60} derivatives are all very similar (Table 3). From C_{60} to monofunctionalized C_{60} derivatives, however, both fluorescence quantum yields and lifetimes increase. The increases in quantum yields are larger than those in lifetimes, corresponding to larger fluorescence radiative rate constants of the C_{60} derivatives than that of C_{60}. The results are consistent with higher molar absorptivities for the 0-0 transitions in the C_{60} derivatives. In a comparison of the C_{60} derivatives with the parent C_{60}, it is obvious that the difference in the fluorescence properties is much less significant among different classes of C_{60} derivatives than between the derivatives and the parent C_{60}.

Photophysical properties of the C_{60}-dimer **21** have also been studied by Sun and coworkers [90]. The dimer may be treated as two identical covalently linked monofunctionalized C_{60} derivatives. While the absorption spectral profile of the C_{60}-dimer is similar to those of monofunctionalized C_{60} derivatives, the molar absorptivity of the C_{60}-dimer at the 0-0 absorption band at ~700 nm is larger than those of the other C_{60} derivatives by approximately a factor of 2 (Figure 1, Table 3). There are also characteristic absorption spectral changes for the dimer in the solvent series of methyl substituted benzenes [90]. The fluorescence spectral profile, quantum yield, and lifetime of the dimer are similar to those of other classes of C_{60} derivatives as well (Figure 2, Table 3). The results are consistent with the notion that the dimer is effectively a pair of monofunctionalized C_{60} derivatives. The fluorescence properties of the C_{60}-dimer are also insensitive to solvent changes [90].

There have been several picosecond transient absorption studies of monofunctionalized C_{60} derivatives [79–81]. Guldi and Asmus examined excited singlet-state absorptions of the monofunctionalized C_{60} derivative **8** [80]. The singlet-singlet absorption spectrum was found to be similar to that of the parent C_{60}. For **8** in an oxygen-free toluene solution, a transient absorption band peaking at 920 nm is formed ~100 ps after the laser excitation at 532 nm. The transient decays with a time constant of 1.54 ns, in good agreement with the excited singlet-state lifetime obtained from fluorescence measurements (Table 2).

Excited singlet-state absorptions in the monofunctionalized pyrrolidino-C_{60} derivatives **15** and **22** were also investigated [79]. For both derivatives, the singlet-singlet absorption spectra are broad, with the maxima at 880 and 890 nm for **15** and **22**, respectively (Figure 3). The rise times of the transients are on the order of 60 ps [79]. Apparently, the transient absorption results of the monofunctionalized C_{60} derivatives are qualitatively similar to those of the parent C_{60}.

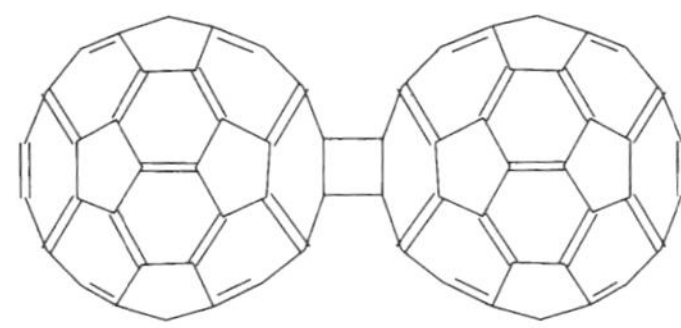

21

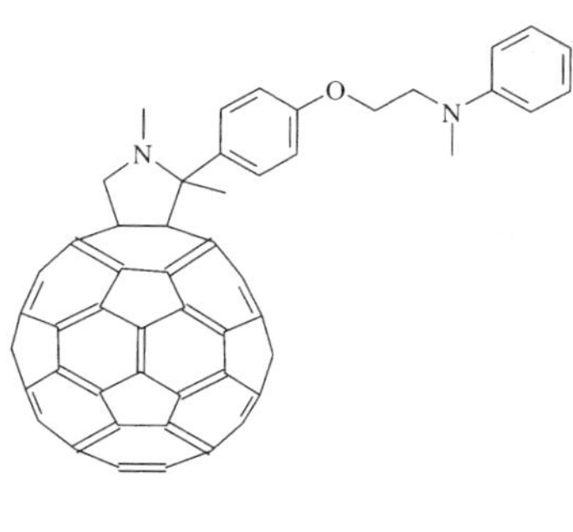

22

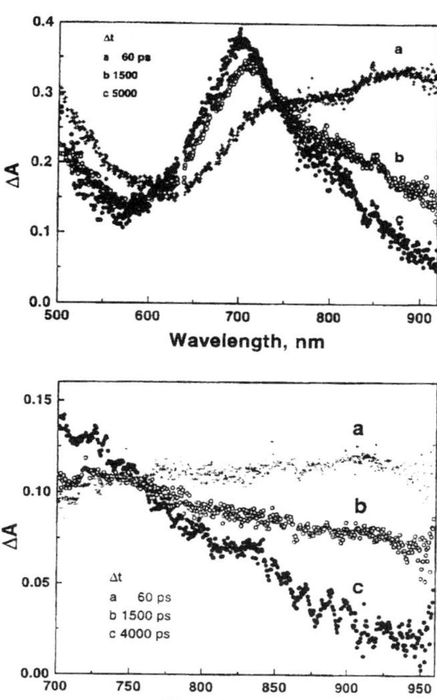

Figure 3 Transient absorption spectra recorded following 355 nm laser pulse (pulse width 18 ps) excitation of toluene solution containing **15** (top) at delay times (a) 60, (b) 1500, and (c) 5000 ps and **22** (bottom) at delay times (a) 60, (b) 1500, and (c) 4000 ps. (*From Ref. 79.*)

3.2   Multiple-Functionalized C_{60} Derivatives

There have been several recent studies of photophysical properties of multiple-functionalized C_{60} derivatives [80, 91–97]. Among structurally well-characterized C_{60} derivatives with multiple cage functionalizations is a series of C_{60}-malonate ester adducts **23–28** [98].

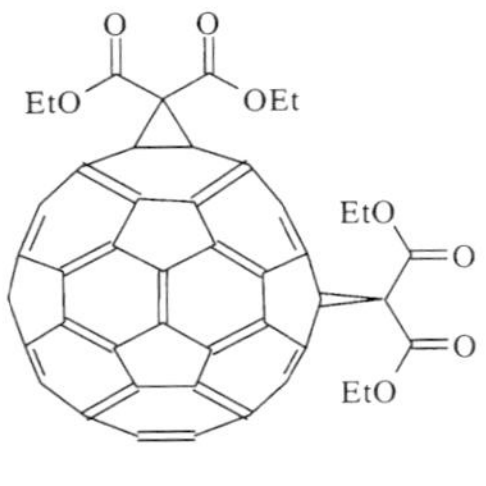

23

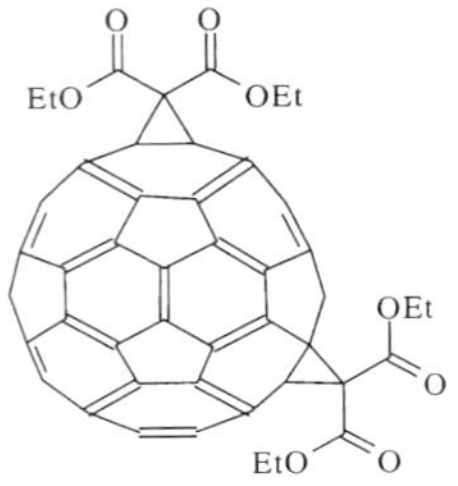

24

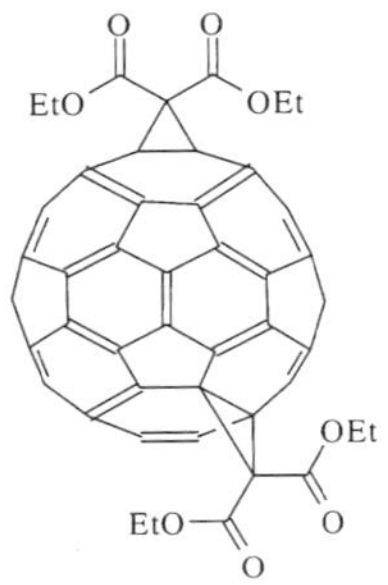

25

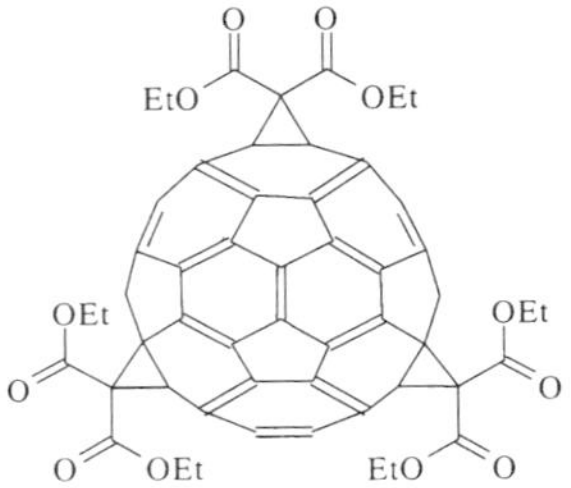

26

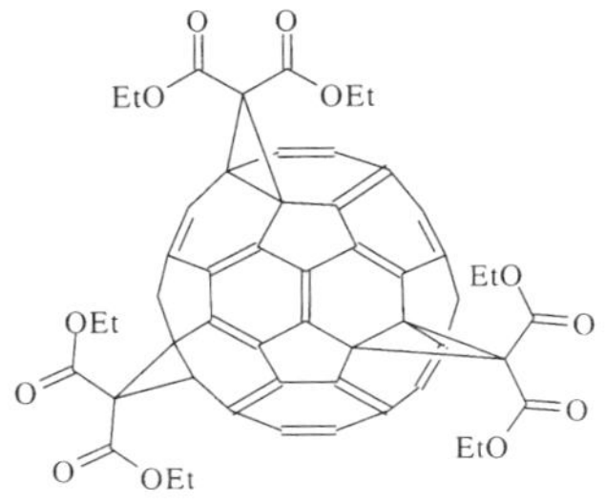

27

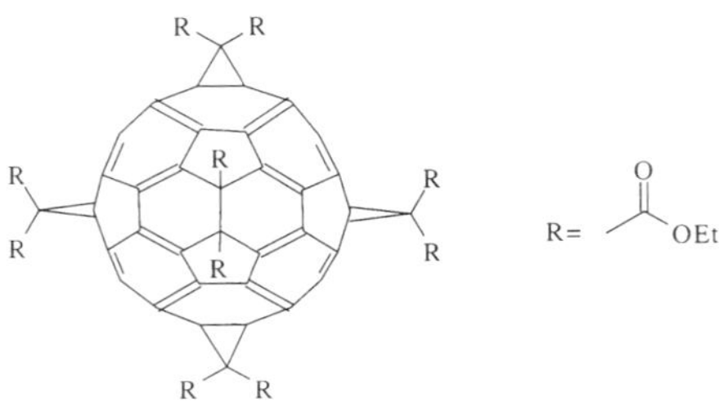

28

Guldi and Asmus studied the photoexcited-state properties of the methano-C_{60} derivatives with different multiple cage functionalizations **23–25** and **27** [80]. As discussed earlier, the singlet transient absorption spectrum of the C_{60}-malonate ester mono-adduct **8** is similar to that of the parent C_{60}, with the absorption maximum at ~920 nm in toluene. As the cage functionalization increases in bis-adducts **23–25** and the tris-adduct **27**, the singlet transient absorption maxima undergo significant blue-shifts. For example, the singlet-singlet absorption maximum of the tris-adduct **27** is 50 nm blue-shifted from that of the mono-adduct **8** [80]. Excited singlet-state lifetimes of the derivatives were obtained from the transient decays (Figure 4, Table 4). For the derivatives **23–25**, and **27**, fluorescence spectra in the ~650–800 nm wavelength region were obtained in methylcyclohexane glass at 77 K [80]. The spectra of bis-adducts **23–25** and the tris-adduct **27** are broader and somewhat red-shifted from that of the mono-adduct **8**. The fluorescence spectral features were attributed to the multiple-cage functionalizations, which perturb the fullerene π-electronic system [80].

Sun and coworkers have also been studying the absorption and emission properties of the C_{60}-malonate ester adducts, including the hexakis-adduct with an octahedral addition pattern **28** [96]. The absorption spectrum of **28** in toluene solution is different from those of the derivatives with less substitution on the cage (Figure 5). Molar absorptivities of the symmetric tris-adduct **26** in the visible region are higher than those of the bis-adduct **23** and the mono-adduct **8** (Figure 5, Table 4).

Fluorescence properties of the multiple-functionalized C_{60} derivatives **23**, **26**, and **28** in room-temperature solution were determined quantitatively [96]. As shown in Figure 6, the fluorescence spectrum of the bis-adduct **23** has a similar profile to that of the mono-adduct. However, the tris-adduct **26** and the hexakis-adduct **28** have much broader fluorescence spectra, which are also blue-shifted from those of the mono- and bis-adducts (Figure 6).

Fluorescence quantum yields of the multiple-functionalized C_{60} derivatives including the hexakis-adduct **28** are somewhat different from those of monofunctionalized methano-C_{60} derivatives (Table 2) [70, 72, 96]. As compared in Table 4 quantitatively, the symmetric tris-adduct **26** is actually somewhat less fluorescent than the mono-adduct. The hexakis-adduct **28** has a higher fluorescence quantum yield of 2.4×10^{-3} in comparison with 1.0×10^{-3} for the mono-adduct and 1.5×10^{-3} for the bis-adduct **23** in room-temperature toluene. The hexakis-addition to the C_{60} cage significantly changes the fullerene π-electronic system, which leaves the cage with eight individual phenyl rings. Although the phenyl rings in **28** are somewhat isolated in comparison with those in the parent C_{60} and the derivatives that are less functionalized, there is still an sp^2 carbon network throughout the fullerene cage in the hexakis-adduct [96]. The absorption and fluorescence spectral properties of **28** are likely a reflection of the fact that the π-electronic system in the hexakis-adduct is very different from those in the par-

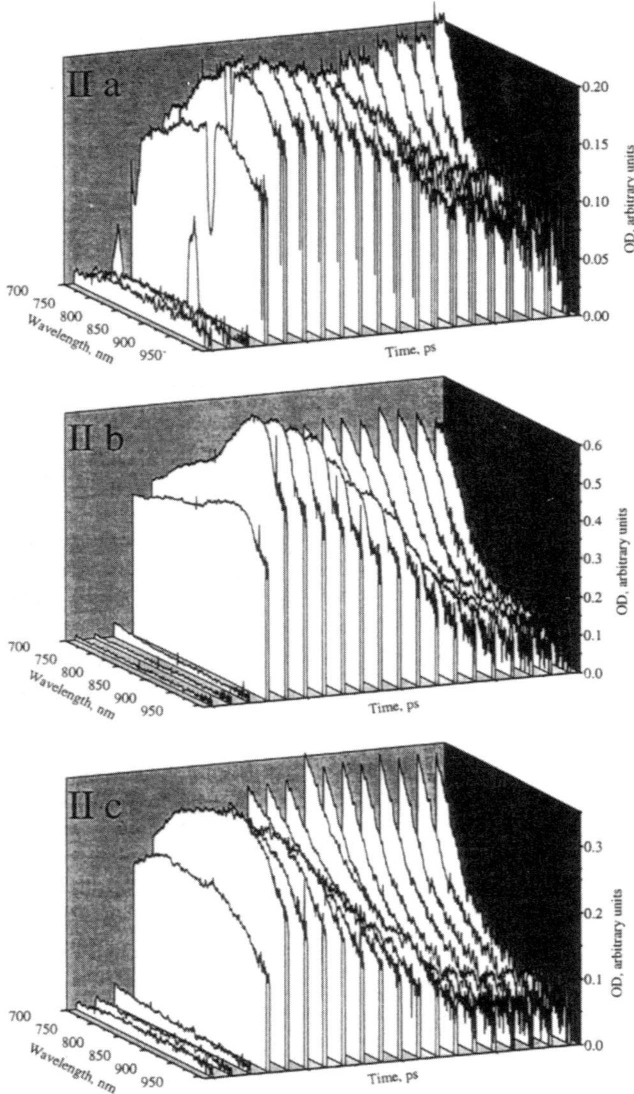

Figure 4 Transient absorbance changes observed following picosecond flash photolysis at 532 nm of bis-adducts **23** (**IIa**), **24** (**IIb**), and **25** (**IIc**) in deaerated toluene solution. Time scale: −50, 0, 50, 100, 200, 300, 400, 500, 750, 1000, 1250, 2000, 2500, 3000, 3500, 4000, 4500, 5000, and 6000 ps. (*From Ref. 80.*)

Table 4 Photophysical Properties of the C_{60}-Malonate Ester Multiple Adducts

Compound	Solvent	$ABS_{0\text{-}0}$ (nm)	ε_{max} ($M^{-1}\ cm^{-1}$)	$FLSC_{0\text{-}0}$ (nm)	Φ_F (10^{-3})	τ_F (ns)	Ref.
C_{60}	Toluene				0.32	1.2	52
Mono **8**	Chloroform	688	195[a]	715	1.0	1.48	70
Mono **8**	Toluene	690		690[b]		1.54	80
Bis **23**	Toluene	481	2800[b]	781	1.5	2.94	96
Bis **23**	Toluene	694		697[b]		2.88	80
Bis **24**	Toluene	697		702[b]		1.96	80
Bis **25**	Toluene	696		700[b]		1.68	80
Tris **26**	Toluene	482	2900	725	0.86	2.45	96
Tris **26**	CH_2Cl_2			726	0.75	2.10	96
Tris **27**	Toluene	694		702[b]		3.12	80
Hexakis **28**	Toluene	385(sh)	6500[c]	680	2.4	2.14	96
Hexakis **28**	CH_2Cl_2	385(sh)	6200[c]	680	2.3	2.03	96

[a] The 0-0 band.

[b] In methylcyclohexane at 77°K.

[c] At the first shoulder.

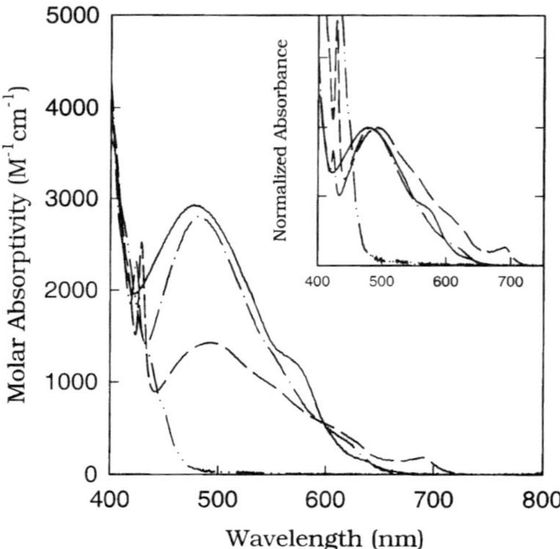

Figure 5 Absorption spectra of the C_{60}-malonate ester mono-adduct **8** (---) and multiple adducts **23** (-•-), **26** (—), and **28** (-••-) in room-temperature solutions. Shown in the inset is a comparison of the normalized absorption spectra. (*From Ref. 96.*)

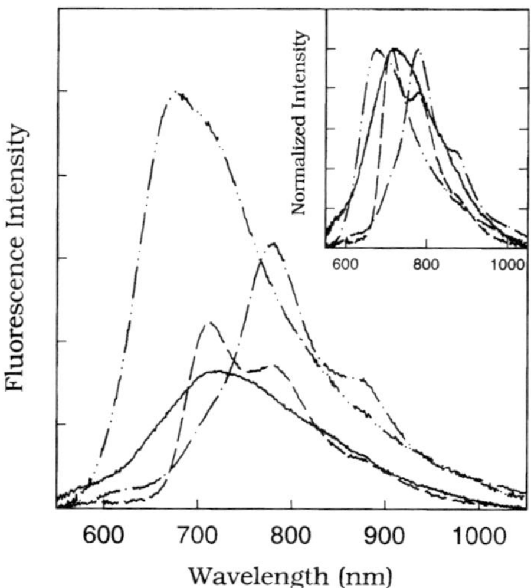

Figure 6 Fluorescence spectra of the C_{60}-malonate ester mono-adduct **8** (---) and multiple adducts **23** (-•-), **26** (—), and **28** (-••-) in room-temperature solutions. Shown in the inset is a comparison of the normalized fluorescence spectra. (*From Ref. 96.*)

ent C_{60} and mono-, bis-, and tris-adducts and also different from those in classical aromatic chromophores [96].

Fluorescence lifetimes of the C_{60}-malonate ester multiple adducts were determined quantitatively [80, 96]. As summarized in Table 4, all the multiple adducts have longer-lived excited singlet states than the mono-adduct. However, changes in the fluorescence lifetime do not correlate with the number of cage functionalizations. The fluorescence lifetime of the hexakis-adduct **28** is almost the same as that of the tris-adduct **26**, but shorter than that of the bis-adduct **23**. The fluorescence radiative rate constants of the bis-adduct **23** and the tris-adduct **26** are in fact rather similar, reflecting their similar 0-0 transition probabilities. The radiative rate constant of the hexakis-adduct is much smaller, consistent with the low absorptivities at the onset of the absorption spectrum (Figure 5).

The excited triplet-state absorption of the hexakis-adduct **28** is also different from those of the parent C_{60} and other C_{60} derivatives [97]. A comparison of the triplet-triplet transient absorption spectra of the mono-adduct **8** [82], tris-adduct **26**, and hexakis-adduct **28** in carefully deoxygenated toluene is shown in Figure 7. The triplet-state absorption maximum of **28** is significantly blue-shifted from those of the other C_{60} derivatives.

The multiple-functionalized C_{60} derivatives, **28** in particular, have different solubility characteristics from the parent C_{60} and monofunctionalized C_{60} deriva-

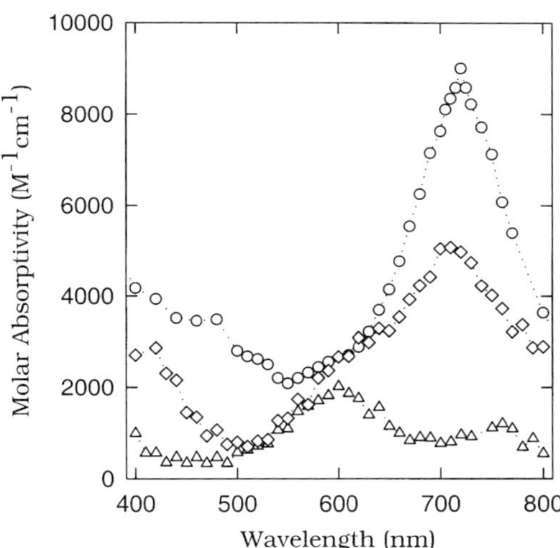

Figure 7   The triplet-triplet absorption spectra of the C_{60}-malonate ester symmetric tris-adduct **26** ($\diamond$) and hexakis-adduct **28** ($\triangle$) are compared with that of the mono-adduct **8** ($\bigcirc$). (*From Refs. 82, 97.*)

tives. As a result, polar solvents such as THF, acetonitrile, and DMF may be used in studies of the absorption and emission properties. However, according to Sun and coworkers the photophysical properties of the C_{60}-malonate ester multiple adducts are insensitive to solvent changes [96].

Other kinds of multiple-functionalized C_{60} derivatives have also been studied. For example, Palit et al. investigated the photophysical properties of $C_{60}(C_6H_5)_5Cl$ [93]. The UV-vis absorption spectrum of the derivative extends to only ~650 nm, and the absorptivities are significantly weaker than those of C_{60} throughout the ~200–650 nm region. The excited singlet-state absorption spectra show broad overlapping bands in the ~450–900 nm region with the spectral maximum at ~880 nm (Figure 8), which is blue-shifted from that of the parent C_{60}. The excited singlet-state lifetime was determined to be ~1.6 ns, comparable to those of other C_{60} derivatives [93].

A photophysical study of $C_{60}H_{18}$ and $C_{60}H_{36}$ was reported by Bensasson et al. [94]. While the study was primarily for the excited triplet-state properties, ground-state absorption spectra of the molecules were reported and discussed. However, there is a possibility that the results were influenced by anthracene contamination due to the synthetic procedure [94]. Recently, Palit et al. also in-

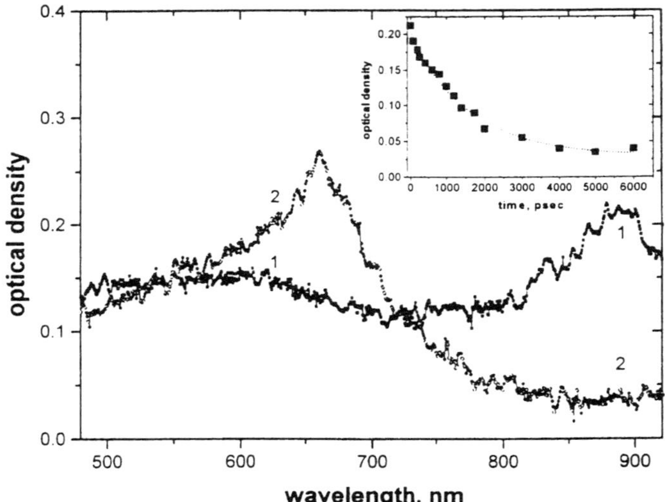

Figure 8 Transient optical absorption spectra obtained on laser flash photolysis (355 nm excitation) of a N_2 saturated solution of $C_{60}(C_6H_5)_5Cl$ in benzene at 0 ps (curve 1) and 6 ns (curve 2) after the laser pulse. Inset shows the decay of the transient absorption monitored at 880 nm. (*From Ref. 93.*)

vestigated the photophysical properties of $C_{60}H_{18}$ and $C_{60}H_{36}$ [92]. The results are somewhat different from those reported by Bensasson et al. The discrepancy was attributed to the different molecular symmetries imparted by the different hydrogenation techniques employed in the syntheses [94].

Mohan et al. investigated the photophysical properties of $C_{60}(OH)_{18}$ in an aqueous solution [91]. It was found on the basis of a deviation from the Beer-Lambert law that there is molecular aggregation at high solution concentrations. The aggregates have a broad absorption spectrum in the visible wavelength region. The fluorescence spectrum of the molecule in water is weak and broad. The singlet-singlet transient absorption spectrum of the molecule is in the $\sim$550–810 nm region without a clearly defined absorption maximum. The excited singlet-state lifetime of the molecule was reported to be $\sim$500 ps [91].

Multiple-functionalized C_{60} derivatives represent an important class of fullerene materials for both fundamental studies and potential applications. It is expected that the photophysical and photoexcited state properties of multiple-functionalized C_{60} derivatives will continue to attract widespread attention.

4 PHOTOINDUCED ELECTRON TRANSFERS IN FULLERENES AND RELATED MATERIALS

It is well established that both ground- and excited-state fullerenes are excellent electron acceptors. As a result, photoexcited states of fullerenes and related materials are quenched effectively by a variety of electron donors. Recently, there have been extensive experimental studies of photoinduced intermolecular [99–122] and intramolecular [123–146] electron transfers in fullerenes and derivatives, and fullerene-based molecular systems.

4.1 Photoinduced Intermolecular Electron Transfers

It has been well established that the excited singlet states of C_{60} and C_{70} are quenched effectively by electron donors such as N,N-dialkylanilines [7, 118–122]. The same electron transfer quenchings have been observed for C_{60} derivatives [70–72, 96]. For monofunctionalized C_{60} derivatives, Sun and coworkers studied intermolecular electron transfers in fluorescence intensity and lifetime quenching experiments [70–72]. The fluorescence quenchings are typically very efficient with diffusion-controlled quenching rate constants. There are also significant contributions from static quenchings, which result in an upward curvature in the Stern-Volmer plots for the quenching of fluorescence intensities. The same kind of static quenching is known to be significant in the electron transfer interactions between the parent fullerene excited states and aromatic amines [7, 119]. The intermolecular electron transfer interactions of the mono-functionalized C_{60}

derivatives with N,N-dialkylamines are strongly dependent on the polarity of the solvent environment, also similar to the parent fullerenes. For example, the quenching of fluorescence emissions of **5** and **8** by N,N-diethylaniline (DEA) results in the formation of emissive exciplexes in room-temperature hexane and toluene, but not in polar solvents such as chloroform or hexane-acetone mixtures [72]. For some derivatives, emissive exciplexes are absent even in toluene, which is more polarizable than a saturated hydrocarbon solvent.

The electron transfer quenching of the excited singlet state of the C_{60}-dimer **21** is similar to those for the monofunctionalized C_{60} derivatives [90]. The results again support the notion that the dimer is effectively a pair of monofunctionalized C_{60} derivatives.

Quenchings of fluorescence intensities and lifetimes of the C_{60}-malonate ester multiple adducts, including the hexakis-adduct **28**, by aromatic amines were also investigated [96]. The results are generally similar to those of the parent C_{60} and monofunctionalized C_{60} derivatives. For example, shown in Figure 9 are

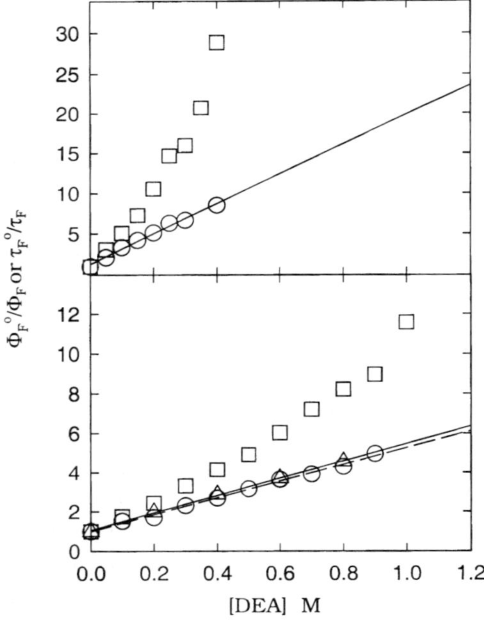

Figure 9 Stern-Volmer plots for the quenching of fluorescence intensity and lifetime of the C_{60}-malonate ester for bis-adduct **23** (top) and hexakis-adduct **28** (bottom) by N,N-diethylaniline in toluene (○) and methylene chloride (□, lifetime: △, and dashed line). (*From Ref. 96.*)

fluorescence quenching results of the bis-adduct **23** and hexakis-adduct **28** in solvents of different polarities. The fluorescence intensities and lifetimes of **23** and **28** are strongly quenched by the electron donor DEA. In room-temperature toluene, the Stern-Volmer plots for fluorescence intensity quenchings are linear for both **23** and **28** (Figure 9). For fluorescence intensity quenchings in a polar solvent methylene chloride, however, there are significant upward curvatures in the Stern-Volmer plots, but the plot for fluorescence lifetime quenching remains linear even in the polar solvent (Figure 9). Thus, the upward curvatures can be attributed to static quenching contributions [96]. Without the static quenching contributions, the dynamic fluorescence quenchings in toluene and methylene chloride are in fact rather similar (Figure 9). All of the quenching rate constants from linear Stern-Volmer plots are close to diffusion-controlled. This is particularly interesting with respect to the hexakis-adduct **28**. Despite the significant changes to the π-electronic structure resulted from the cage multiple functionalizations, the photoexcited hexakis-adduct exhibits the same electron accepting properties as the parent C_{60} and other C_{60} derivatives [96].

Guldi and Asmus also reported that intermolecular electron transfers involving C_{76} and C_{78} may be used as models for studying the Marcus inverted region [67].

4.2 Photoinduced Intramolecular Electron Transfers

For C_{60} derivatives containing amino groups, there are possibilities for photoinduced intramolecular electron transfers [70, 71, 73, 86]. Experimentally, a common feature for the electron transfers is a strong solvent dependence of the observed photophysical properties.

Nakamura et al. investigated the strong solvent polarity dependence of fluorescence properties of a C_{60}-o-quinodimethane derivative with an NMe_2 moiety [86]. In the nonpolar solvent cyclohexane, the fluorescence lifetime is 1.19 ns, similar to those of other o-quinodimethane adducts [86] and mono-functionalized C_{60} derivatives (Table 3). However, the lifetime becomes considerably shorter in the polar solvent benzonitrile (63 ps). The significant quenching of fluorescence lifetime with increasing solvent polarity was attributed to intramolecular electron transfer from the NMe_2 moiety to the photoexcited fullerene cage by a through-bond mechanism [86].

Sun and coworkers recently reported a systematic investigation of amino-C_{60} derivatives **11** and **12** [71]. The experimental results suggest strongly that the sensitivity of observed photophysical properties to changes in solvent polarity are due to photoinduced intramolecular electron transfers. There is also a significant difference between the amino-C_{60} derivatives **11** and **12** with respect to the formation of an emissive charge transferred state [71].

Fluorescence spectra of the compound **11** in solvents of different polarities show only small changes, in sharp contrast to the wide variations in fluorescence quantum yield (Table 5) [71]. The results suggest that observed fluorescence intensities in different solvents are due only to emissions from the vertical excited singlet state of the molecule. The strong fluorescence quenching for the compound **11** in polar solvents was attributed to the intramolecular n (amino unit) to π^* (C_{60} moiety) electron transfer. The involvement of n-orbital electrons in the electron transfer process is made evident by the acidification effect (Table 5). For example, the fluorescence quantum yield of **11** in a toluene-acetonitrile (40%, v/v) mixture is less than 15% of that in neat toluene, but a small amount of trifluoroacetic acid (TFA) added to the solvent mixture essentially eliminates the fluorescence quenching. Since TFA only has a minor volume fraction in the solvent mixture, its effect as a solvent component other than acidity on molecular absorption and emission properties is insignificant. The acidification effect on observed fluorescence intensities of **11** is most likely associated with the protonation of amino groups in the molecule, which effectively shuts off the n-π^* electron transfer process. The acidification has only a minor effect on the absorption and fluorescence properties of the pyrrolidino-C_{60} derivative **10**, which undergoes no photoinduced intramolecular electron transfer so that the electronic transitions in the C_{60} derivative are dictated by the monofunctionalized fullerene cage [71].

The fluorescence properties of **12** are even more solvent-sensitive [71]. For example, fluorescence quenchings are already significant for **12** in polarizable solvents such as toluene and CS_2 and even more so in polar solvents and solvent mixtures. A significant difference of **12** from **11** is that **12** has a second emissive excited state [71]. The second emission observed for **12** in a polar or polarizable solvent environment is red-shifted from the fluorescence band of the vertical excited singlet state (Figure 10). Because it can essentially be eliminated when the solution is acidified in the presence of TFA, the second emission was assigned to an intramolecular n-π^* charge transferred state [71].

The photoexcited state properties of the amino-C_{60} derivatives were explained in terms of the schematic energy diagram in Figure 11 [71]. For **12**, the electron transfer is already a major excited-state decay process even in nonpolar but more polarizable solvents such as toluene and CS_2. The radiative process is apparently a competitive decay pathway of the n-π^* charge transferred state in **12**. However, there is a longer-lived component in the observed fluorescence decays of both **11** and **12** in polar solvents (Table 5). The longer-lived emissions were assigned to the regenerated (delayed) vertical excited singlet state due to back electron transfer in **11** and to both the delayed vertical excited state and the emissive charge transferred state in **12** (Figure 11).

Similarly strong solvent polarity dependence of fluorescence properties was found in the C_{60} derivative **14** (Table 6). Photoinduced intramolecular electron transfer from the tertiary amine unit to the fullerene cage is most likely responsible for the fluorescence results [73].

Table 5 Photophysical Properties of the Amino-Fullerene Derivatives

Solvent	Dielectric Constant	11			12		
		Φ_F (10^{-4})	τ_{F1} (ns)	τ_{F2} (ns)	Φ_F (10^{-4})	τ_{F1} (ns)	τ_{F2} (ns)
Hexane	1.89	11	1.4		10	1.3	
Toluene	2.38	9.1	1.3		1.9	0.32	1.7
Carbon Disulfide	2.64	7.0	1.1		5.7	1.0	1.4
Chloroform	4.81	8.5	1.2		3.9	0.44	1.6
THF	7.58	2.0	0.17	1.5	<0.5	<0.1	
Dichloromethane	8.93	5.3	0.54	1.4	<0.5	<0.1	
Dichloromethane+ 1% (v/v) TFA		8.3	1.3				
Toluene+40% (v/v) Acetonitrile		1.3					
Toluene+40% (v/v) Acetonitrile+1% (v/v) TFA		9.0	1.5				
MCH	2.02				10	1.3	
MCH+10% (v/v) Acetone					1.3	0.32	1.7
MCH+10% (v/v) Acetone +1% (v/v) TFA					8.3	1.4	

Source: From Ref. 71.

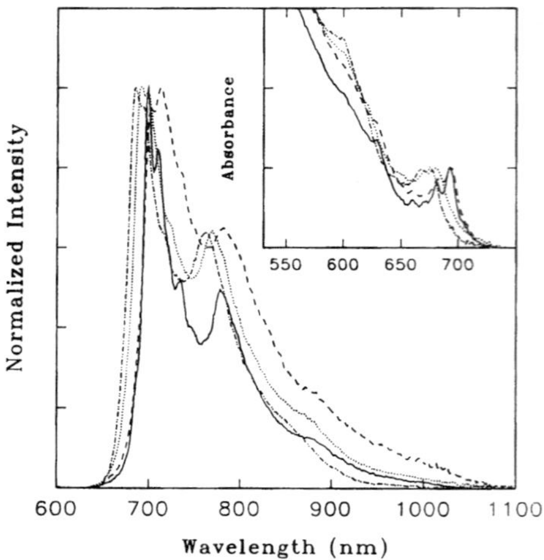

Figure 10 Absorption (inset) and fluorescence spectra of **12** in MCH (—) and mixtures of MCH-10% acetone (---), MCH-10% acetone-2% TFA (•••), and MCH-2% TFA (-••-). (*From Ref. 71.*)

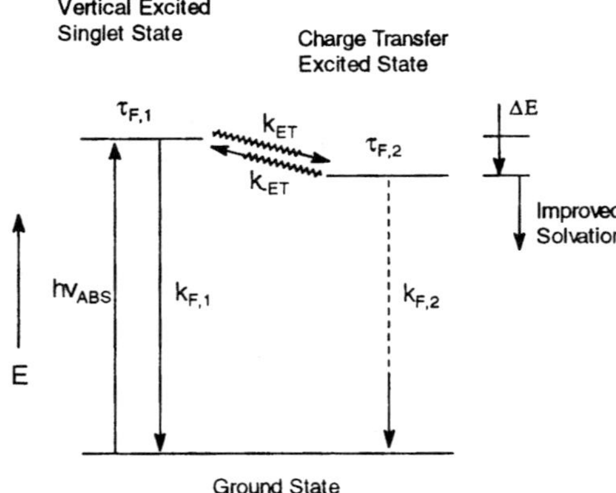

Figure 11 Schematic energy diagram for the intramolecular n-π^* electron transfer in amino-fullerene molecules. (*From Ref. 71.*)

Table 6 Fluorescence Parameters of **14** under Different Solvent Conditions

Solvent	Dielectric Constant	Φ_F ($\times 10^{-4}$)	τ_{F1} (ns)	τ_{F2} (ns)
Hexane	1.89	12	1.25	
Toluene	2.38	5.1	0.35	0.85
o-Xylene	2.57	5.8		
1,2,4-Trimethylbenzene	2.26	8.1		
1,2,3,5-Tetramethylbenzene	2.29	8.6		
Carbon Disulfide	2.64	6.9		
Chloroform	4.81	4.0		
THF	7.58	0.92		
Methylene Chloride	8.93	1.5	0.18	1.05
Methylene Chloride+1% TFA		11	1.3	
Dichlorobenzene	9.93	2.6		
Hexane+5% (v/v) Methylene Chloride		11		
Hexane+10% (v/v) Methylene Chloride		9.2		
Hexane+20% (v/v) Methylene Chloride		6.7		
Hexane+30% (v/v) Methylene Chloride		5.3		
Hexane+40% (v/v) Methylene Chloride		4.2		

Source: From Ref. 73.

The amino-C_{60} derivatives and related molecules are effectively redox dyads with some of the simplest spacers. They may be used as models for more complex redox systems, particularly with respect to effects of solvent environment on photoinduced intramolecular electron transfer processes.

There have been many recent reports on the preparation of fullerene-based redox dyads and triads for fundamental studies of photoinduced intramolecular electron transfers, modeling long-distance electron transfers in biological systems, and other purposes [79, 123–146]. The results have demonstrated that the fullerene cage is an excellent electron acceptor moiety in intramolecular redox systems. For example, Gust and coworkers investigated photoinduced energy and electron transfers in covalently linked porphyrin-fullerene dyads **29** using time-resolved fluorescence and transient absorption techniques [126]. The photoexcited state processes of the dyads with free-base and zinc porphyrins are strongly solvent-dependent. In toluene solution, the excited singlet state of the porphyrin moiety decays in ~20 ps through singlet-singlet energy transfer to the fullerene moiety. The fullerene excited singlet state thus formed undergoes intersystem crossing to the excited triplet state, which reaches equilibrium with the porphyrin excited triplet state [126]. In a more polar solvent benzonitrile, however, photoinduced electron transfer from the excited singlet porphyrin to the acceptor fullerene becomes a significant competing process to the singlet-singlet energy transfer. In addition, the fullerene excited singlet state from the energy transfer is also quenched by the porphyrin moiety through electron transfer. Thus, the overall yields for the formation of porphyrin-fullerene charge-separated state upon photoexcitation of the dyads in a polar solvent environment are close to unity. The charge transferred state has a lifetime of 290 ps in the dyad with free-base porphyrin and 50 ps in the zinc analog [126].

29

Williams et al. investigated the solvent dependence of photoinduced electron transfers in two donor–rigid spacer–C_{60} dyads [137]. For the dyad containing a 3-σ-bond bridge **30**, the photoexcited state intramolecular charge separation occurs very fast, followed by as fast charge recombination. The charge separation and recombination processes are both independent of the solvent polarity. However, in the dyad with an 11-σ-bond spacer **31**, photoinduced charge separation occurs only in a polar solvent environment [137]. The charge recombination in **31** becomes considerably slower. The results of **31** in polar solvents were explained such that because of the special symmetry properties, the fullerene π system may have very strong electronic coupling with the hydrocarbon bridge to allow fast photoinduced charge separation, while at the same time the electronic coupling relevant to the charge recombination remains relatively small [137].

30

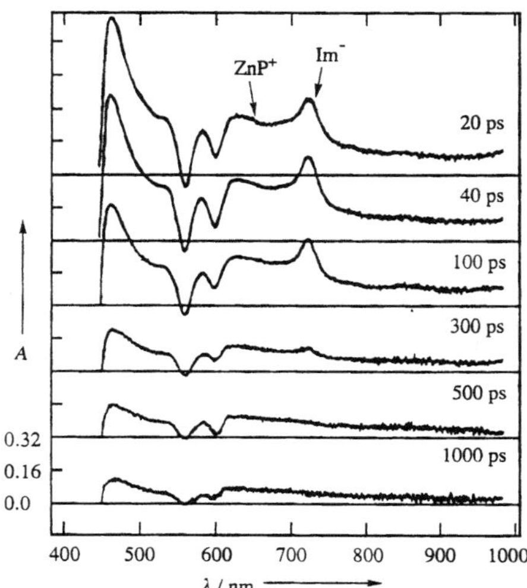

31

Photophysical properties of four zinc porphyrin-C_{60} dyads were studied systematically by Imahori et al. [131]. There are photoinduced charge separation and subsequent charge recombination in the dyads regardless of the spacers between the two chromophores. For the dyads in solution, the charge separation occurs from both excited singlet porphyrin and C_{60} moieties [131].

Imahori et al. also studied a zinc porphyrin (ZnP)-pyromellitimide (Im)-C_{60} triad **32** [130]. Upon photoexcitation of the triad in dioxane with a laser pulse at 590 nm, the initial electron transfer yields the $ZnP^{\bullet+}$-$Im^{\bullet-}$-C_{60} radical ion pair (Figure 12). The charge separated species $ZnP^{\bullet+}$-Im-$C_{60}^{\bullet-}$ is formed with an estimated rate constant of 1.6×10^{10} s^{-1}. The overall quantum yield for $ZnP^{\bullet+}$-Im-$C_{60}^{\bullet-}$ is 0.46 [130]. In polar solvents, the formation of the charge separated spe-

Figure 12 Absorption spectra of **32** in dioxane at different times (resolution in picosecond range) after excitation at 590 nm. The delay times between excitation and measurement are indicated on the right; A = absorbance. (*From Ref. 130.*)

cies becomes faster, but the charge recombination of the reverse process is even faster. As a result, the yield for $ZnP^{\bullet+}\text{-Im-}C_{60}^{\bullet-}$ is lower in a more polar solvent environment [130].

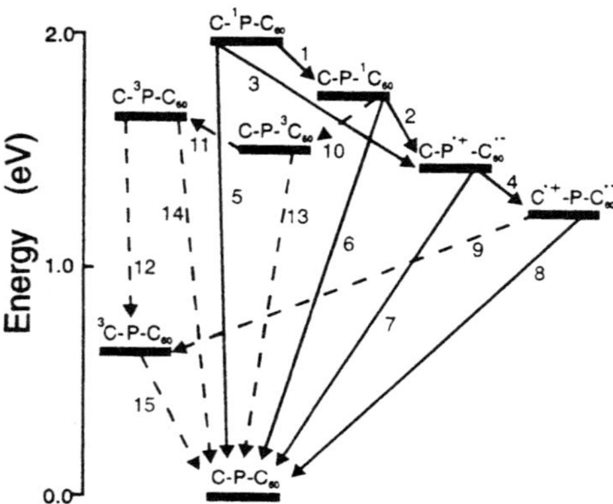

32

Liddell et al. prepared a molecular triad **33** that contains a C_{60} fullerene cage with a covalently linked diarylporphyrin (P) and a carotenoid polyene (C) moieties [124]. Upon photoexcitation, the triad in a room-temperature 2-methyl-THF solution undergoes electron transfer to form first a $C\text{-}P^{\bullet+}\text{-}C_{60}^{\bullet-}$ radical ion pair and then evolves into a $C^{\bullet+}\text{-}P\text{-}C_{60}^{\bullet-}$ charge transferred species with an overall quantum yield of 0.14. The $C^{\bullet+}\text{-}P\text{-}C_{60}^{\bullet-}$ ion pair state decays with a time constant of 170 ns to yield the carotenoid triplet state through charge recombination (Figure 13). Similar processes were observed for the triad **33** in a glassy matrix at 77 K [124]. The $C^{\bullet+}\text{-}P\text{-}C_{60}^{\bullet-}$ charge transferred species is formed with a quantum

Figure 13 Transient states of the triad **33** and their relevant interconversion pathways. Spin multiplicities for radical pair states are not included. (*From Ref. 124.*)

yield of ~0.10 in the low-temperature glassy medium, which again decays to form mainly the carotenoid excited triplet state through charge recombination. The initial photoinduced electron transfer seems not quantitative for the triad in the glassy matrix at 77 K because there is excited triplet C_{60} from the direct intersystem crossing process [124]. The photochemistry of the triad **33** was also studied at 20 K using time-resolved EPR spectroscopy [123]. The spin-polarized EPR signal of the charge transferred species $C^{\bullet+}$-P-$C_{60}^{\bullet-}$ was observed and simulated. There is a weak exchange interaction between the electrons in the ion pair ($J = 1.2G$) [123]. The $C^{\bullet+}$-P-$C_{60}^{\bullet-}$ state decays with a lifetime of 1.2 µs to produce the carotenoid triplet in a high yield. The spin polarization of $^3C^*$-P-C_{60} is characteristic of a triplet formed by the charge recombination of a singlet-derived radical pair. The kinetics for the decay of the $^3C^*$-P-C_{60} to the ground state was also examined [123].

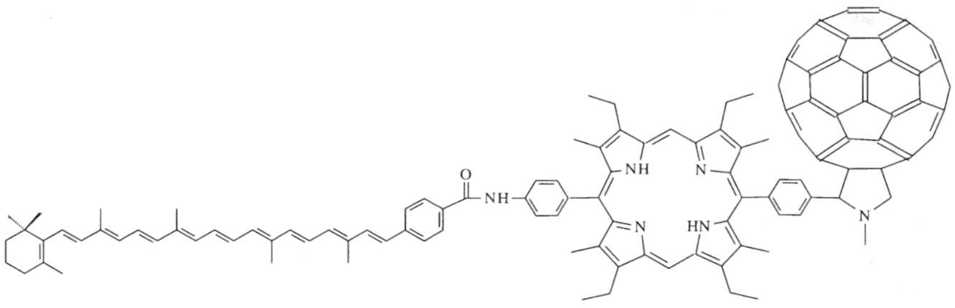

33

By preparing a series of redox dyads using ferrocene as a donor and fullerene as an acceptor, Guldi et al. investigated the spacer dependence of the photoinduced intramolecular electron transfers in the dyads [141]. Two mechanisms were suggested for intramolecular quenchings of the fullerene excited singlet state by the ferrocene moiety. One is the through-bond electron transfer and the other is through the formation of a transient intramolecular exciplex. The two mechanisms apply to the ferrocene-C_{60} dyads of different spacers [141].

Baran et al. synthesized two novel classes of porphyrin-fullerene hybrids with open crown-ether spacers [138]. UV-vis absorption and fluorescence properties of the hybrids were studied. In the presence of alkali metal cations, hybrid **34** forms a complex in which porphyrin-fullerene intramolecular interactions in both ground and excited states are enhanced. Because of the stronger intramolecular interactions, the absorption spectra of the hybrids in the presence of K^+ are red-shifted from those of the uncomplexed hybrids. The complexation also changes the fluorescence properties of the hybrids. The fluorescence intensities due to the porphyrin moiety in the hybrids become substantially lower in the presence of low-concentration K^+ [138].

34

There is a different class of redox active molecular systems based on fullerenes. These systems take advantage of the two important fullerene properties. One is the ability for a fullerene cage to accept multiple electrons (up to 6 [19]) and the other is the possibility to carry out multiple additions to a fullerene cage in a controllable fashion. Recently, Sun and coworkers prepared a series of multiple-functionalized C_{60} derivatives that may be described as fullerene-(donor)$_n$ redox active molecular systems, where the subscript n varies from 0 to 12. For example, the redox molecular systems **35–40** using the N,N-dimethylaniline (DMA) unit as electron donor are studied for their photophysical and electron transfer properties [144].

35: $R_1 = R_2 = $

36: $R_1 = $ $R_2 = $

37: $R_1 = R_2 = $

38

39

40

$D = $

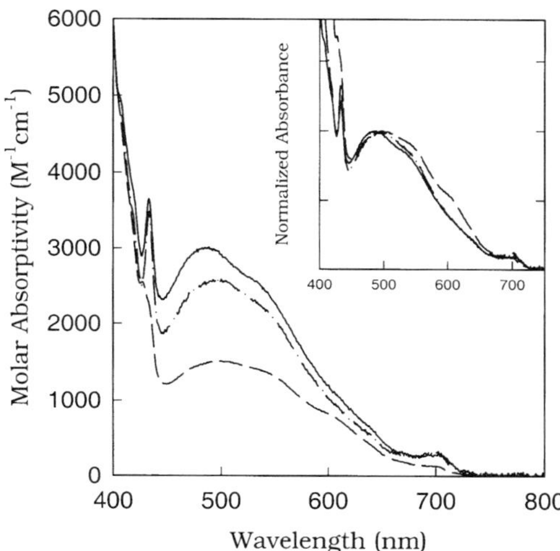

Figure 14 Absorption spectra of the fullerene-(donor)$_n$ molecules **35** (---), **36** (-•-), and **37** (—) in room-temperature solutions. Shown in the inset is a comparison of the normalized absorption spectra. (*From Ref. 144.*)

The absorption spectra of **35–37** are shown in Figure 14. There is a clear trend that the absorptivities increase with the number of electron donating DMA units in the molecules. It is known that there are ground-state C_{60}-DMA charge transfer complexes in a solution of C_{60} in neat DMA and that the complexes absorb more strongly in the visible than the parent C_{60} [7, 118–122]. Thus, the absorptivity increases in the C_{60}-(DMA)$_n$ molecules of larger n values may be attributed to absorption contributions from the ground-state intramolecular charge transfer complexes.

The fluorescence spectra of **35–37** in room-temperature toluene have also been obtained. As shown in Figure 15, the spectra of **35** and **36** are essentially the same; they are also similar to those of other methano-C_{60} derivatives [70, 72, 144]. The fluorescence quantum yields of **35** and **36** in toluene are quite similar as well (Table 7). For **37** in toluene, however, the fluorescence spectrum is different, showing an additional broad emission band at longer wavelengths (Figure 15). The new emission band may be assigned to intramolecular fullerene cage–DMA exciplexes in **37** [144]. Thus, the observed fluorescence quantum yield of **37** in toluene is a combination of emission contributions from both the excited singlet state and intramolecular exciplexes in **37**. The portion assignable to the excited singlet state of **37** is lower than the fluorescence yield of **35**, which may be

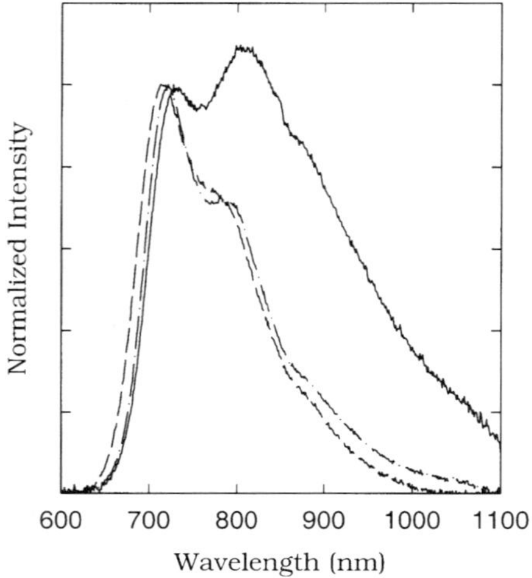

Figure 15 Normalized fluorescence spectra of the fullerene-(donor)$_n$ molecules systems **35** (---), **36** (-•-), and **37** (—) in room-temperature solutions. (*From Ref. 144.*)

explained in terms of the excited state quenching through electron transfer in **37**. With the emission due to exciplexes included, the observed total fluorescence yield of **37** is in fact slightly higher than that of **35** in toluene (Table 7).

Fluorescence properties of **35–37** were also measured in polar solvents such as methylene chloride [144]. The fluorescence spectrum and quantum yield of **35** in methylene chloride are similar to those in toluene, showing no solvent polarity dependence (Figure 15). For **36** and **37** in methylene chloride, however, there are no detectable fluorescence emissions. The strong solvent polarity effects on excited singlet state electron transfer processes in **36** and **37** are similar to

Table 7 Photophysical Properties of the Fullerene-(DMA)$_n$ Molecules

Compound	Solvent	ABS$_{max}$ (nm)	ε_{max} (M^{-1}cm^{-1})	FLSC$_{0-0}$ (nm)	Φ_F (10^{-3})
C$_{60}$	Toluene				0.32
35 ($n = 0$)	Dichlorobenzene	497	1500	713	1.2
36 ($n = 1$)	Toluene	498	2600	721	1.6
37 ($n = 2$)	Toluene	489	3000	727	1.4

Source: From Ref. 144.

those in the C_{60}-DMA intermolecular electron transfer interactions, though the effects are apparently more dramatic in the intramolecular systems [144].

Fullerene cages are three-dimensional molecular species capable of undergoing multiple functionalizations with multiple electron donors. The results of **35–37** provided the initial evidence for the notion that fullerene molecules are ideal building blocks for (donor)$_n$-acceptor type three-dimensional redox molecular systems for a number of potential technological applications.

ACKNOWLEDGMENT

Financial support from the National Science Foundation (CHE-9320558 and CHE-9727506), the Department of Energy through the DoE/EPSCoR cooperative agreement, and the Center for Advanced Engineering Fibers & Films, a National Science Foundation Engineering Research Center at Clemson University, is gratefully acknowledged.

REFERENCES

1. Foote, C. S., in *Topics in Current Chemistry: Electron Transfer I* (J. Mattay, ed.), Springer-Verlag, Berlin, 1994, p. 347.
2. Foote, C. S., in *Light-Activated Pest Control, ACS Symp. Ser. 616* (J. R. Heitz, K. R. Downum, eds.), American Chemical Society, Washington D.C. 1995, p. 17.
3. Foote, C. S., in *Physics and Chemistry of the Fullerenes* (K. Prassides, ed.), Kluwer Academic Publishers, Dordrecht, 1994, p. 79.
4. Gust, D.; Moore, T. A.; Moore, A. L. *Res. Chem. Intermed.* **1997**, *23*, 621.
5. Imahori, H.; Sakata, Y. *Adv. Mater.* **1997**, *9*, 537.
6. Leach, S. K., in *Physics and Chemistry of the Fullerenes* (K. Prassides, ed.), Kluwer Academic Publishers, Dordrecht, 1994, p. 117.
7. Sun, Y.-P., in *Molecular and Supramolecular Photochemistry, vol. 1* (V. Ramamurthy, and K. S. Shanze, eds.) Marcel Dekker, New York, 1997, p. 325.
8. Tutt, L. W.; Kost, A. *Nature* **1992**, *356*, 225.
9. Tutt, L.; Boggess, T. F. *Prog. Quantum Electron.* **1993**, *17*, 299.
10. Perry, J. W., in *Nonlinear Optics of Organic Molecules and Polymers* (H. S. Nalwa and S. Miyata, eds.), CRC Press, New York, 1997, p. 813.
11. Van Stryland, E. W.; Hagan, D. J.; Xia, T.; Said, A. A., in *Nonlinear Optics of Organic Molecules and Polymers* (H. S. Nalwa and S. Miyata, eds.) CRC Press, New York, 1997, p. 841.
12. *Materials for Optical Limiting—Proc. Mater. Res. Soc. Symp, vol. 374* (R. Crane, K. Lewis, E. Van Stryland, and M. Khoshnevisan, eds.), Materials Research Society, Pittsburgh, PA, 1995.
13. *Nonlinear Optical Liquids and Power Limiters—Proc. SPIE, vol. 3146* (C. M. Law-

son, ed.), Society of Photo-Optical Instrumentation Engineers, Bellingham, WA, 1997.

14. *Materials for Optical Limiting II—Proc. Mater. Res. Soc. Symp., vol. 479* (R. Sutherland, R. Pachter, P. Hood, D. Hagan, K. Lewis, and J. Perry, eds.), Materials Research Society, Pittsburgh, PA, 1998.

15. McLean, D. G.; Sutherland, R. L.; Brant, M. C.; Brandelik, D. M. *Opt. Lett.* **1993**, *18*, 858.

16. Riggs, J. E.; Sun, Y.-P. *J. Phys. Chem. A*, **1999**, *103*, 485.

17. Sun, Y.-P.; Riggs, J. E. *Int. Rev. Phys. Chem.* **1999**, *18*, 43.

18. Imahori, H.; Hagiwara, K.; Akiyama, T.;Aoki, M.; Taniguchi, S.; Okada, T.; Shirakawa, M.; Sakata, Y.; *Chem. Phys. Lett.* **1996**, *263*, 545.

19. Xie, Q.; Perez-Cordero, E.; Echegoyen, L. *J. Am. Chem. Soc.* **1992**, *114*, 3978.

20. (a) Lawson, G. E.; Kitaygorodskiy, A.; Ma, B.; Bunker, C. E.; Sun, Y.-P. *J. Chem. Soc., Chem. Commun.* **1995**, 2225. (b) Liou, K. F.; Cheng, C. H. *J. Chem. Soc., Chem. Commun.* **1996**, 14. (c) Lawson, G. E.; Guduru, R.; Kitaygorodskiy, A.; Sun, Y.-P. *J. Org. Chem.* submitted.

21. Prato, M.; Maggini, M. *Acc. Chem. Res.* **1998**, *31*, 519.

22. Martin, N.; Sanchez, L.; Illescas, B.; Perez, I. *Chem. Rev.* **1998**, *98*, 2527.

23. Prato, M. *J. Mater. Chem.* **1997**, *7*, 1097.

24. *Science of Fullerenes and Carbon Nanotubes* (M. S. Dresselhaus, G. Dresselhaus, and P. C. Eklund, eds.), Academic Press, San Diego, 1996.

25. Diederich, F.; Thilgen, C. *Science* **1996**, *271*, 31.

26. *The Chemistry of Fullerenes* (R. Taylor, ed.) World Scientific, Singapore, 1995.

27. Hirsch, A., *The Chemistry of the Fullerenes*, Thieme, Stuttgart, 1994.

28. (a) Miller, E. K.; Lee, K.; Hasharoni, K.; Hummelen, J. C.; Wudl, F.; Heeger, A. J. *J. Chem. Phys.* **1998**, *108*, 1390. (b) Miller, E. K.; Lee, K.; Cornil, J.; Pei, Q.; Wudl, F.; Heeger, A. J.; Bredas, J. L. *Synth. Met.* **1997**, *84*, 631. (c) Hasharoni, K.; Keshavarz, K. M.; Sastre, A.; Gonzalez, R.; Bellavia, L. C.; Greenwald, Y.; Swager, T.; Wudl, F.; Heeger, A. J. *J. Chem. Phys.* **1997**, *107*, 2308. (d) Kraabel, B.; Hummelen, J. C.; Vacar, D.; Moses, D.; Sariciftci, N. S.; Heeger, A. J.; Wudl, F. *J. Chem. Phys.* **1996**, *104*, 4267.

29. Brabec, C. J.; Dyakonov, V.; Sariciftci, N. S. *J. Chem. Phys.* **1998**, *109*, 1185.

30. Vacar, D.; Maniloff, E. C.; McBranch, D. W.; Heeger, A. J. *Phys. Rev. B* **1997**, *56*, 4573.

31. Itaya, A.; Suzuki, I.; Tsuboi, Y.; Miyasaka, H. *J. Phys. Chem. B* **1997**, *101*, 5118.

32. Kakimoto, M.; Kashihara, H.; Kashiwagi, T.; Takiguchi, T. *Macromolecules* **1997**, *30*, 7816.

33. Yoshino, K.; Tada, K.; Fujii, A.; Hosoda, K.; Kawabe, S.; Kajii, H.; Hirohata, M.; Hidayat, R.; Araki, H.; Zakhidov, A. A.; Sugimoto, R.; Iyoda, M.; Ishikawa, M.; Masuda, T. *Fullerene Sci. Tech.* **1997**, *5*, 1359.

34. Khairullin, I. I.; Chen, Y.-H.; Hwang, L.-P. *Chem. Phys. Lett.* **1997**, *275*, 1.

35. Gao, J.; Hide, F.; Wang, H. *Synth. Met.* **1997** *84*, 979.

36. Bruening, J.; Friedman, B. *J. Chem. Phys.* **1997**, *106*, 9634.

37. Qian, J.; Xu, C.; Qian, S.; Peng, W. *Chem. Phys. Lett.* **1996**, *257*, 563.

38. Bouchtalla, S.; Deronzier, A.; Janot, J.-M.; Moutet, J.-C.; Seta, P. *Synth. Met.* **1996**, *82*, 129.

39. Akashi, K. Y. T.; Morita, S.; Yoshida, M.; Hamaguchi, M.; Tada, K.; Fujii, A.; Kawai, T.; Uto, S.; Ozaki, M.; Onoda, M.; Zakhidov, A. A. *Synth. Met.* **1995**, *70*, 1317.

40. Nath, S.; Pal, H.; Palit, D. K.; Sapre, A. V.; Mittal, J. P. *J. Phys. Chem. B* **1998**, *102*, 10158.

41. Negri, F.; Orlandi, G. *J. Chem. Phys.* **1998**, *108*, 9675.

42. Sassara, A.; Zerza, G.; Chergui, M. *J. Phys. Chem. A* **1998**, *102*, 3072.

43. Sassara, A.; Zerza, G.; Chergui, M.; Negri, F.; Orlandi, G. *J. Chem. Phys.* **1997**, *107*, 8731.

44. Ahn, J. S.; Suzuki, K.; Iwasa, Y.; Otsuka, N.; Mitani, T. *J. Lumin.* **1998**, *76*, 201.

45. Ahn, J. S.; Suzuki, K.; Iwasa, Y.; Mitani, T. *J. Lumin.* **1997**, *72*, 464.

46. Wang, D.; Zuo, J.; Zhang, Q.; Luo, Y.; Ruan, Y.; Wang, Z. *J. Appl. Phys.* **1997**, *81*, 1413.

47. Ichida, M.; Sakai, M.; Yajima, T.; Nakamura, A. *J. Lumin.* **1997**, *72*, 499.

48. Argentine, S. M.; Kotz, K. T.; Rudalevige, T.; Zaziski, D.; Francis, A. H.; Zand, R.; Schleuter, J. A. *Res. Chem. Intermed.* **1997**, *23*, 601.

49. Fedorov, A.; Berberan-Santos, M. N.; Lefevre, J.-P.; Valeur, B. *Chem. Phys. Lett.* **1997**, *267*, 467.

50. Werner, A. T.; Byrne, H. J.; Roth, S. *Fullerene Sci. Technol.* **1996**, *4*, 757.

51. Palit, D. K.; Mittal, J. P. *Fullerene Sci. Technol.* **1995**, *3*, 643.

52. Sun, Y.-P.; Ma, B. *J. Chem. Soc. Perkin Trans. 2* **1996**, 2157.

53. Riggs, J. E.; Bunker, C. E.; Sun, Y.-P., unpublished results.

54. (a) Salazar, F. A.; Fedorov, A.; Berberan-Santos, M. N. *Chem. Phys. Lett.* **1997**, *271*, 361. (b) Berberan-Santos, M. N.; Garcia, J. M. M. *J. Am. Chem. Soc.* **1996**, *118*, 9391.

55. (a) Kim, D.; Lee, M.; Suh, Y. D.; Kim, S. K. *J. Am. Chem. Soc.* **1992**, *194*, 446. (b) Lee, M.; Song, O.-K.; Seo, J.-C.; Kim, D.; Suh, Y. D.; Jin, S. M.; Kim, S. K. *Chem. Phys. Lett.* **1992**, *196*, 325.

56. (a) Palit, D. K.; Sapre, A. V.; Mittal, J. P.; Rao, C. N. R. *Chem. Phys. Lett.* **1992**, *195*, 1. (b) Williams, R. M.; Verhoeven, J. W. *Chem. Phys. Lett.* **1992**, *194*, 446.

57. Tanigaki, K.; Ebbesen, T. W.; Kuroshima, S. *Chem. Phys. Lett.* **1991**, *185*, 189.

58. Ebbesen, T. W.; Tanigaki, K.; Kuroshima, S. *Chem. Phys. Lett.* **1991**, *181*, 501.

59. Wasielewski, M. R.; O'Neil, M. P.; Lykke, K. R.; Pellin, M. J.; Gruen, D. M. *J. Am. Chem. Soc.* **1991**, *113*, 2774.

60. (a) Reindl, S.; Penzkofer, A.; Gratz, H. *J. Photochem. Photobiol. A* **1998**, *115*, 89. (b) Watanabe, A.; Ito, O.; Watanabe, M.; Saito, H.; Koishi, M. *J. Phys. Chem.* **1996**, *100*, 10518.

61. Bensasson, R. V.; Bienvue, E.; Janot, J.-M.; Land, E. J.; Leach, S.; Seta, P. *Chem. Phys. Lett.* **1998**, *283*, 221.

62. Kallay, M.; Nemeth, K.; Surjan, P. R. *J. Phys. Chem. A* **1998**, *102*, 1261.

63. Fujisuka, M.; Watanabe, A.; Ito, O.; Yamamoto, K.; Funasaka, H. *J. Phys. Chem. A* **1997**, *101*, 4840.

64. Guldi, D. M.; Liu, D.; Kamat. P. V. *J. Phys. Chem. A* **1997**, *101*, 6195.

65. Guldi, D. M. *J. Phys. Chem. B* **1997**, *101*, 9600.

66. Kamat, P. V.; Sauve, G.; Guldi, D. M.; Asmus, K.-D. *Res. Chem. Intermed.* **1997**, *23*, 575.

67. (a) Guldi, D. M.; Asmus, K.-D. *J. Am. Chem. Soc.* **1997**, *119*, 5744. (b) Guldi, D. M.; Asmus, K.-D. *J. Am. Chem. Soc.* **1997**, *119*, 9588.

68. Fujitsuka, M.; Watanabe, A.; Ito, O. *J. Phys. Chem. A* **1997**, *101*, 7960.

69. (a) Bunker, C. E.; Rollins, H. W.; Sun, Y.-P. *J. Chem. Soc., Perkin Trans. 2* **1996**, 1307. (b) Sun, Y.-P., in *Fullerenes, Recent Advances in the Chemistry and Physics of Fullerenes and Related Materials* (K. M. Kadish and R. S. Ruoff, eds.), Electrochemical Society, 1995, p. 510.

70. Sun, Y.-P.; Lawson, G. E.; Riggs, J. E.; Ma, B.; Wang, N.; Moton, D. K. *J. Phys. Chem. A* **1998**, *102*, 5520.

71. Sun, Y.-P.; Ma, B.; Bunker, C. E. *J. Phys. Chem. A* **1998**, *102*, 7580.

72. Ma, B.; Bunker, C. E.; Guduru, R.; Zhang, X.; Sun, Y.-P. *J. Phys. Chem. A* **1997**, *101*, 5626.

73. Sun, Y.-P.; Guduru, R.; Rollins, H. W.; Guo, Z., manuscript in preparation.

74. Higashida, S; Nishiyama, K.; Yusa, S.-I.; Morishima, Y.; Janot, J.-M.; Seta, P.; Imahori, H.; Kaneda, T.; Sakata, Y. *Chem. Lett.* **1998**, 381.

75. Lee, S.; Park Y. S.; Park, Y. W.; Li, Y.; Zhu, D.; Tada, K.; Kawai, T.; Yoshino, K. *Solid State Commun.* **1998**, *105*, 345.

76. Bensasson, R. V.; Bienvenue, E.; Fabre, C.; Janot, J.-M.; Land, E. J.; Leach, S.; Leboulaire, V.; Rassat, A.; Roux, S.; Seta, P. *Chem. Eur. J.* **1998**, *4*, 270.

77. Bensasson, R. V.; Bienvenue, E.; Janot, J.-M.; Leach, S.; Seta, P.; Schuster, D. I.; Wilson, S. R.; Zhao, H. *Chem. Phys. Lett.* **1995**, *245*, 566.

78. Higashida, S.; Nishiyama, K.; Yusa, S.-I.; Morishima, Y.; Janot, J.-M.; Seta, P.; Imahori, H.; Kaneda, T.; Sakata, Y. *Chem. Lett.* **1998**, 381.

79. Thomas, K. G.; Biju, V.; George, M. V.; Guldi, D. M.; Kamat, P. V. *J. Phys. Chem. A* **1998**, *102*, 5341.

80. Guldi, D. M.; Asmus, K. D. *J. Phys. Chem. A* **1997**, *101*, 1472.

81. Guldi, D. M. *J. Phys. Chem. A* **1997**, *101*, 3895.

82. Guldi, D. M.; Hungerbuehler, H.; Asmus, K.-D. *J. Phys. Chem.* **1995**, *99*, 9380.

83. Luo, C.; Fujitsuka, M.; Watanabe, A.; Ito, O.; Gan. L.; Huang, Y.; Huang, C.-H. *J. Chem. Soc., Faraday Trans.* **1998**, *94*, 527.

84. Zhou, D.-J.; Gan, L.-B.; Luo, C.-P.; Huang, C.-H. *Chin. J. Chem.* **1997**, *15*, 35.

85. Zhou, D.; Gan, L.; Tan, H.; Luo, C.; Huang, C.; Yao, G.; Zhang, B. *J. Photochem. Photobiol. A* **1996**, *99*, 37.

86. Nakamura, Y.; Minowa, T.; Hayashida, Y.; Tobita, S.; Shizuka, H.; Nishimura, J. *J. Chem. Soc., Faraday Trans.* **1996**, *92*, 377.

87. Williams, R. M.; Zwier, J. M.; Verhoeven, J. W. *J. Am. Chem. Soc.* **1995**, *117*, 4093.

88. Glenis, S.; Cooke, S.; Chen, X.; Labes, M. M. *Synth. Met.* **1995**, *70*, 1313.

89. Andeson, J. L.; Yi-Zhong, A.; Rubin, Y.; Foote, C. S. *J. Am. Chem. Soc.* **1994**, *116*, 9763.

90. Ma, B.; Riggs, J. E.; Sun, Y.-P. *J. Phys. Chem. B* **1998**, *102*, 5999.

91. Mohan, H.; Palit, D. K.; Mittal, J. P.; Chiang, L. Y.; Asmus, K. D.; Guldi, D. M. *J. Chem. Soc., Faraday Trans.* **1998**, *94*, 359.

92. Palit, D. K.; Mohan, H.; Mittal, J. P. *J. Phys. Chem. A* **1998**, *102*, 4456.

93. Palit, D. K.; Mohan, H.; Birkett, P. R.; Mittal, J. P. *J. Phys. Chem. A* **1997**, *101*, 5418.

94. Bensasson, R. V.; Hill, T. J.; Land, E. J.; Leach, S.; McGarvey, D. J.; Truscott, T. G.; Ebenhoch, J.; Gerst, M.; Ruchart, C. *Chem. Phys.* **1997**, *215*, 111.
95. Kadish, K. M.; Gao, X.; Van, C. E.; Hirasaka, T.; Suenobu, T.; Fukuzumi, S. *J. Phys. Chem. A* **1998**, *102*, 3898.
96. Sun, Y.-P.; Guduru, R.; Rollins, H. W.; Bunker, C. E., manuscript in preparation.
97. Riggs, J. E.; Sun, Y.-P., manuscript in preparation.
98. (a) Lamparth, I.; Maichle-Mossmer, C.; Hirsch, A. *Angew. Chem. Int. Ed. Engl.* **1995**, *34*, 1607. (b) Hirsch, A.; Lamparth, I.; Grosser, T. *J. Am. Chem. Soc.* **1994**, *116*, 9385. (c) Hirsch, A.; Lamparth, I.; Karfunkel, H. R. *Angew. Chem. Int. Ed. Engl.* **1994**, *33*, 437.
99. (a) Nojiri, T.; Watanabe, A.; Ito, O. *J. Phys. Chem. A* **1998**, *102*, 5215. (b) Nojiri, T.; Alam, M. M.; Konami, H.; Watanabe, A.; Ito, O. *J. Phys. Chem. A* **1997**, *101*, 7943. (c) Ito, O.; Sasaki, Y.; Watanabe, A.; Hoffmann, R.; Siedschlag, C.; Mattay, J. *J. Chem. Soc., Perkin Trans.* **1997**, *2*, 1007. (d) Sasaki, Y.; Fujitsuka, M.; Watanabe, A.; Ito, O. *J. Chem. Soc., Faraday Trans.* **1997**, *93*, 4275. (e) Alam, M. M.; Watanabe, A.; Ito, O. *J. Photochem. Photobiol. A.: Chem.* **1997**, *104*, 59. (f) Ito, O. *Res. Chem. Intermed.* **1997**, *23*, 389.
100. Hudhomme, P.; Boulle, C.; Rabreau, J. M.; Cariou, M.; Jubault, M.; Gorgues, A. *Synth. Met.* **1998**, *94*, 73.
101. Ghosh, H. N.; Palit, D. K.; Sapre, A. V.; Mittal, J. P. *Chem. Phys. Lett.* **1997**, *265*, 365.
102. Stasko, A.; Brezova, V.; Biskupic, S.; Dinse, K.-P.; Gross, R.; Baumgarten, M.; Guegel, A.; Belik, P. *J. Electroanal. Chem.* **1997**, *423*, 131.
103. Sun, Y.-P.; Bunker, C. E.; Liu, B. *Chem. Phys. Lett.* **1997**, *272*, 25.
104. Zhang, G. P.; Sun, X.; George, T. F.; Pandey, L. N. *J. Chem. Phys.* **1997**, *106*, 6398.
105. Yee, K. A.; Han, K. R.; Kim, C.-H.; Pyun, C.-H. *J. Phys. Chem. A* **1997**, *101*, 5692.
106. Guldi, D. M. *Res. Chem. Intermed.* **1997**, *23*, 653.
107. Michaeli, S.; Meiklyar, V.; Endeward, B.; Mobius, K.; Levanon, H. *Res. Chem. Intermed.* **1997**, *23*, 505.
108. Brezova, V.; Gugel, A.; Rapta, P.; Stasko, A. *J. Phys. Chem.* **1996**, *100*, 16232.
109. Janssen, R. A. J.; Jansen, J. F. G. A.; Van Haare, J. A. E. H.; Meijer, E. W. *Adv. Mater.* **1996**, *8*, 494.
110. Tronel, E.; Miquel, G.; Vanel, P. Seta, P. *Chem. Phys. Lett.* **1998**, *285*, 294.
111. Bourdelande, J. L.; Font, J.; Gonzalez-Moreno, R.; Nonell, S. *J. Photochem. Photobiol. A* **1998**, *115*, 69.
112. Crooks, E. R.; Eastoe, J.; Beeby, A. *J. Chem. Soc., Faraday Trans.* **1997**, *93*, 4131.
113. Konarev, D. V.; Semkin, V. N.; Lyubovskaya, R. N.; Graja, A. *Synth. Met.* **1997**, *88*, 225.
114. Takahashi, K.; Etoh, K.; Tsudo, Y.; Yamaguchi, T.; Komura, T.; Ito, S.; Murata K.; *J. Electroanal. Chem.* **1997**, *426*, 85.
115. Alam, M. M.; Sato, M.; Watanabe, A.; Akasaka, T.; Ito, O. *J. Phys. Chem. A.* **1998**, *102*, 7447.
116. Fukuzumi, S.; Suenobu, T.; Hirasaka, T.; Arakawa, R.; Kadish, K. M. *J. Am. Chem. Soc.* **1998**, *120*, 9220.

117. Fukuzumi, S.; Suenobu, T.; Patz, M.; Hirasaka, T.; Itoh, S.; Fujitsuka, M.; Ito, O. *J. Am. Chem. Soc.* **1998**, *120*, 8060.

118. Sibley, S. P.; Campbell, R. L.; Silber, H. B. *J. Phys. Chem.* **1995**, *99*, 5274.

119. Sun, Y.-P.; Bunker, C. E.; Ma, B. *J. Am. Chem. Soc.* **1994**, *116*, 9692.

120. Seshardi, R.; D'Souza, F.; Krishnan, V.; Rao, C. N. R. *Chem. Lett.* **1993**, 217.

121. Wang, Y. *J. Phys. Chem.* **1992**, *96*, 764.

122. Sension, R. J.; Szarka, A. Z.; Smith, G. R.; Hochstrasser, R. M. *Chem. Phys. Lett.* **1991**, *185*, 179.

123. Carbonera, D.; Di Valentin, M.; Corvaja, C.; Agostini, G.; Giacometti, G.; Liddell, P. A.; Kuciauskas, D.; Moore, A. L.; Moore, T. A.; Gust, D. *J. Am. Chem. Soc.* **1998**, *120*, 4398.

124. Liddell, P. A.; Kuciauskas, D.; Sumida, J. P.; Nash, B.; Nguyen, D.; Moore, A. L.; Moore, T. M.; Gust, D. *J. Am. Chem. Soc.* **1997**, *119*, 1400.

125. Gust, D.; Moore, A. L.; Moore, T. A. *Res. Chem. Intermed.* **1997**, *23*, 621.

126. Kuciauskas, D.; Lin, S.; Seely, G. R.; Moore, A. L.; Moore, T. A.; Gust, D.; Drovetskaya, T.; Reed, C. A.; Boyd, P. D. W. *J. Phys. Chem.* **1996**, *100*, 15926.

127. Imahori, H.; Cardoso, S.; Tatman, D.; Lin, S.; Noss, L.; Seely, G. R.; Sereno, L.; Chessa De Silber, J.; Moore, T. A.; Moore, A. L.; Gust, D. *Photochem. Photobiol.* **1995**, *62*, 1009.

128. Liddell, P. A.; Sumida, J. P.; Macpherson, A. N.; Noss, L.; Seely, G. R.; Clark, K. N.; Moore, A. L.; Moore, T. M.; Gust, D. *Photochem. Photobiol.* **1994**, *60*, 537.

129. Higashida, S.; Imahori, H.; Kaneda, T.; Sakata, Y. *Chem. Lett.* **1998**, 605.

130. Imahori, H.; Yamada, K.; Hasegawa, M.; Taniguchi, S.; Okada, T.; Sakata, Y. *Angew. Chem. Int. Ed. Engl.* **1997**, *36*, 2626.

131. Imahori, H.; Hagiwara, K.; Aoki, M.; Akiyama, T.; Taniguchi, S.; Okada, T.; Shirakawa, M.; Sakata, Y. *J. Am. Chem. Soc.* **1996**, *118*, 11771.

132. Imahori, H.; Sakata, Y. *Chem. Lett.* **1996**, 199.

133. Imahori, H.; Hagiwara, T.; Akiyama, T.; Taniguchi, S.; Okada, T.; Sakata, Y. *Chem. Lett.* **1995**, 265.

134. Bell, T. D. M.; Smith, T. A.; Ghiggino, K. P.; Ranasinghe, M. G.; Shephard, M. J.; Paddon-Row, M. N. *Chem. Phys. Lett.* **1997**, *268*, 223.

135. Ranasinghe, M. C.; Oliver, A. M.; Rothenfluh, D. F.; Salek, A.; Paddon-Row, M. M. *Tetrahedron Lett.* **1996**, *37*, 4797.

136. Lawson, J. M.; Oliver, A. M.; Rothenfluh, D. F.; An, Y.-Z.; Ellis, G. A.; Ranasinghe, M. G.; Khan, S. I.; Franz, A. G.; Ganapathi, P. S.; Shepherd, M. J.; Paddon-Row, M.; Rubin, Y. *J. Org. Chem.* **1996**, *61*, 5032.

137. Williams, R. M.; Koeberg, M.; Lawson, J. M.; An, Y.-Z.; Rubin, Y.; Paddon-Row, M.; Verhoeven, J. W. *J. Org. Chem.* **1996**, *61*, 5055.

138. Baran, P. S.; Monaco, R. R.; Khan, A. U.; Schuster, D. I.; Wilson, S. R. *J. Am. Chem. Soc.* **1997**, *119*, 8363.

139. Safonov, I. G.; Baran, P. S.; Schuster, D. I. *Tetrahedron Lett.* **1997**, *38*, 8133.

140. Jensen, A. W.; Wilson, S. R.; Schuster, D. I. *Bioorg. Medicinal Chem.* **1996**, *4*, 767.

141. (a) Guldi, D. M.; Maggini, M.; Scorrano, G.; Prato, M. *J. Am. Chem. Soc.* **1997**, *119*, 974. (b) Guldi, D. M.; Maggini, M.; Scorrano, G.; Prato, M. *Res. Chem. Intermed.* **1997**, *23*, 561.

142. Sariciftci, N. S.; Wudl, F.; Heeger, A. J.; Maggini, M.; Scorrano, G.; Prato, M.; Bourassa, J.; Ford, P. C. *Chem. Phys. Lett.* **1995**, *247*, 510.
143. Maggini, M.; Dono, A.; Scorrano, G.; Prato, M. *J. Chem. Soc., Chem. Commun.* **1995**, 845.
144. Huang, W.; Guo, Z.; Sun, Y.-P., manuscript in preparation.
145. Drovetskaya, T.; Reed, C. A.; Boyd, P. *Tetrahedron Lett.* **1995**, *36*, 7971.
146. Linssen, T. G.; Durr, K.; Hanack, M.; Hirsch, A. *J. Chem. Soc., Chem. Commun.* **1995**, 103.

4
Optical Studies of Fullerene Triplet States

R. Bruce Weisman
Rice University
Houston, Texas

1 INTRODUCTION

In the gas phase or in solution, optically excited fullerenes lose their excess energy in ways typical of individual molecules influenced relatively weakly by their environments. Traditional molecular photophysics therefore provides the proper tools and concepts for describing the behavior of electronically energized fullerenes. A useful starting point in such descriptions is a Jablonski diagram, as shown in Figure 1. Here, each horizontal line marks the energy of a molecular electronic state plus the corresponding zero-point vibrational energy. This type of diagram, which is appropriate for compounds composed of low-Z atoms, sorts a molecule's electronic states into "singlets," with no net electron spin, and "triplets," with net electron spin of 1. For typical compounds having closed-shell ground states, such as those discussed here, these two categories include all electronic states that are important in photophysics. A molecule in S_0, the singlet ground state, can efficiently absorb light at an appropriate visible or ultraviolet wavelength to generate an excited singlet state. Similar absorption to a triplet state is far less probable because of the spin-conserving optical selection rule.

Once promoted high into the manifold of excited singlets, a molecule will typically undergo rapid radiationless decay to S_1, the lowest-lying excited singlet state. This process converts the difference in electronic state energies into excess vibrational excitation, which is then drained into the surroundings on a picosecond time scale in typical condensed phase environments. The resulting vibra-

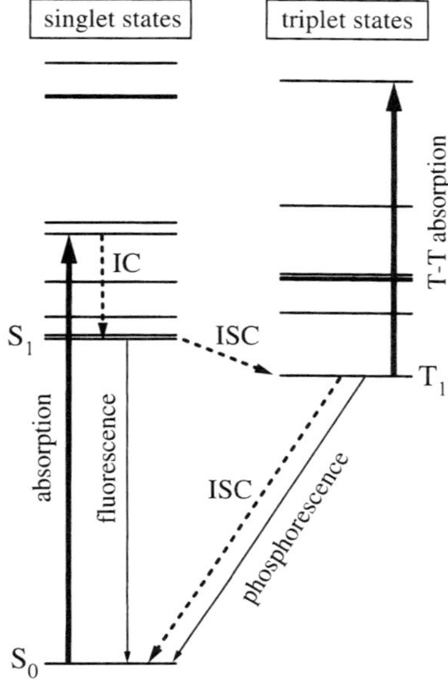

Figure 1 Schematic electronic energy-level diagram showing radiative and nonradiative transitions.

tionally thermalized S_1 state may relax further by fluorescence emission to S_0 or by secondary radiationless decay to S_0 ("internal conversion") or to a triplet state ("intersystem crossing"). $S_1 \rightarrow S_0$ internal conversion is omitted for clarity from Figure 1 because that channel is slow compared to $S_1 \rightarrow T_n$ intersystem crossing for C_{60}, C_{70}, and many of their derivatives. A triplet state formed through intersystem crossing quickly relaxes to T_1, the triplet of lowest energy. Unimolecular electronic decay of T_1 to S_0 then normally occurs only through two slow channels: phosphorescent photon emission, which must overcome the selection rule for spin conservation, and $T_1 \rightarrow S_0$ intersystem crossing, which is slowed both by spin conservation and by the large energy gap between the two states. In the most common fullerenes, fluorescence and phosphorescence compete poorly against the nonradiative steps and therefore have very small quantum yields. The dominant decay sequence after optical excitation is therefore internal conversion from S_n to S_1, followed by intersystem crossing and internal conversion from S_1 through T_n to T_1, and finally intersystem crossing from T_1 back to S_0. With a high quantum yield for formation and a lifetime orders of magnitude longer than other excited states, T_1 acts as a kinetic bottleneck and a centrally important metastable state for fullerene photophysics and photochemistry. This state, commonly referred to simply as the ''triplet,'' commands attention both for its intrinsic scientific interest and for its importance in potential photoapplications.

Although the small rate constants for spin-forbidden radiative decay can make triplet states difficult to monitor through optical emission methods, strong spin-allowed electronic absorptions within the triplet manifold allow observation of T_1 populations through induced absorption spectroscopy. As suggested by Figure 1, the most common wavelengths for such $T_n \leftarrow T_1$ absorption studies fall in the region of ground-state transparency to the red of the $S_1 \leftarrow S_0$ transition. Induced optical absorption may be measured as a function of probe wavelength to find a sample's triplet-triplet absorption spectrum, or as a function of delay after pulsed optical excitation to measure T_1 population kinetics. This chapter will summarize kinetic, spectroscopic, and thermodynamic findings about fullerene triplet states based mainly on such transient optical absorption methods. The emphasis will be on results from the author's laboratory.

In addition to the unimolecular triplet decay channels of phosphorescence and intersystem crossing, there are bimolecular channels that may become kinetically significant or even dominant under various conditions. The scheme below shows several of these decay processes.

$$
\begin{aligned}
\text{Phosphorescence:} \qquad & T_1 \xrightarrow{k_{rad}} S_0 \\[4pt]
\text{Intersystem crossing:} \qquad & T_1 \xrightarrow{k_1} S_0 \\[4pt]
\text{Oxygen quenching:} \qquad & T_1 + O_2 \xrightarrow{k_{O_2}} S_0 + O_2^* \\[4pt]
\text{Triplet-triplet annihilation:} \qquad & T_1 + T_1 \xrightarrow{k_{TT}} S_0 + T_1 \\[4pt]
\text{Energy transfer:} \qquad & T_1 + S_0' \underset{k_{et}'}{\overset{k_{et}}{\rightleftharpoons}} S_0 + T_1' \\[4pt]
\text{Self-quenching:} \qquad & T_1 + S_0 \xrightarrow{k_{sq}} 2S_0 \\[4pt]
\text{Electron transfer:} \qquad & T_1 + \text{donor} \xrightarrow{k_{red}} D_0^{\bullet-} + \text{donor}^+
\end{aligned}
\tag{1}
$$

The symbol $D_0^{\bullet-}$ in the electron transfer equation denotes the doublet ground state of the radical anion. When sample molecules are immobilized in a solvent of high viscosity, such as a low-temperature glass, crystal, or matrix, the bimolecular contributions to triplet-state decay become negligible relative to unimolecular components. However, in fluid solvents, molecular mobilities may be large enough to let excited sample molecules encounter quenching species during their unimolecular lifetime. This leads to accelerated depopulation of the triplet state. Such quenching by oxygen, impurities, ground-state sample molecules, or other added solutes normally gives pseudo-first-order rather than second-order kinetic components. As a result, the observed triplet decay kinetics will appear accurately first-order, but with a rate constant that is inflated over the true unimolecular rate constant by contributions from bimolecular quenching. By contrast, the triplet-

triplet annihilation channel obeys true second-order decay kinetics. Although it is possible in principle to separate first- and second-order kinetic components in a sample displaying concurrent decays of both types, in practice the findings will have large uncertainties unless data of very high quality are available.

It is challenging, then, to extract a fluid sample's basic unimolecular and bimolecular rate constants from measurements of triplet population kinetics. To determine the unimolecular or "intrinsic" decay constant (whose inverse is the intrinsic lifetime), one should first reduce the concentrations of dissolved oxygen, solvent impurities, and solute impurities to negligible levels. The sample concentration should also be kept as low as possible to suppress self-quenching from ground-state molecules. In addition, the concentration of excited states should be minimized to reduce the amount of second-order decay from triplet-triplet annihilation. These considerations point to the benefits of instrumentation that can measure very small levels of induced absorption. Using such instruments, one may keep ground- and excited-state concentrations low enough to measure nearly pure intrinsic triplet decays in many fullerene samples. By contrast, with less sensitive equipment one may instead have to isolate first-order components from mixed-order kinetic data for several sample concentrations and then extrapolate the first-order rate constant to zero concentration. Accurate, sensitive induced absorption instruments are also very valuable for measuring $T_n \leftarrow T_1$ spectra that are reliable enough for quantitative shape analyses.

2 EXPERIMENTAL METHODS

Induced absorption measurements are never easy because one must accurately determine small changes in a large quantity, the probe intensity. To measure triplet kinetics with high sensitivity, our laboratory has constructed an apparatus that is customized for the characteristics of fullerene samples. Fullerene transient spectra show the presence of $T_n \leftarrow T_1$ bands in the red and near-infrared spectral regions, with many peak positions in the 650–1050 nm range. This region includes the output wavelengths of commercially available continuous-wave diode lasers, some of which have nearly Gaussian beam profiles and excellent intensity stability. Although they are fixed in wavelength, such diode lasers make nearly ideal sources of probe light for measurements of fullerene induced absorption kinetics. A second area of instrument customization involves detection bandwidth. We find that the intrinsic lifetimes of fullerene triplet states vary considerably, as discussed below, but are no shorter than several microseconds. It is therefore useful to limit the time resolution of the apparatus to approximately 0.5–1 μs in order to suppress noise and optimize sensitivity. Another useful design feature is collinear or near-collinear alignment of the excitation and probing beams, allowing greater sensitivity through longer sample path lengths. A final consideration is the need to suppress systematic errors as well as photometric random noise. Small induced absorptions can be reliably measured only after one has eliminated such problems as optical and electronic cross-talk between excitation and probing lasers, thermal lensing artifacts, and spurious transients arising from optical elements instead of the sample.

Figure 2 shows a schematic diagram of an apparatus constructed for fixed-wavelength measurements of triplet kinetics. In studies of fluid samples near room temperature, solutions are sealed in a cylindrical spectrophotometer cell (path length of 0.5, 1.0, or 5 cm) that has been fused to a tube used for freeze-pump-thaw degassing and a fitting for attachment to a vacuum line. Thermostatted water is circulated through a jacket of the cell to control the sample temperature. Nanosecond-scale 355 or 532 nm excitation pulses from a small Q-switched Nd:YAG laser system (New Wave Research, Mini-Lase II-20) are partially reflected from an uncoated fused silica window used as a beam combiner and directed into the sample cell. The 676 nm output of a continuous diode laser (PTI model ACM15) passes through the beam-combining window and traverses the sample cell collinearly with the excitation beam. The probing beam is angularly

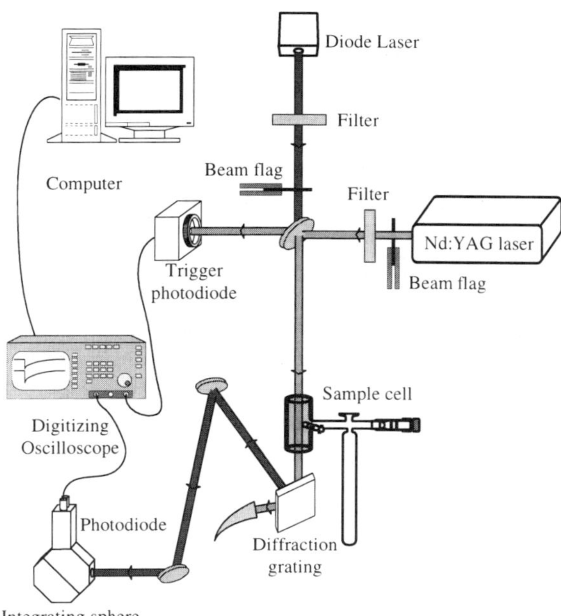

Figure 2 Schematic diagram of an apparatus designed to measure induced absorption kinetics at a single wavelength with high accuracy and precision.

separated from the residual excitation beam by a diffraction grating and then mildly focused through a long-pass filter into the aperture of a 2.5 cm integrating sphere. A biased 1 cm^2 silicon photodiode (Hamamatsu S3590-01) mounted in the sphere at 90° to the probing axis generates a current proportional to the probe beam intensity. The function of the integrating sphere is to suppress artifactual changes in the measured intensity caused by transient refraction of the probe beam as it passes through the excited sample volume.

The photodiode's output is connected through a 40 cm coaxial cable to the input of a digitizing oscilloscope (Hewlett-Packard 54504A) that is externally terminated with a 2.6 kΩ resistor. To acquire transient data, the oscilloscope is set for DC coupling, 20 MHz bandwidth limiting, a sensitivity of 7 mV per division, and a vertical offset of ~700 mV to cancel the probe pedestal signal. The oscilloscope's time base is triggered from the output of a separate photodiode that detects a portion of the excitation pulse. Between 256 and 2048 traces are averaged for each of the four combinations of opened and closed flags in the excitation and probe beams. These four 500-point waveforms are transmitted through a GPIB interface to a laboratory computer, where appropriate subtractions give a waveform, denoted $S(t)$, representing only excitation-induced changes in probe intensity after correction for dark current, excitation-induced luminescence, and electronic noise pickup in the detector or probe laser. Letting $S°$ represent the average value of this waveform before arrival of the excitation pulse, the induced absorbance is computed as $\Delta A(t) = \log_{10}(S°/S(t))$. This function, which is proportional to the concentration of a transient species, is used for subsequent kinetic analysis.

We use a custom-written program to deduce kinetic parameters from induced absorbance traces. This program incorporates a selection of kinetic models, including simple first-order, concurrent first- and second-order, multiexponential, and energy-transfer schemes. These models are defined through differential rate equations and then integrated for selected initial concentrations and rate constants to find the concentrations of all relevant species as a function of time. Using specified absorptivities of these species, a simulated induced absorbance trace is calculated and then convoluted with the experimental instrument response function for comparison with experimental data. A Levenberg-Marquardt nonlinear least-squares fitting routine adjusts any selected parameters to obtain an optimized kinetic simulation.

Transient spectra of fullerene samples are measured with a separate custom-built apparatus that differs from the one described above by using for the probing source a 20 W tungsten-halogen lamp coupled with a monochromator (ISA Triax 190). Because this configuration gives a probe intensity lower than in the fixed-wavelength apparatus, an amplified photodiode detector is required. For the visible region we use a Hamamatsu S-3887 silicon photodetector; for the near infrared, a Terahertz Technologies TIA-500 InGaAs unit. Both detectors are adjusted for 1 MHz bandwidths. Their outputs are recorded by a digitizing oscilloscope (Tektronix model TDS-430A) whose vertical offset and sensitivity are adjusted at each probing wavelength by a GPIB-interfaced control computer. Transient spectral scans are obtained by acquiring averaged kinetic traces at each probe wavelength before advancing the monochromator to the next spectral position. To correct for the effects of slow variations in the excitation beam's energy, induced kinetic traces at different wavelengths are normalized to the corresponding excitation energies. Transient spectra are then constructed by combining specified kinetic slices from the set of probe wavelengths covered in the scan.

3 TRIPLET KINETICS

3.1 Unimolecular Processes

C_{60}

The kinetics of triplet formation in optically excited C_{60} has been deduced from three types of measurements, all of which should give the same rate constant. These measurements are S_1 fluorescence decay [1], $S_n \leftarrow S_1$ absorption decay [2, 3], and $T_n \leftarrow T_1$ absorption appearance [2–4]. Although the various reported values are not entirely consistent, it seems clear that the S_1 lifetime of C_{60} in room-temperature solution (and therefore the characteristic time constant for triplet appearance) is 1.2 ns [1]. The quantum yield of fluorescence emission is approximately 3×10^{-4} in a variety of solvents [5]. In contrast to this very small probability for radiative relaxation from S_1, the quantum yield for $S_1 \rightarrow T_1$ nonradiative decay is very large. Reported values for this triplet formation quantum yield include 0.96 ± 0.04 [6], 1.0 [7], 0.93 ± 0.07 [8], 0.88 ± 0.15 [9], and 0.98 [10]. It seems likely that the true value is very close to unity. We therefore conclude that the S_1 state of C_{60} decays nonradiatively to form the lowest triplet state with an efficiency near 100% and a rate constant of 8.3×10^8 s^{-1}.

Phosphorescence emission from the T_1 state of C_{60} is so weak that it initially escaped detection, even in a cryogenic sample [11]. It has since been observed in a frozen solvent containing heavy atoms and in rare gas matrices [12, 13]. Spectroscopic and theoretical analyses both indicate that the lowest triplet state of C_{60} has T_{2g} symmetry [9, 13–16]. The weak phosphorescence reflects a low radiative rate for $T_1 \rightarrow S_0$ emission, which is both spin and orbitally forbidden, plus comparatively rapid $T_1 \rightarrow S_0$ nonradiative decay. The intrinsic triplet lifetimes of C_{60} and most other fullerenes are therefore entirely controlled by the rate of intersystem crossing to the ground state.

Published values for the triplet lifetime of C_{60} in room-temperature fluid solution range at least from 37 to 285 μs [2, 6, 7, 9, 10, 17–20]. The wide discrepancies among these values likely reflect the presence of bimolecular decay channels such as self-quenching or unrecognized triplet-triplet annihilation, or the effects of impurities in solute or solvent, or uncertainties in the kinetic data. Figure 3 shows transient absorption kinetics measured in a 4 μM solution of C_{60} in toluene using the apparatus described in Section 2. We believe these data to be reliable because of the high purity of the solute and solvent, the use of 12 freeze-pump-thaw cycles to remove dissolved oxygen, the low sample concentration, the mild excitation conditions and resulting absence of second-order decay component, and the high signal-to-noise ratio. Our data imply an exponential lifetime of 142 μs. We obtain the same lifetime (within 3 μs) from measurements on a variety of dilute C_{60} solutions made independently by different individuals using three separate apparatus. Work described in a later section shows that the self-quenching component of this decay is 0.21%/μM. Including this small correction, we conclude that the intrinsic triplet lifetime of C_{60} in room-temperature toluene solution is reliably determined to be 143 $\pm$ 3 μs, corresponding to a rate constant k_1 of 7000 s^{-1}.

Because interactions between solute and solvent may influence nonradiative decay processes, it is relevant to examine the solvent dependence of the intrinsic triplet lifetime [21, 22]. We have found no significant variation in the C_{60} triplet lifetime for solutions in benzene, benzene-d_6, toluene, the three isomers

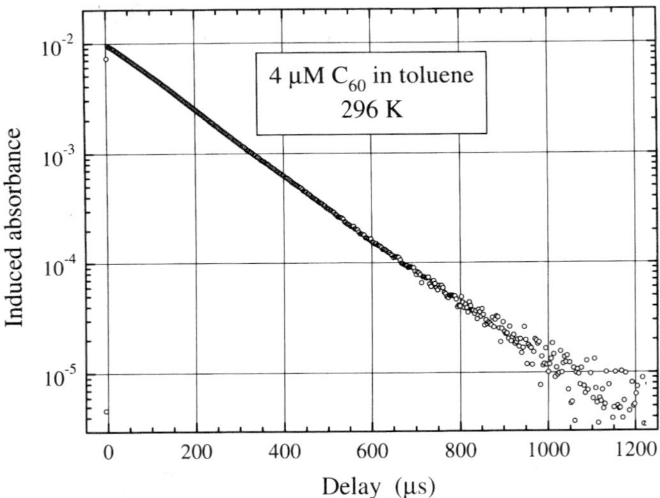

Figure 3 Semilog plot of induced absorbance vs. delay after excitation for a 4 μM solution of C_{60} in toluene at 296 K. The excitation wavelength was 532 nm and the probing wavelength was 750 nm. The nearly single-exponential decay corresponds to a lifetime of 142 μs.

of xylene, ethylbenzene, and hexane. In solvents containing atoms from lower rows of the periodic table, triplet lifetimes are shortened through the well-known external heavy atom effect. We measure intrinsic C_{60} lifetimes of 103 µs in carbon disulfide, 109 µs in dichloromethane, 126 µs in chloroform, 127 µs in carbon tetrachloride, and 131 µs in chlorobenzene. The heavy-atom effect is stronger for bromine-containing solvents, reducing the triplet lifetime to 66 µs in bromo-benzene and 43 µs in bromoform.

Temperature is another important environmental parameter that may affect triplet decay constants. In the case of fullerenes in fluid solution, it is essential to distinguish the temperature dependence of intrinsic decay from those of bimo-lecular processes such as self-quenching [23]. Of course, in a rigid glass all bimo-lecular channels will be suppressed, leaving only intrinsic decay. Figure 4 shows the triplet-state kinetics measured in a sample of C_{60} dissolved in a glass com-posed of methylcyclohexane and isopentane at 77 K. The data are well fit by a single exponential component of 4220 s^{-1} (237 µs lifetime). The C_{60} intrinsic rate constant k_1 thus varies by less than a factor of 2 between 77 and 296 K. This weak sensitivity to temperature has been confirmed up to 348 K by careful measurements made on fluid C_{60} solutions [23].

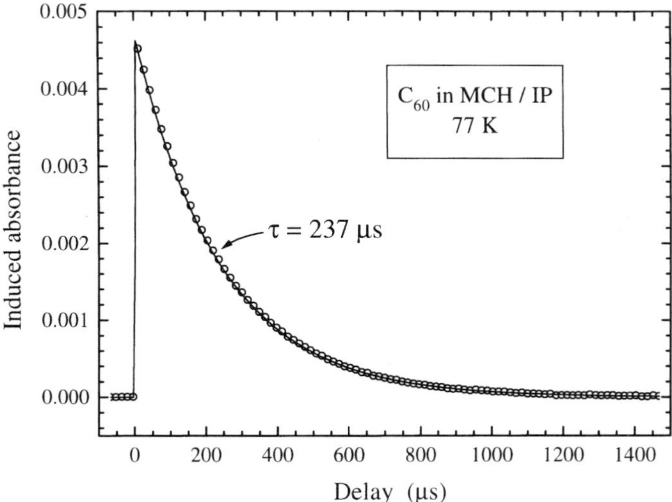

Figure 4 Triplet decay kinetics measured at 676 nm following 532 nm excitation of a solid solution of C_{60} in 1:1 methylcyclohexane and isopentane at 77 K. The open circles are data points and the solid curve is a computed simulation based on simple first-order decay at a rate constant of 4220 s^{-1}, corresponding to an exponential lifetime of 237 µs. For plotting clarity, only 25% of the data points are drawn.

The nature of temperature-dependent triplet decay in C_{60} has been further explored through experiments on high-temperature vapors. Using a sublimation cell with optical windows, we have measured triplet decay in samples of C_{60} vapor plus argon at approximately 1000 K [24]. Figure 5 illustrates the resulting kinetic data, which show a distinctly nonexponential shape and much more rapid decay than was found at lower temperatures. We interpret the complex kinetic shape as a superposition of many first-order components covering a range of rate constants. These components reflect the wide span of molecular vibrational energy contents present in a thermalized sample at high temperatures. The solid curve through the data is a simulation computed using the predicted distribution of C_{60} energy contents combined with a simple exponential dependence of triplet decay constant on vibrational energy content. This simulation specifies two parameters that describe the variation of k_1 with energy content. The heavy solid line in Figure 6 plots this dependence over the range of energy contents found at 1000 K. When this line is extrapolated to lower energies, as shown by its dashed continuation, it accounts surprisingly well for independent kinetic measurements made in a molecular beam [25], in room-temperature solution [26], and in low-temperature glasses [27, 28]. We therefore conclude that vibrational energy content, rather than temperature, is the quantity that most directly controls

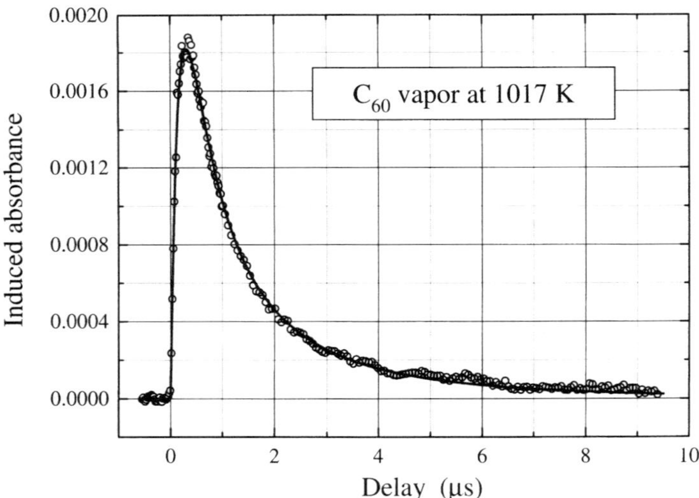

Figure 5 Triplet decay kinetics measured at 758 nm following excitation of a sample containing C_{60} vapor plus 100 Torr of argon at 1017 K. The solid curve shows a kinetic simulation computed as described in the text.

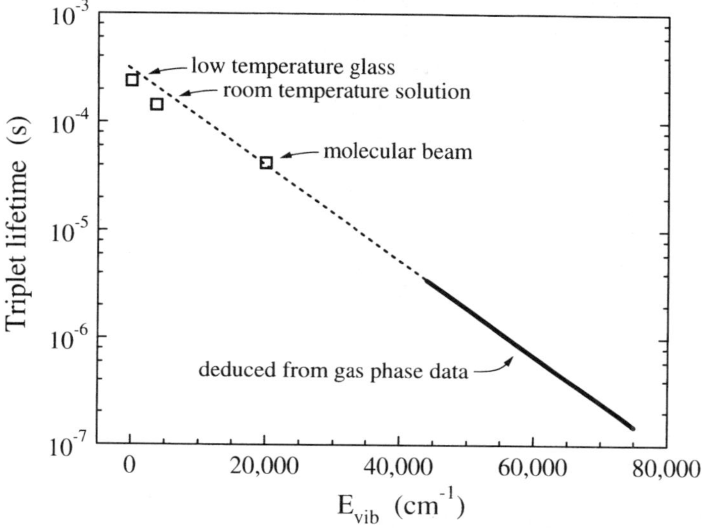

Figure 6 C_{60} intrinsic triplet lifetime as a function of vibrational energy content per molecule. The heavy line is deduced from the fit shown in the previous figure and the dashed line is an extrapolation to lower energy contents.

the intrinsic triplet decay of molecular C_{60} in gaseous, liquid, and solid environments. The mild temperature dependence found near and below room temperature reflects the small changes in average vibrational energy content throughout this range.

C_{60} Derivatives

Chemical derivatization strongly alters the intrinsic triplet lifetime of C_{60}. The simplest derivative is the dihydride $C_{60}H_2$, first synthesized in 1993 [29]. In this and other "[6,6]-closed dihydrofullerenes," the derivatization sites are two adjacent carbon atoms at the fusion of six-membered rings. Figure 7 shows the triplet kinetics measured in this laboratory for a dilute solution of $1,2\text{-}C_{60}H_2$ in toluene at 296 K. The accurate simulation drawn through the data points is first-order decay with a rate constant of 22,700 s^{-1}, corresponding to an exponential lifetime of 44.0 μs. We identify this value as the intrinsic triplet lifetime based on its negligible variation with sample concentration in this dilute regime and the absence of second-order kinetic components. Thus derivatization to form the most stable dihydride accelerates the intrinsic triplet decay of C_{60} by a factor slightly greater than 3 [30].

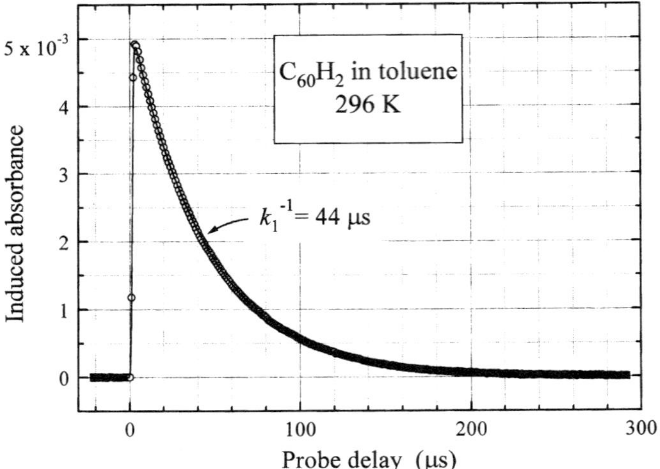

Figure 7 Triplet decay kinetics measured at 676 nm following 532 nm excitation of a dilute solution of $C_{60}H_2$ in toluene at 296 K. The open circles are data points and the solid curve is a computed simulation based on simple first-order decay at a rate constant of 22,700 s^{-1}. The corresponding exponential lifetime is 44.0 μs.

Similar measurements on the analogous [6,6]-closed derivative 1,2-$C_{60}(CH_3)_2$ gave an intrinsic triplet lifetime of 45.3 μs ($k_1 = 22,100$ s^{-1}). In the *ortho*-xylyl adduct of C_{60}, which has a six-membered ring external to the fullerene, the corresponding triplet decay constant is slightly reduced to 20,200 s^{-1}. N,N'-dimethyl-1,2-ethylenediamine C_{60} is another derivative with an external six-membered ring, but in this compound the adjacent fullerene carbons are bonded to nitrogen atoms and the external ring is twisted, imparting some degree of torsional strain to the fullerene cage. We find the triplet decay constant of this compound to be significantly lower, 17,700 s^{-1}. The malonic acid diethylester derivative of C_{60} has a single carbon atom bridging a fullerene [6,6] site to give a three-membered external ring. For this compound the triplet decay constant has a still lower value of 12,700 s^{-1}, corresponding to an exponential lifetime of 79 μs, the longest yet found for a C_{60} adduct. Luo et al. [10] have reported a triplet lifetime of 31 μs for the *p*-dimethylaniline-substituted *N*-methylpyrrolidine adduct of C_{60} in room-temperature cyclohexane solution. However, as the lifetime presented for C_{60} in the same study is only 37 μs, it seems that bimolecular decay processes dominate these kinetic findings.

$C_{60}O$ is another [6,6]-closed adduct, with the oxygen atom bridging two carbon atoms to form an epoxide structure [31]. The triplet-state decay of this compound is quite anomalous. In hexane solution at 296 K, the $C_{60}O$ triplet decays with a rate constant near 235,000 s^{-1}, giving an intrinsic lifetime of only 4.3 µs, by far the shortest yet found for any fullerene derivative. Variable temperature kinetic studies have provided insight into the nature of this rapid decay [28, 32]. We find that the $C_{60}O$ triplet lifetime varies strongly with temperature between ~180 and 300 K, with an apparent Arrhenius activation energy of approximately 20 kJ mol^{-1}. Below 180 K, however, the temperature dependence becomes very weak, as is typical of C_{60} over the full temperature range. We suggest that the low-temperature variation reflects normal radiationless decay, whereas above 180 K a new unimolecular decay channel, probably involving reversible breaking of an epoxide C—O or C—C bond, becomes kinetically dominant.

Recently, triplet lifetime measurements have been extended to a different type of fullerene derivative: C_{60} molecules containing a single endohedral krypton atom. In a frozen hydrocarbon solution at 77 K, the exponential lifetime of $Kr@C_{60}$ was found to be 210 µs, indicating a 12% acceleration in intrinsic triplet decay as a result of krypton incorporation [33]. This observation is the first example of the "endohedral heavy atom effect," in which electronic interactions between the encapsulated atom and the surrounding moiety allow the krypton atom's strong spin-orbit coupling to be transmitted to the fullerene's π electrons. As a result, spin-changing processes (such as $T_1 \rightarrow S_0$ radiationless decay) are enhanced. The small kinetic acceleration found in this system points to weak electronic coupling between the krypton atom and the surrounding fullerene. The effects induced by rare gas incorporation should increase monotonically in going from helium through xenon.

C_{70}

Although C_{70}, the second most abundant fullerene, has not been investigated as thoroughly as C_{60}, a reasonable amount has been learned about its photophysics. Time-resolved fluorescence measurements have shown that the S_1 lifetime of C_{70} in room temperature solution is 0.66 ns [1]. As in the case of C_{60}, the dominant decay channel of S_1 C_{70} is intersystem crossing to the triplet state, for which the quantum yield of formation has been reported to be very close to one [34–36]. Fluorescence emission is a minor decay channel, with a quantum yield near 5×10^{-4} reported in room-temperature solutions [37]. Combining these data, we can state that S_1 C_{70} decays nonradiatively with a rate constant of 1.3×10^9 s^{-1} to produce the lowest triplet state with near total efficiency.

Phosphorescence is stronger from C_{70} than from C_{60} [11]. Analysis of emission spectra from a 77 K hexane matrix gives a value of 12,695 cm^{-1} for the T_1 origin and 15,400 cm^{-1} for the S_1 origin [38]. The symmetry of the lowest C_{70} triplet state has been deduced to be $^3A_2'$ or $^3E_1'$ through differing vibronic analyses of phosphorescence spectra [39–41]. Recent entropic data (described in the Section 5 below) supports the $^3A'_2$ assignment. Despite the increased phosphorescence quantum yield of C_{70} relative to C_{60}, decay of the lowest triplet state is strongly dominated by $T_1 \rightarrow S_0$ radiationless decay in both molecules.

The intrinsic triplet lifetime of C_{70} in fluid solution has proven more challenging to measure than that of C_{60}. The earliest reports, giving 130 and 250 μs [6, 20] were followed by more sensitive measurements that found a far longer lifetime of at least 2.2 ms [26]. A subsequent study from this laboratory under even more carefully controlled conditions set a revised lower limit of 12 ms for the intrinsic lifetime of C_{70} in room-temperature toluene [42]. Figure 8 shows representative decay data and a simulation based on concurrent first- and second-order decays. Because such a long-lived triplet is readily quenched by impurities, the room-temperature experiments should still be viewed as providing lower limits rather than a definitive lifetime value.

The C_{70} triplet lifetime in low-temperature rigid media, where second-order decay processes are suppressed, was first measured as 53 ms for a 77 K sample through measurement of phosphorescence decay [11]. This value agrees rather well with a lifetime of 50.0 ms obtained in our laboratory from transient absorption kinetics in a hydrocarbon glass at the same temperature [28], as shown in Figure 9. It therefore appears that the intrinsic triplet decay constant of C_{70} in-

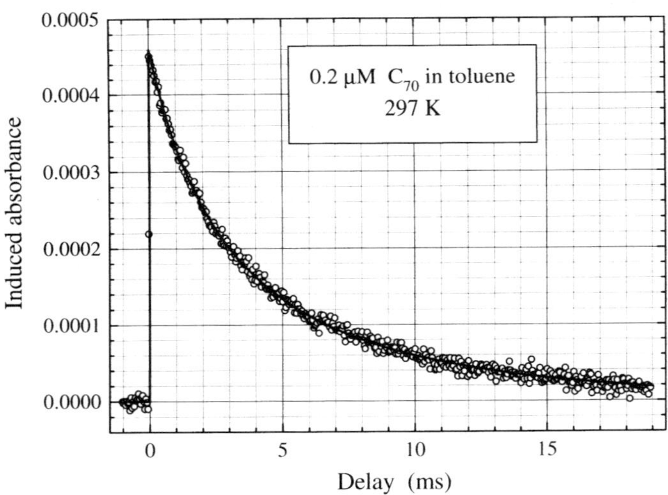

Figure 8   Triplet decay kinetics measured at 672 nm following 532 nm excitation of a 0.2 μM solution of C_{70} in toluene at 297 K. The open circles are data points and the solid curve is a computed simulation based on concurrent first- and second-order decay. The first-order decay constant of 110 s^{-1} has contributions from self- and impurity quenching.

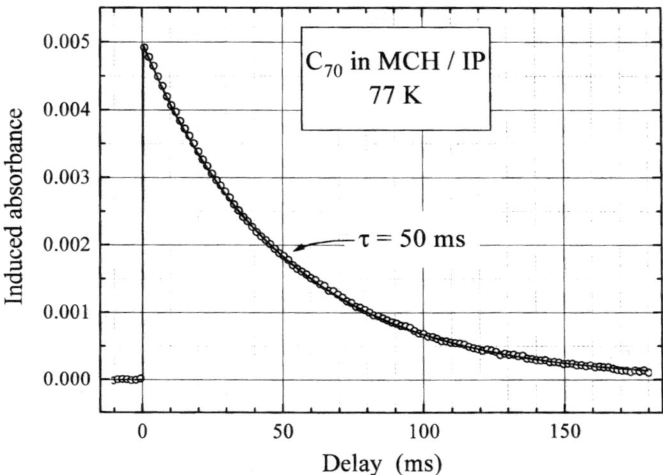

Figure 9 Triplet decay kinetics measured at 676 nm following 532 nm excitation of C_{70} dissolved in a methylcyclohexane/isopentane glass at 77 K. The open circles are data points and the solid curve is a computed simulation based on first-order decay with a rate constant of 20.0 s^{-1}, corresponding to an intrinsic exponential lifetime of 50.0 ms. For clarity, only 25% of the data points are plotted.

creases by less than a factor of 4 as the temperature is raised from 77 to 296 K. This mild temperature dependence appears similar to that found for C_{60}, although the C_{70} data are sparser and do not yet extend above room temperature. Little is currently known about the solvent dependence of C_{70} intrinsic decay in fluid solution.

C_{70} Derivatives

The photophysical properties of C_{70} derivatives remain nearly unexplored. Preliminary results from this laboratory indicate that the intrinsic lifetime of 1,2-$C_{70}H_2$ in room-temperature toluene solution is at least 1.8 ms, or approximately 7 times shorter than that of C_{70} under similar conditions. Of course, this value and other fluid solution triplet lifetimes on the millisecond time scale remain uncertain for the reasons discussed earlier. In the case of $C_{70}O$, we have available some recent triplet decay data that are free of bimolecular interference. In a rigid glass at 85 K, the $C_{70}O$ triplet lifetime is 14.7 ms, or approximately 3 times shorter than that of underivatized C_{70} under similar conditions. Based on these preliminary findings, it therefore appears that simple derivatization of C_{70} may shorten the intrinsic triplet lifetime by a factor not far from 3, as is found in C_{60}

systems. However, as will be discussed below, the two fullerene families differ in their $T_n \leftarrow T_1$ spectral shifts on derivatization.

Higher Fullerenes

More limited studies of triplet photophysics have been reported for the less abundant fullerenes beyond C_{70}. In the case of the D_2-symmetry isomer of C_{76}, the quantum yield of triplet formation was measured to be only 0.03 or 0.05 [43, 44]. Phosphorescence spectroscopy gave an estimate of 12,400 cm^{-1} for the T_1 energy [44], indicating a small S_1-T_1 gap of 0.15 eV [43]. A reliable value for the C_{76} intrinsic triplet lifetime is not yet available. One group has reported the lifetime to be 7.1 μs for a 100 μM solution of C_{76} in room-temperature toluene [45]. However, this value may require revision, because the accompanying decay trace suggests a lifetime closer to 20 μs and it is also unclear whether bimolecular channels were accounted for. Another study found a triplet lifetime near 3 μs, influenced by bimolecular quenching processes, and ~ 100 μs under other conditions [44], while a third group presented a value of 5.6 μs at high excitation intensity [43]. For the C_{2v}' isomer of C_{78}, a triplet quantum yield of 0.12 has been reported, along with a T_1 energy and kinetics similar to C_{76} [44]. The dominant (D_2) isomer of C_{84} appears also to have a small quantum yield for triplet formation, and a triplet lifetime estimated from energy transfer studies to be less than 100 μs [46].

3.2 Bimolecular Processes

Triplet-Triplet Annihilation

It is well known that the loss of triplet population through annihilating encounters between two triplet-state molecules causes a second-order component in decay kinetics. Fullerenes in fluid solution have been found to undergo such triplet-triplet annihilation with large rate constants typical of many organic compounds [47]. To extract k_{TT}, the rate constant for second-order decay, one must know the absolute concentration of the fullerene triplet state. In optical experiments this requires knowledge of the triplet's molar absorptivity, a quantity that tends to be difficult to determine. As a result, numerical values for k_{TT} often contain uncertainties in the 50% range. Reported k_{TT} values also depend on whether the annihilation reaction is assumed to deactivate one or two triplet molecules. Results from our laboratory assume the net reaction shown in Eq. (1), in which two triplets initially form one ground-state singlet and one highly excited singlet, which subsequently repopulates T_1 to give a net loss of one triplet molecule. The k_{TT} values defined in this way are found to be 5×10^9 M^{-1} s^{-1}, or approximately one-half of the diffusion-limited rate constant, for C_{60} and C_{70} in room-temperature toluene solution. Other fullerenes are likely to show similar rate constants for triplet-triplet annihilation.

Oxygen Quenching

It is well known that organic molecules in triplet states are readily deactivated by encounters with ground state O_2. In this energy-transfer quenching process, the oxygen becomes excited from its $^3\Sigma_g^-$ ground state to its low-lying $^1\Delta_g$ singlet excited state, 7882 cm^{-1} higher in energy [48]. It is not surprising that common fullerene triplet states, whose energies lie significantly above 7882 cm^{-1}, are efficiently quenched by oxygen. For C_{60}, reported k_{O_2} values include 1.9×10^9 M^{-1} s^{-1} and 1.4×10^9 M^{-1} s^{-1} [6, 19]. These are somewhat larger than the k_{O_2} values found for C_{70} under similar conditions, 9.4×10^8 M^{-1} s^{-1} and 9.7×10^8 M^{-1} s^{-1} [19, 34]. The rate constants for oxygen quenching of fullero-1,2,5-triphenylpyrrylodine and a related C_{60} derivative have been reported to be 2.1×10^9 M^{-1} s^{-1} [49]. The similarity between this value and that for C_{60} suggests that derivatization will often have little effect on the oxygen quenching of fullerene triplets.

Energy Transfer

Triplet-state deactivation can also occur by energy transfer to species other than oxygen. One category of special interest is the transfer of triplet energy between different fullerene species in a mixed solution. Such mixtures may be prepared either deliberately or inadvertently, if a sample is contaminated with small amounts of other fullerenes. The first and best characterized example of fullerene energy transfer is the C_{60}/C_{70} system, in which the two species are close enough in triplet energy that transfer in both directions must be considered [26]. Here the most important kinetic processes can be described using a simple model:

$$^3C_{70} + C_{60} \underset{k'_{ET}}{\overset{k_{ET}}{\rightleftharpoons}} C_{70} + {}^3C_{60} \qquad (2)$$

$$\downarrow k_1 \qquad\qquad\qquad \downarrow k'_1$$

$$C_{70} \qquad\qquad\qquad C_{60}$$

Rate constants k_1 and k'_1 may be taken to include pseudo-first-order processes such as self-quenching as well as intrinsic triplet decay. In a mixed sample, C_{70} will be preferentially excited by 532 nm light, because it has a ground-state absorptivity at that wavelength $\sim$9 times greater than C_{60}. The resulting triplet population will then tend to flow from C_{70} into C_{60} if sample concentrations are high enough to allow energy transfer to compete kinetically against deactivation. The differing triplet-triplet spectra of C_{60} and C_{70} allow this energy flow to be monitored by time-resolved absorption probing at many wavelengths in the visible

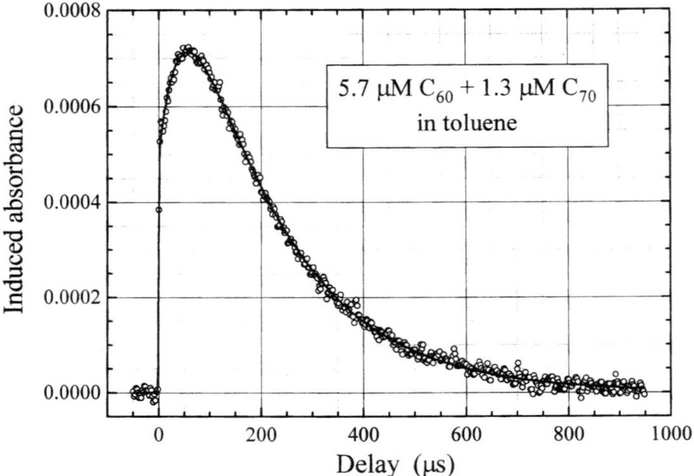

Figure 10   Induced absorption kinetics measured at 672 nm following 532 nm excitation of a sample containing 5.7 µM C_{60} and 1.3 µM C_{70}. The solid line is a kinetic simulation computed as described in the text.

and near-infrared. As an example, Figure 10 shows a characteristic kinetic shape in which the slow rise reveals initial transfer of energy from C_{70} to C_{60} (which has stronger triplet absorption at the probing wavelength), and the later sections reflect the decay of a preequilibrated pool of triplet energy. The solid curve drawn through the data is a kinetic simulation that has been computed using the model of Eq. (2) and known values for k_1 and k_1'. The fitting process reveals the values of both k_{ET} and k_{ET}'. In the case of the C_{60}/C_{70} system in room-temperature toluene solution, both of these rate constants are near $2.2 \times 10^9 \, M^{-1} \, s^{-1}$, or approximately 20% of the diffusion-limited value [26].

The similar forward and reverse rate constants for energy transfer in the C_{60}/C_{70} system reflect the small free-energy difference between triplet excitation of these two species. Consequently, energy becomes partitioned between them in a ratio given by the triplet exchange equilibrium constant, $K^T \equiv e^{-\Delta G_T^\circ/RT}$, multiplied by the ground-state concentration ratio of the two fullerenes. So long as energy exchange is much more rapid than deactivation, the triplet excitation will form a pool whose decay constant is a weighted average of the constants of the individual solutes. The $\sim$100-fold difference in decay constants between C_{60} and C_{70} permits the pooled triplet energy decay to be varied widely by adjustment of ground-state concentrations. It is also important to note that contamination of

C_{60} by C_{70} lengthens rather than shortens the apparent triplet lifetime, whereas contamination of C_{70} by C_{60} has the opposite effect.

When the two fullerene solutes differ in triplet free energy by significantly more than RT, one will observe irreversible transfer behavior instead of energy pooling. An example is the effect of $C_{60}O$ on excited C_{60} [32]. Here the $C_{60}O$ acts as a strong quencher because its triplet state lies lower than that of C_{60} and has a much shorter intrinsic lifetime. Figure 11 shows the effect of added $C_{60}O$ on the triplet decay constant of a C_{60} solution. The slope gives the bimolecular quenching constant, $2.9 \pm 0.6 \times 10^9$ M^{-1} s^{-1}. It seems likely that this large value, which is similar to the energy transfer rate constants deduced in the C_{60}/C_{70} system, may be typical for interfullerene triplet energy transfer. One possible cause of the surprisingly discrepant triplet lifetime measurements described in Section 3.1 may be the undetected presence of small amounts of quenching fullerene derivatives such as $C_{60}O$.

Self-Quenching

The final category of bimolecular processes to be discussed here is "self-quenching," the deactivation of triplet-state molecules through encounters with ground-state molecules of the same species. Note that self-quenching differs from the

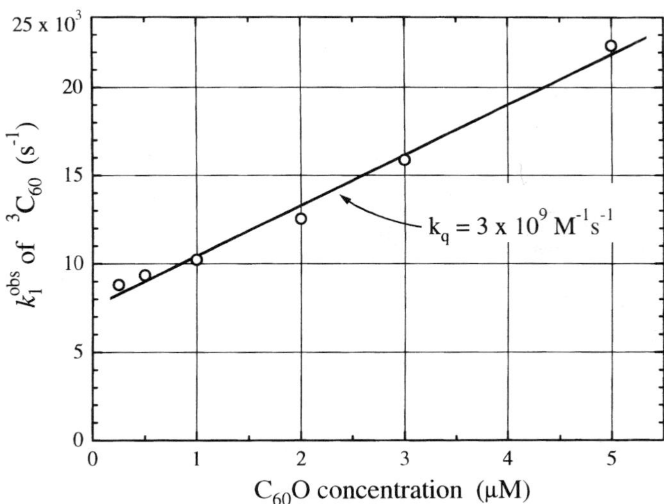

Figure 11 Apparent first-order decay constant for induced absorption in a room-temperature hexane solution containing 30 μM C_{60} plus varying amounts of $C_{60}O$.

simple exchange of triplet energy between equivalent molecules, a process that, although surely as efficient as those discussed above, cannot be observed by normal transient absorption methods because it leaves the triplet population unchanged. The rate constants for fullerene self-quenching are high enough both to complicate triplet lifetime measurements and also to motivate investigations of the process's underlying mechanism.

The self-quenching rate constant, k_{sq}, is readily measured as the slope of a plot of observed first-order decay constant vs. sample concentration in simple fluid solutions. However, at least two experimental problems can cause inaccuracies. First, if quenching impurities are present in the solute, one will observe true first-order kinetics and a linear self-quenching plot but the deduced k_{sq} value will be artificially high. Second, any second-order decay components will be more prominent at increased sample concentrations. If these components are present but not recognized and the decays are mistakenly fit as purely first-order, one will also find an inflated value of k_{sq}. When evaluating discrepant values of self-quenching constants, one should therefore normally presume the lowest to be the most reliable.

The self-quenching constant for C_{60} in benzene solution at 296 K was reported from this laboratory as $1.5 \times 10^7 \, M^{-1} \, s^{-1}$ [26], a value in excellent agreement with the more recent data for toluene solutions shown in Figure 12. This

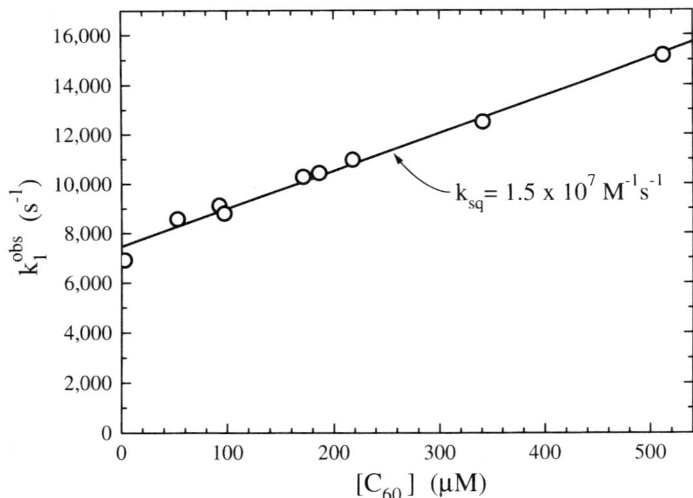

Figure 12   First-order triplet decay constant as a function of sample concentration for solutions of C_{60} in toluene at 296 K.

result is significantly lower than the k_{sq} value of 2×10^8 M^{-1} s^{-1} reported for C_{60} by two other groups [19, 20]. Our study of C_{60} triplet decay kinetics as a function of temperature and concentration revealed that k_{sq} is strongly temperature-dependent [23]. Figure 13 shows the decay constants measured in that study as a function of both sample concentration and temperature. To generate the mesh surface through these data we have fit to a functional form that is based on transition state kinetics for two independent processes, unimolecular intrinsic decay and bimolecular self-quenching:

$$k_1^{obs}([C_{60}], T) = C_1 T \exp\left(\frac{-\Delta H_1^{\ddagger}}{RT}\right) + C_{sq} T \exp\left(\frac{-\Delta H_{sq}^{\ddagger}}{RT}\right) \tag{3}$$

Even though the temperature dependence of intrinsic decay follows the molecular vibrational energy content rather than an Arrhenius or transition state form, the first term in Eq. (3) provides a convenient empirical description over the required temperature range. The small value deduced for $\Delta H_1^{\ddagger}$, 0.7 ± 0.5 kJ mol^{-1}, reflects the mild temperature dependence of intrinsic triplet decay near room temperature. By contrast, the much larger value of $\Delta H_{sq}^{\ddagger}$, 10.1 ± 0.9 kJ mol^{-1} or approximately 840 cm^{-1} per molecule, signifies a substantial dependence of the self-quenching constant on temperature. This variation is illustrated in Figure 14.

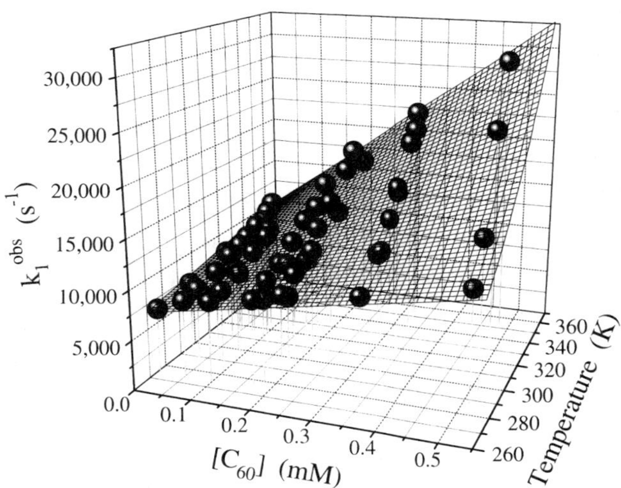

Figure 13 Observed first-order decay constant for C_{60} in toluene as a function of ground-state concentration and sample temperature. The mesh surface shows a fit calculated using the model described in the text.

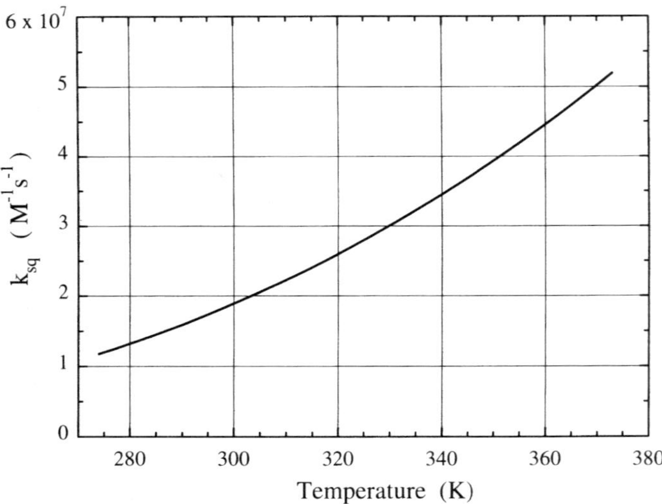

Figure 14 Triplet self-quenching constant as a function of temperature for C_{60} in toluene solution. The curve is computed from the parameters of the fit in Figure 13.

Self-quenching is also significant in C_{70} solutions. However, accurate determinations of k_{sq} are much more difficult for C_{70} because of its far longer intrinsic triplet lifetime and the resulting kinetic sensitivity to impurities. The value of k_{sq} for C_{70} in room-temperature aromatic solvents has been variously reported as 6×10^8 [20], 8.5×10^7 [26], 1.2×10^8 [42], and 2.3×10^9 M^{-1} s^{-1} [50]. It is now clear that the 1993 and 1995 reports from this laboratory were dominated by sample impurity effects. Our most recent results point to a lower value for k_{sq} near 3×10^7 M^{-1} s^{-1}, or only about twice that of C_{60}.

Although the fullerene self-quenching rate constants measured to date are two orders of magnitude below those for thermoneutral energy transfer, self-quenching must be considered a relatively efficient process, and one whose mechanism remains in doubt. An earlier report presented evidence for triplet excimer formation in C_{70} solutions and proposed a self-quenching mechanism based on rapid deactivation of the excimers [42]. However, it has since been found that the kinetic anomalies interpreted as excimer formation actually resulted from energy transfer to impurities [51]. In addition, the temperature dependence of self-quenching in C_{60} and C_{70} is not easily reconciled with a triplet excimer mechanism. Further studies are needed to clarify the nature of fullerene self-quenching.

4 TRIPLET SPECTRA

C_{60}

Induced electronic absorption spectra attributable to the triplet state have been observed for a number of fullerenes and derivatives. Although the first report of the C_{60} triplet spectrum revealed features only in the 400–500 nm range [6], subsequent work has consistently shown the presence of strong, characteristic absorption peaks near 345 and 750 nm in aromatic solvents [2–4, 8, 9, 17, 19, 20, 52]. There is poorer agreement among investigators concerning the absolute molar absorptivity scale of the C_{60} $T_n \leftarrow T_1$ spectrum. Because it lies in a region free of ground-state absorption, the peak near 750 nm has received the most attention. Early published values of ε_T or $\varepsilon_T\phi_T$ for this peak include 6100 [17], 7000 [4], 15000 [2], 12000 [20], 16100 [8], and 20200 [9] M^{-1} cm^{-1}. Although these determinations were not all in the same solvent, it is likely that nearly all of the variation reflects experimental uncertainties rather than solvent effects. Two more recent reports gave peak ε_T values of 18,800 [10] and 19,500 [52] M^{-1} cm^{-1}. The latter value was obtained using a new photometric method that allows measurement of $\varepsilon_T\phi_T$ on an absolute basis, without comparison to a reference sample, with an estimated error limit of 7% [52]. Based on the close agreement of the three most recent determinations, it seems reasonable to accept 19500 ± 1400 M^{-1} cm^{-1} as the reliable value of ε_T at the 750 nm peak for C_{60} in a room-temperature aromatic solvent. Figure 15 shows the visible portion of the room-temperature triplet-triplet absorption spectrum, corrected for ground-state depletion and calibrated in absorptivity units. An uncalibrated 77 K spectrum is plotted in Figure 16 along with three Lorentzian components whose superposition accurately simulates the measured data.

Spectroscopic assignment of the triplet-triplet absorption spectrum is hampered by the difficulty of performing accurate excited-state quantum calculations on systems the size of C_{60} [53]. As an example, an attempt to account for the C_{60} triplet spectrum based on CNDO/S-CI computations was not successful [16]. However, in a pioneering early analysis, the main absorption features were convincingly assigned as transitions between $^3T_{2g}$, the lowest triplet state, and higher-lying triplets of G_u symmetry [9]. It is hoped that advanced methods recently found useful in describing fullerene ground state spectra can soon be extended to excited-state spectroscopy [54].

C_{60} Derivatives

Chemical derivatization of C_{60} significantly changes the triplet absorption spectrum. Such spectra have been reported for the 4-hydroxycyclohexane adduct (a [6,6]-dihydrofullerene) [55], for $1,2\text{-}C_{60}H_2$ [30, 56], for $1,2\text{-}C_{60}(CH_3)_2$ [30], for

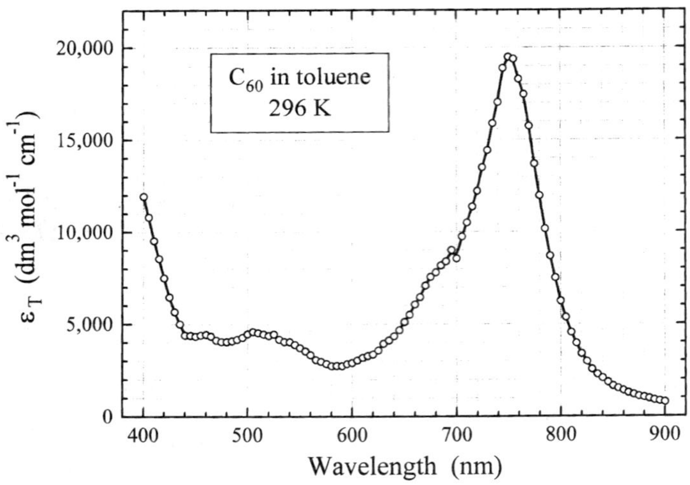

Figure 15 Triplet-triplet absorption spectrum of C_{60} in toluene solution, after correction for ground-state absorption and conversion to absolute absorptivity units. The value of ϕ_T was presumed to be 1.00.

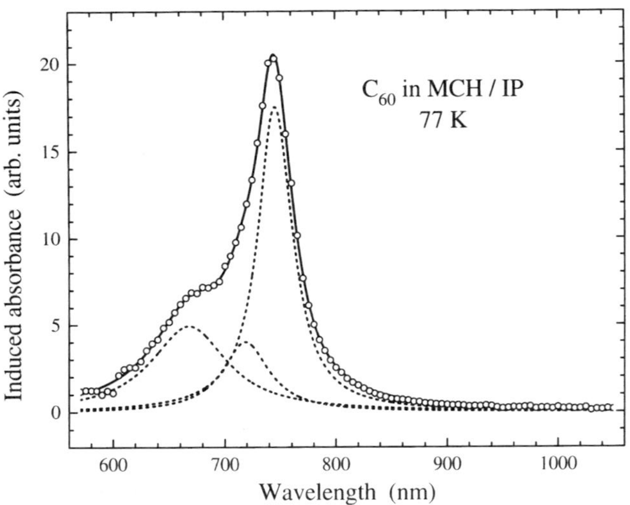

Figure 16 Induced spectrum of C_{60} in a 77 K glass of 1:1 methylcyclohexane:isopentane. The circles are measured data; dotted curves show Lorentzian components that add to give the simulation drawn as a solid curve.

1,2,3,4,-$C_{60}H_4$ [56], and for 1-methylsuccinate-4-methyl-cyclohexadiene-2,3-C_{60} [56]. Other [6,6]-adducts whose $T_n \leftarrow T_1$ spectra have been measured include several *N*-methylpyrrolidinofullerenes [10], $C_{60}O$ [30], and *o*-xylyl-C_{60} [30]. In addition, a series of mono-, bis-, and tris-adducts of C_{60} with malonic acid diethyl ester have been examined for systematic variations in their triplet spectra [57, 58]. In all of these C_{60} derivatives, the prominent 750 nm band is blue-shifted to wavelengths between 650 and 720 nm, and a shoulder is commonly seen between 810 and 830 nm. As an example, Figure 17 displays the (uncorrected) transient spectrum in this region for 1,2-$C_{60}(CH_3)_2$ in room-temperature toluene solution. Fitting to two Lorentzians clarifies the position of the unresolved shoulder band. Relative to pristine C_{60}, the peak molar absorptivity of the red triplet bands is reportedly reduced by ~50% in dihydrofullerenes [56], and by 15–25% in *N*-methylpyrrolidinofullerenes [10].

We have recently uncovered an empirical correlation in C_{60} derivatives between the spectral position of the dominant red triplet-triplet absorption band and the intrinsic decay constant of the T_1 state [28]. Figure 18 shows this correlation as a graph of the first quantity vs. the second for C_{60} and a number of its [6,6]-derivatives. It seems likely that the pattern in Figure 18 reflects in part the influence of geometrical strain induced by derivatization on molecular orbitals

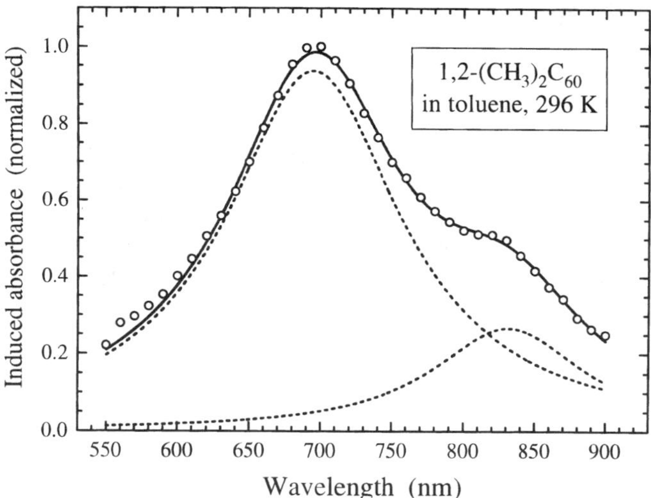

Figure 17 Induced absorption spectrum of 1,2-$C_{60}(CH_3)_2$ in toluene solution at 296 K, in a region of weak ground-state absorption. Excitation was at 532 nm. Open circles show measured data points and the solid curve shows its simulation as a superposition of two Lorentzian components, drawn as dotted curves.

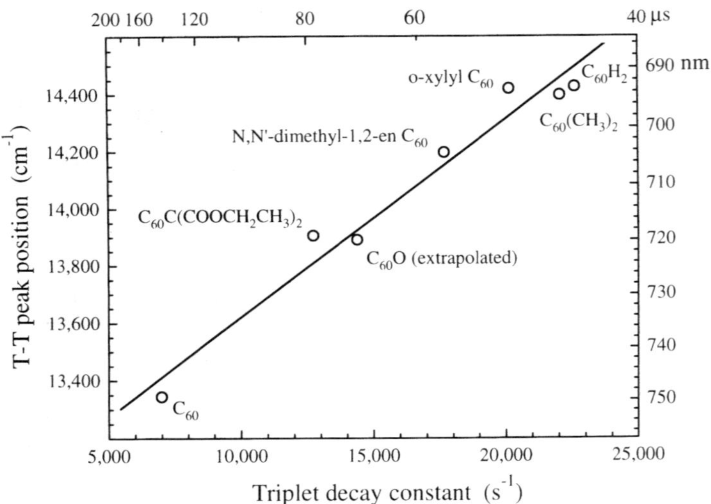

Figure 18 Correlation plot for C$_{60}$ and some of its derivatives showing the frequency of the most intense red band of the $T_n \leftarrow T_1$ absorption spectrum on the y axis and the intrinsic decay constant of the T_1 state on the x axis. All data are for toluene solutions at ambient temperature. (The decay constant for C$_{60}$O is extrapolated from lower-temperature data to estimate the nonphotochemical decay component.)

occupied in the triplet state. Although quantum calculations have had some success in accounting for ground-state spectral features of fullerene derivatives in terms of orbital shifts and symmetry-breaking caused by derivatization [59], a similar interpretation of derivatization effects in triplet-triplet spectra is not yet available. We believe that the accumulating body of data on triplet spectra, kinetics, and related properties will prove essential in gaining a proper understanding of fullerene triplet electronic structures.

C$_{70}$ and Its Derivatives

The triplet-triplet absorption spectrum of C$_{70}$ is broader than that of C$_{60}$, with bands extending from the ultraviolet into the near-infrared [19, 20, 34, 35, 60]. In room-temperature toluene solution, the strongest low-energy $T_n \leftarrow T_1$ transition of C$_{70}$ peaks near 970 nm, or 220 nm to the red of the prominent C$_{60}$ band. The molar absorptivity at this 970 nm peak has been reported to be 3800 M^{-1} cm^{-1} [35], or approximately one-fifth that of the C$_{60}$ band. Figure 19 shows the induced absorbance spectrum of C$_{70}$ in a frozen saturated hydrocarbon glass at 77 K [28]. The sharp structure at wavelengths below 660 nm results from the superposition

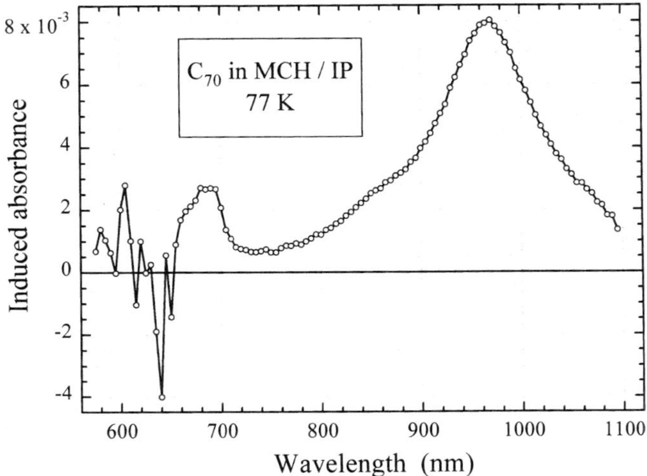

Figure 19   Induced absorbance spectrum of C_{70} in a solid isopentane/methylcyclohexane solution at 77 K. The sample was excited at 532 nm.

of broad, positive-going triplet-state absorption and narrow, negative-going ground-state depletion bands. It is clear that at 77 K these ground-state bands are far narrower than the $T_n \leftarrow T_1$ features, whose persistent widths and near-Lorentzian shapes likely reflect femtosecond-scale relaxation of T_n via internal conversion to lower-lying triplet states.

As of this writing, triplet-triplet absorption spectroscopy of C_{70} derivatives remains at a very early stage. Results from our laboratory on $C_{70}H_2$ show a dominant $T_n \leftarrow T_1$ absorption band at 1050 nm, red-shifted by 80 nm compared to that of the parent C_{70}. We have found a smaller red shift, to 1010 nm, in the main $T_n \leftarrow T_1$ band of $C_{70}O$. It is notable that these spectral shifts on derivatization occur in the opposite direction from those in the C_{60} family. Correlations between spectral position and triplet decay kinetics for C_{70} derivatives have not yet been explored.

Higher Fullerenes

The $T_n \leftarrow T_1$ absorption spectra of fullerenes larger than C_{70} are not very well characterized, in part because these compounds have rather low quantum yields of triplet formation. For C_{76}, transient peaks have been reported near 540, 620, 840, and 1400 nm [43, 45]. Not yet available, however, is a $T_n \leftarrow T_1$ spectrum corrected for the broad and substantial ground-state depletion over most of this range. The triplet molar absorptivity at 840 nm has been measured as 5730 M^{-1} cm^{-1} for C_{76} in room-temperature benzene solution [43]. No reliable triplet spectral data seem to have been measured for C_{78}. Finally, there is one report of a difference spectrum arising from the triplet state of C_{84} [46]. These data show maxima at 310 and 345 nm plus weaker features extending to ~700 nm. The largest molar absorptivity was estimated to be 4430 M^{-1} cm^{-1} at 310 nm.

5 TRIPLET THERMODYNAMICS

If one views the triplet state of a molecule as a metastable species with well-defined thermodynamic properties, then it may be characterized by measuring its enthalpy and entropy relative to a reference substance. Triplet-state energies, which are equivalent to enthalpies, are most commonly found by estimating the T_1-S_0 energy gap from absorption or phosphorescence spectra or photoacoustic measurements. Spectroscopic values of this sort may differ somewhat from thermodynamic values for transitions that involve small diagonal Franck-Condon factors, large Stokes shifts, Jahn-Teller distortions, or Herzberg-Teller vibronic coupling. The triplet-state entropies of molecules in solution are measured far less often than energies. If solvation of the sample and reference substances is nearly equal, the entropy difference between their triplets will reflect thermal occupations of spin, vibrational, and electronic states. In certain cases, knowledge of triplet-state entropies can therefore help to clarify the orbital degeneracy of T_1 and the presence of nearby excited triplet states.

We have recently developed a method to find the relative thermodynamic properties of fullerene triplet states in solution by measuring equilibrium constants for triplet energy transfer between different solutes [28, 61]. As described earlier in the "Energy Transfer" section, rate constants for such transfer between fullerenes of similar triplet energy are quite large (more than 20% of the diffusion limit), allowing energy transfer to kinetically dominate deactivation in many mixed solutions. This situation leads to the rapid formation of a preequilibrated pool of triplet energy, whose balance is determined by the equilibrium constant for triplet energy exchange, K^T, and the ratio of ground-state concentrations. Measurement of K^T as a function of temperature allows one to extract the enthalpy and entropy changes for the transfer of triplet energy from one species to the other.

To find the value of K^T, the induced spectrum of a mixed fullerene sample is measured as a function of delay after excitation to find the asymptotic spectral shape that reflects triplet preequilibration. This spectrum is then analyzed as a superposition of the pure-component induced spectra to obtain the ratio of triplet

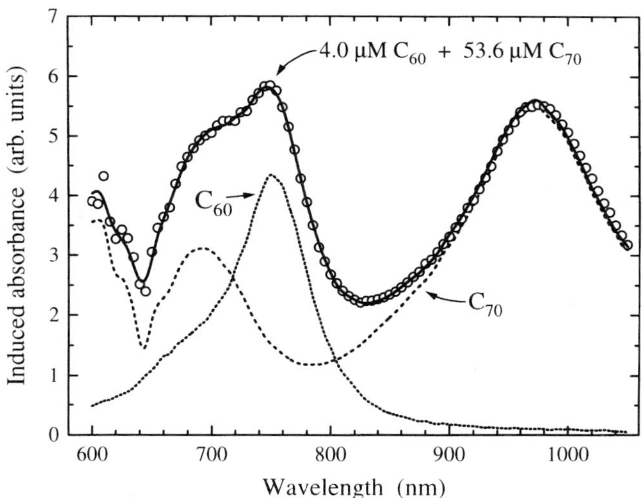

Figure 20 Superposition analysis of the induced spectrum in a pre-equilibrated mixture containing 3.97 μM C_{60} plus 53.6 μM C_{60} in toluene at 23.8°C. The dotted and dashed curves show induced spectra of pure C_{60} and C_{70}; open circles are data points for the mixture; and the solid curve is the best fit found by combining the pure component spectra.

concentrations at equilibrium. Figure 20 illustrates such an analysis for a mixture of C_{60} and C_{70}. By repeating this determination as a function of temperature, we have constructed the van't Hoff plots shown in Figure 21 for C_{60}/C_{70} and $1,2$-$C_{60}(CH_3)_2/C_{70}$ mixtures. The $\Delta H°$ results show that the T_1-S_0 energy gaps of C_{60} and C_{70} in toluene solution are equal to each other, within an experimental uncertainty of 0.2 kJ mol^{-1} (16 cm^{-1}). However, derivatization of C_{60} to form $1,2$-$C_{60}(CH_3)_2$ lowers this gap by approximately 3.4 kJ mol^{-1} (285 cm^{-1}). The entropy changes caused by interfullerene energy transfer likely reflect differences in electronic orbital degeneracies between donor and acceptor. This view is supported by our experimental finding that the entropy of $T_1 C_{60}$ (which is threefold orbitally degenerate) exceeds that of T_1 $1,2$-$C_{60}(CH_3)_2$ (which has no orbital degeneracy) by $1.18 \pm 0.13R$, implying a degeneracy ratio of 3.25 ± 0.4. Apparently, Jahn-Teller splitting of the degenerate T_1 state of C_{60} is not large compared to kT at room temperature. Most interestingly, our entropy data show that the orbital degeneracy of $T_1 C_{70}$ is intermediate between 1 and 2. This result implies that the lowest C_{70} triplet is the nondegenerate $^3A_2'$ state, and that the $^3E_1'$ state theoretically predicted to be nearby or lower [62, 63] lies higher by ~600 cm^{-1}.

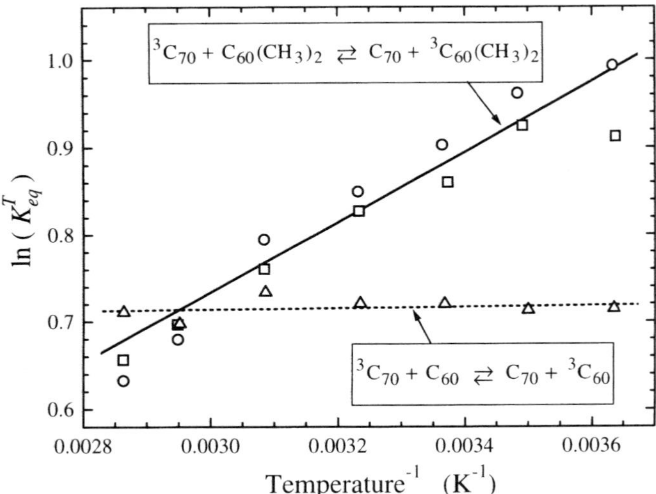

Figure 21 Van't Hoff plot of equilibrium constant data for the exchange of triplet energy between C_{70} and C_{60} ($\triangle$), and between C_{70} and 1,2-$C_{60}(CH_3)_2$ with composition ratios of 5:1 ($\bigcirc$) and 12:1 ($\square$). The straight lines show linear best fits to the data.

The relative thermodynamic measurements on fullerene triplet states thus complement spectroscopic and computational studies of their properties.

6 CONCLUSIONS

Of the many electronically excited states of fullerene molecules, one holds special importance. T_1, the lowest-lying triplet state, is typically formed with substantial efficiency following light absorption and persists far longer than any other electronically excited state. These properties give triplet states key roles in most processes or applications involving fullerene photochemistry or photophysics. A central challenge is measuring the persistence of these fullerene triplet states in a variety of environments, quantitatively isolating the various factors that contribute to their decay, and finally understanding these decay processes in terms of basic molecular structure.

Progress along this path is hampered by the difficulty of disentangling the various channels for triplet-state deactivation in fluid solution. Because fullerene phosphorescence emission is weak, $T_n \leftarrow T_1$ transient absorption is the optical method of choice for monitoring triplet-state kinetics. However, many such attempts to measure the "intrinsic" or unimolecular triplet decay constant are seriously inaccurate because of unrecognized bimolecular contributions from triplet-triplet annihilation, self-quenching, or impurity quenching. As a result, the literature contains many discrepant reports of fullerene triplet-state lifetimes. It is nevertheless possible to deduce valid unimolecular and bimolecular rate constants by employing a sensitive apparatus, using purified samples, and measuring kinetics over a range of sample concentrations and excitation energies.

Reliably determined intrinsic triplet lifetimes in room-temperature fluid solution are found to vary widely among fullerenes, from a low of 4.3 μs for $C_{60}O$ to at least 12 ms for C_{70}. With the exception of $C_{60}O$, whose anomalously rapid decay suggests a photochemical channel, chemical derivatization to form [6,6]-closed adducts shortens the 143 μs intrinsic lifetime of C_{60} to values in the range of 40–80 μs. Incorporation of a krypton atom within the C_{60} cage accelerates the triplet decay through the endohedral heavy atom effect. The C_{70} triplet lifetime is apparently also shortened by derivatization, but members of the C_{70} family retain longer-lived triplet states than C_{60} or its adducts. Reliable intrinsic triplet lifetimes are not yet available for the higher fullerenes.

Below 350 K, there is only a weak temperature dependence to the intrinsic triplet decay of C_{60}, C_{70}, and, probably, their adducts. At 77 K, the lifetimes are 237 μs for C_{60} and 50.0 ms for C_{70}. This temperature variation has been traced from 0 to 1000 K for C_{60} and found to follow a simple exponential function of the molecule's vibrational energy content, which varies slowly at or below room temperature but much more rapidly near 1000 K. The only significant influence of solvent on C_{60} triplet lifetime appears to involve external heavy atom effects.

Bimolecular deactivation of fullerene triplets in room-temperature aromatic solvents can occur by triplet-triplet annihilation, with rate constants near 5×10^9 M^{-1} s^{-1}, or by oxygen quenching, with rate constants in the range of $1–2 \times 10^9$ M^{-1} s^{-1}. Self-quenching of triplet-state fullerenes by ground states of the same species occurs with a rate constant of 1.5×10^7 M^{-1} s^{-1} for C_{60} and 3×10^7 M^{-1} s^{-1} for C_{70}, as measured in room-temperature toluene solutions. The self-quenching constant of C_{60} increases strongly with temperature and can be described by an Arrhenius activation enthalpy of approximately 10 kJ mol^{-1}. Excimer models of self-quenching cannot easily account for this temperature variation. Further work will be needed to clarify the molecular mechanism of self-quenching.

Energetically allowed transfer of triplet excitation between fullerene solutes occurs with rate constants of approximately 2 to 3×10^9 M^{-1} s^{-1}. These high values often permit the rate of energy exchange in a mixed solution of fullerenes to exceed the rates of deactivation, allowing formation of a preequilibrated pool of triplet excitation. By spectroscopically determining the equilibrium constant that controls such pooling, it has been possible to find the relative energies and entropies of fullerene triplet states in fluid solution. Entropy results indicate that

Table 1 Property Comparison for Triplet-State Fullerenes

Compound	T_1 intrinsic decay constant at 296 K (s^{-1})	T_1 intrinsic decay constant at 77 K (s^{-1})	Self-quenching constant[a] ($M^{-1}\ s^{-1}$)	T-T peak wavelength (nm)	T-T shoulder wavelength (nm)
C_{60}	7,000	4,220	1.5×10^7	749.4 (745.2[b])	679 (666, 717[b])
$Kr@C_{60}$		4,760		747.6[b]	667, 723[b]
$C_{60}C(COOCH_2CH_3)_2$	12,700			719.2	631, 812
N,N'-dimethyl-1,2-en C_{60}	17,700			704.4	857
$ortho$-xylyl C_{60}	20,200			693.6	826
$1,2$-$C_{60}(CH_3)_2$	22,100			694.6	831
$1,2$-$C_{60}H_2$	22,700			693.2	823
$C_{60}O$	235,000	6,670		720.0	676, 782
C_{70}	<85	20.0	3×10^7	970	
$1,2$-$C_{70}H_2$[c]	<550			1050	
$C_{70}O$[c]	6,500	74		1010	

[a] In toluene at 296 K.

[b] In saturated hydrocarbon glass at 77 K.

[c] Preliminary results.

the lowest triplet of C_{70} is orbitally nondegenerate. Information obtained by this method can complement that found from phosphorescence emission spectroscopy.

The triplet-triplet absorption spectrum of C_{60} is significantly perturbed by derivatization to form [6,6]-adducts. The prominent 750 nm C_{60} peak is blue-shifted in the adducts to between 650 and 720 nm, a shoulder can be seen near 820 nm in many adduct spectra, and the peak molar absorptivity is reduced. In C_{70}, triplet-triplet absorption shows a strong band at 970 nm, which seems to shift substantially to the red on simple derivatization. Among C_{60} and its [6,6]-adducts, the spectral position of the red triplet-triplet absorption peak correlates strongly with the intrinsic triplet decay constant. Table 1 summarizes some of the fullerene triplet state kinetic and spectral parameters presented in this chapter. The patterns emerging from these properties suggest that photophysical data will provide clues vital to gaining a fundamental understanding of the triplet electronic manifolds of fullerenes and their derivatives.

ACKNOWLEDGMENTS

The author is grateful to his co-workers Meg Fraelich, Tom Etheridge, Kevin Ausman, Angelo Benedetto, Rebecca Brown, David Samuels, and Sergei Bachilo for their invaluable research contributions, to Angelo Benedetto for assistance in preparing this manuscript, and to the National Science Foundation and the Robert A. Welch Foundation for grant support.

REFERENCES

1.   D. Kim, M. Lee, Y. D. Suh, and S. K. Kim, *J. Am. Chem. Soc.* **114**, 4429 (1992).
2.   T. W. Ebbesen, K. Tanigaki, and S. Kuroshima, *Chem. Phys. Lett.* **181**, 501 (1991).
3.   M. Lee, O.-K. Song, J.-C. Seo, D. Kim, Y. D. Suh, S. M. Jin, and S. K. Kim, *Chem. Phys. Lett.* **196**, 325 (1992).
4.   R. J. Sension, C. M. Phillips, A. Z. Szarka, W. J. Romanow, A. R. McGhie, J. P. McCauley, Jr., A. B. Smith, III, and R. M. Hochstrasser, *J. Phys. Chem.* **95**, 6075 (1991).
5.   B. Ma and Y.-P. Sun, *J. Chem. Soc. Perkin Trans.* 2 2157 (1996).
6.   J. W. Arbogast, A. P. Darmanyan, C. S. Foote, Y. Rubin, F. N. Diederich, M. M. Alvarez, S. J. Anz, and R. L. Whetten, *J. Phys. Chem.* **95**, 11 (1991).
7.   M. Terazima, N. Hirota, H. Shinohara, and Y. Saito, *J. Phys. Chem.* **95**, 9080 (1991).
8.   L. Biczok, H. Linschitz, and R. I. Walter, *Chem. Phys. Lett.* **195**, 339 (1992).
9.   R. V. Bensasson, T. Hill, C. Lambert, E. J. Land, S. Leach, and T. G. Truscott, *Chem. Phys. Lett.* **201**, 326 (1993).

10. C. Luo, M. Fujitsuka, A. Watanabe, O. Ito, L. Gan, Y. Huang, and C.-H. Huang, *J. Chem. Soc. Faraday Trans.* **94**, 527 (1998).

11. M. R. Wasielewski, M. P. O'Neil, K. R. Lykke, M. J. Pellin, and D. M. Gruen, *J. Am. Chem. Soc.* **113**, 2774 (1991).

12. Y. Zeng, L. Biczok, and H. Linschitz, *J. Phys. Chem.* **96**, 5237 (1992).

13. A. Sassara, G. Zerza, and M. Chergui, *Chem. Phys. Lett.* **261**, 213 (1996).

14. F. Negri, G. Orlandi, and F. Zerbetto, *Chem. Phys. Lett.* **144**, 31 (1988).

15. I. Laszlo and L. Udvardi, *J. Mol. Struct.* (Theochem) **183**, 271 (1989).

16. R. R. Surjan, L. Udvardi, and K. Nemeth, *J. Mol. Struct.* (Theochem) **311**, 55 (1994).

17. Y. Kajii, T. Nakagawa, S. Suzuki, Y. Achiba, K. Obi, and K. Shibuya, *Chem. Phys. Lett.* **181**, 100 (1991).

18. G. L. Closs, P. Gautam, D. Zhang, P. J. Krusic, S. A. Hill, and E. Wasserman, *J. Phys. Chem.* **96**, 5228 (1992).

19. D. K. Palit, A. V. Sapre, J. P. Mittal, and C. N. R. Rao, *Chem. Phys. Lett.* **195**, 1 (1992).

20. N. M. Dimitrijevic and P. V. Kamat, *J. Phys. Chem.* **96**, 4811 (1992).

21. H. T. Etheridge, M. R. Fraelich, and R. B. Weisman, *Proc.-Electrochem. Soc.* **94-24** (Recent Advances in the Chemistry and Physics of Fullerenes and Related Materials), 803 (1994).

22. K. D. Ausman and R. B. Weisman, *Res. Chem. Intermediates* **6**, 431 (1997).

23. K. D. Ausman and R. B. Weisman, *Proc.-Electrochem. Soc.* **96-10** (Recent Advances in the Chemistry and Physics of Fullerenes and Related Materials), 276 (1996).

24. H. T. Etheridge, III, R. D. Averitt, N. J. Halas, and R. B. Weisman, *J. Phys. Chem.* **99**, 11306 (1995).

25. R. E. Haufler, L.-S. Wang, L. P. F. Chibante, C. Jin, J. Conceicao, Y. Chai, and R. E. Smalley, *Chem. Phys. Lett.* **179**, 449 (1991).

26. M. R. Fraelich and R. B. Weisman, *J. Phys. Chem.* **97**, 11145 (1993).

27. D. J. van den Heuvel, I. Y. Chan, E. J. J. Groenen, J. Schmidt, and G. Meijer, *Chem. Phys. Lett.* **231**, 111 (1994).

28. K. D. Ausman, A. F. Benedetto, D. A. Samuels, and R. B. Weisman, *Proc.-Electrochem. Soc.* **98-8** (Recent Advances in the Chemistry and Physics of Fullerenes and Related Materials), 281 (1998).

29. C. C. Henderson and P. A. Cahill, *Science* **259**, 1885 (1993).

30. K. D. Ausman, A. F. Benedetto, D. A. Samuels, R. J. Brown, and R. B. Weisman, *Proc.-Electrochem. Soc.* **97-14** (Recent Advances in the Chemistry and Physics of Fullerenes and Related Materials), 111 (1997).

31. K. M. Creegan, J. L. Robbins, W. K. Robbins, J. M. Millar, R. D. Sherwood, P. J. Tindall, D. M. Cox, A. B. Smith, J. P. McCauley, D. R. Jones, and R. T. Gallagher, *J. Am. Chem. Soc.* **114**, 1103 (1992).

32. A. F. Benedetto and R. B. Weisman, *Chem. Phys. Lett.*, in press (1999).

33. K. Yamamoto, M. Saunders, A. Khong, R. J. Cross, M. Grayson, M. L. Gross, A. F. Benedetto, and R. B. Weisman, *J. Am. Chem. Soc.*, **121**, 1591 (1999).

34. J. W. Arbogast and C. S. Foote, *J. Am. Chem. Soc.* **113**, 8886 (1991).

35. R. V. Bensasson, T. Hill, C. Lambert, E. J. Land, S. Leach, and T. G. Truscott, *Chem. Phys. Lett.* **206**, 197 (1993).

36. M. N. Berberan-Santos and J. M. M. Garcia, *J. Amer. Chem. Soc.* **118**, 9391 (1996).
37. Y. P. Sun and C. E. Bunker, *J. Phys. Chem.* **97**, 6770 (1993).
38. K. Palewska, J. Sworakowski, H. Chojnacki, E. C. Meister, and U. P. Wild, *J. Phys. Chem.* **97**, 12167 (1993).
39. S. M. Argentine, K. T. Kotz, and A. H. Francis, *J. Am. Chem. Soc.* **117**, 11762 (1995).
40. J. B. M. Warntjes, I. Holleman, G. Meijer, and E. J. J. Groenen, *Chem. Phys. Lett.* **261**, 495 (1996).
41. A. Sassara, G. Zerza, and M. Chergui, *J. Phys. Chem. A* **102**, 3072 (1998).
42. H. T. Etheridge, III, and R. B. Weisman, *J. Phys. Chem.* **99**, 2782 (1995).
43. R. V. Bensasson, E. Bienvenue, J.-M. Janot, E. J. Land, S. Leach, and P. Seta, *Chem. Phys. Lett.* **283**, 221 (1998).
44. D. M. Guldi, D. Liu, and P. V. Kamat, *J. Phys. Chem. A* **101**, 6195 (1997).
45. M. Fujitsuka, A. Watanabe, O. Ito, K. Yamanoto, and H. Funasaka, *J. Phys. Chem. A* **101**, 4840 (1997).
46. G. Sauve, P. V. Kamat, and R. S. Ruoff, *J. Phys. Chem.* **99**, 2162 (1995).
47. N. J. Turro, *Modern Molecular Photochemistry*, Benjamin/Cummings, Menlo Park, CA, 1978.
48. D. R. Kearns, *Chem. Rev.* **71**, 395 (1971).
49. K. G. Thomas, V. Miju, M. V. George, D. M. Guldi, and P. V. Kamat, *J. Phys. Chem. A* **102**, 5341 (1998).
50. D. K. Palit and J. P. Mittal, *Fullerene Science and Technology* **3**, 643 (1995).
51. K. D. Ausman, Ph.D. Dissertation, Rice University, 1998.
52. D. A. Samuels and R. B. Weisman, *Chem. Phys. Lett.* **295**, 105 (1998).
53. J. Cioslowski, *Electronic Structure Calculations on Fullerenes and Their Derivatives*, Oxford University Press, New York, 1995.
54. R. Bauernschmitt, R. Ahlrichs, F. H. Hennrich, and M. M. Kappes, *J. Am. Chem. Soc.* **120**, 5052 (1998).
55. X. Zhang, J. Anderson, and C. S. Foote, *Proc.-Electrochem. Soc.* (Recent Advances in the Chemistry and Pysics of Fullerenes and Related Materals) **94-24**, 797 (1994).
56. R. V. Bensasson, E. Bienvenue, J.-M. Janot, S. Leach, P. Seta, D. I. Schuster, S. R. Wilson, and H. Zhao, *Chem. Phys. Lett.* **245**, 566 (1995).
57. D. M. Guldi, H. Hungerbuhler, and K.-D. Asmus, *J. Phys. Chem.* **99**, 9380 (1995).
58. D. M. Guldi and K.-D. Asmus, *J. Phys. Chem. A* **101**, 1472 (1997).
59. Q. Teng, J. Feng, C. Sun, and M. Zerner, *Int. J. Quantum Chem.* **55**, 35 (1995).
60. K. Tanigaki, T. W. Ebbesen, and S. Kuroshima, *Chem. Phys. Lett.* **185**, 189 (1991).
61. K. D. Ausman and R. B. Weisman, *J. Am. Chem. Soc.*, **121**, 1110 (1999).
62. J. Baker, P. W. Fowler, P. Lazzeretti, M. Malagoli, and R. Zanasi, *Chem. Phys. Lett.* **184**, 182 (1991).
63. P. R. Surjan, K. Nemeth, and M. Kallay, *J. Mol. Struct.* (*Theochem*) **398**, 293 (1997).

5
The Electronic Third-Order Nonlinear Optical Properties of C_{60} and C_{70} Films

F. P. Strohkendl
University of Southern California
Los Angeles, California

Zakya H. Kafafi
U.S. Naval Research Laboratory
Washington, D.C.

1 INTRODUCTION

The rigid cage structure of fullerenes coupled with extended delocalized electronic π-bonds leads to large optical nonlinearities [1–11]. Nevertheless, the nonlinearities of the fullerenes, due to their three-dimensional character, are found to be below those of organic linear π-conjugated polymers [12]. The fullerenes have, however, the advantage that they do not exhibit intrinsic absorption in the important telecommunications window near 1.5 μm. This is due to the absence of $C-H$, $N-H$ and $O-H$ bonds which give rise to vibrational overtone absorption in organic materials [5]. The fullerenes are an interesting model for the understanding of electronic structure and nonlinear optical properties, as they are at once simple due to their high symmetry and elementally pure nature and complicated due to the large number of atoms involved. As a consequence, a large number of theoretical studies have been devoted to computing the excited-state electronic spectra (e.g., Refs. 13–17) and nonlinear optical properties (e.g., Refs. 12, 18–24) of these molecules. Our interest in the nonlinear optical properties of the fullerenes is twofold: (1) Nonlinear optical spectra reveal excited state levels which do not appear in the linear absorption spectrum; (2) We want to estimate the potential of these materials for all optical switching and frequency shifting.

The importance of the fullerenes from a fundamental point of view, as well as their applications potential, has spawned a large number of nonlinear optical measurements. Such measurements, which were based on different nonlinear optical techniques, yielded widely varying results [4]. Some measurements yielded very large values for the optical nonlinearity whereas others did not. Certain measurement techniques yielded resonances which were not apparent with others. The reason for these varying and sometimes contradictory results lies in the relative complexity of the third-order nonlinear optical susceptibility, often simply referred to as "$\chi^{(3)}$." The absence of common reference standards and clear conventions as well as ambiguities in relating experimental data to the underlying electronic structure have further complicated the issue. Some experimental techniques, such as third-harmonic generation, are assured that they measure the purely electronic nonlinearity but the nature of their resonances is not unique. Other techniques, such as degenerate four-wave mixing, do not necessarily measure the electronic $\chi^{(3)}$, but if they do, the nature of their resonances is unambiguous.

In the present chapter we outline first some of the basic properties of $\chi^{(3)}$ and detail some of the experimental precautions which have to be taken in order to observe the purely electronic $\chi^{(3)}$. Such considerations greatly reduce the number of measurements which we need to consider. We also comment on the different resonance structures exhibited by different nonlinear optical techniques and their relation to the underlying electronic structure. We discuss some of the relevant facts about the electronic structure of C_{60} and C_{70}, and then proceed to compare actual nonlinear optical spectra. The result of this investigation is the identification of the intrinsic two-photon states of C_{60} and C_{70} below the linear-absorption edge, the corresponding two-photon absorption coefficients, and the nonresonant electronic $\chi^{(3)}$, a quantity important for nonlinear optical switching as well as for comparison to theoretical computations.

2 THE ELECTRONIC THIRD-ORDER NONLINEAR OPTICAL SUSCEPTIBILITY

Third-order optical nonlinearities can be based on a variety of fundamental mechanisms, such as ac and dc electric field–induced changes in the polarizability of the electronic cloud, excited-state population, structural relaxation involving nuclear motion, molecular rotation, heating. These processes can be distinguished by the time it takes for the nonlinear-optical material to revert back to its ground state after excitation. The fastest types of nonlinearities are based on the nonlinear polarizability of the electronic charge density cloud. In the condensed phase these have relaxation times on the order of a few optical cycles. Such nonlinearities are often referred to as electronic or ''instantaneous'' since they result from a very short integration time over the action of the optical driving field. Other nonlinearities based on processes with sufficiently slow relaxation times are the result of the accumulation of the action of the optical driving field during the whole duration of an optical pulse. Such nonlinearities depend on the pulse energy and we refer to them as ''energy-dependent'' in contrast to the instantaneous or ''intensity-dependent'' effects. The presence of energy-dependent effects even in nonresonant regions of the linear absorption spectrum are apparent in pulse-duration-dependent damage-thresholds which in C$_{60}$ and C$_{70}$ are near 100 GW/cm^2 for 100 fs pulses [6, 9] and near 10 GW/cm^2 for 30 ps pulses [4]. Degenerate-four-wave-mixing (DFWM) experiments done at 10 GW/cm^2 in C$_{60}$ with a non-resonant electronic nonlinearity of 10^{-12} esu cause a refractive index modulation with an amplitude of only 0.02%; i.e., it is small compared to index-modulations achievable with energy-dependent effects. Experience has shown that the electronic DFWM nonlinearity is essentially impossible to observe with nanosecond pulses due to the unfavorable peak-power to pulse-energy ratio. Already with 30 ps pulses near 1 μm we find an asymmetric tail [4] in time resolved DFWM a signature of some contribution by energy-dependent effects. In general, nonlinear optical experiments done with short optical pulses favor intensity-dependent effects, whereas long-pulse experiments favor energy-dependent effects.

In contrast to DFWM third harmonic generation (THG) shows little sensitivity to nonelectronic effects and meaningful results can be derived with nanosecond pulses. This advantage of THG is due to the large frequency shift between pump and signal beams. However, in comparison to DFWM the interpretation of the THG spectra may be ambiguous at times, as is explained below. Also, essentially all switching applications based on $\chi^{(3)}$ are directly related to DFWM while THG gives switching related properties only in its low frequency limit. Here we are interested in the purely electronic, i.e., intensity-dependent, nonlinearities of C$_{60}$ and C$_{70}$.

The polarization density in isotropic bulk samples, $\vec{P}_{dc}$, which is induced by an applied static or dc electric field $\vec{E}_{dc}$ can be expressed by a Taylor expansion:

$$\vec{P}_{dc} = \overleftrightarrow{\chi}^{(1)}(0, 0) \cdot \vec{E}_{dc} + \overleftrightarrow{\chi}^{(3)}(0, 0, 0, 0) : \vec{E}_{dc}\vec{E}_{dc}\vec{E}_{dc} + \cdots \tag{1}$$

We assume here for conceptual purposes that $\overleftrightarrow{\chi}^{(1)}(0, 0)$ and $\overleftrightarrow{\chi}^{(3)}(0, 0, 0, 0)$ are the dc or ''zero-frequency'' limits of the *electronic* parts of the linear and third-order nonlinear optical susceptibility tensors, respectively. (In general one will have to take nuclear contributions to these susceptibilities into account under dc or low-frequency conditions.) We make this assumption since we are here only interested in the determination of the electronic part of $\chi^{(3)}$.

The frequency-dependent third-order susceptibility tensor $\overset{\leftrightarrow}{\chi}^{(3)}(-\omega_4, \omega_3, \omega_2, \omega_1)$ has four frequency arguments, where ω_1, ω_2, and ω_3 are frequencies of monochromatic driving fields of the form $\mathrm{Re}(\vec{E}_{\omega q}e^{-i\omega_q t})$, $q = 1, 2, 3$, while ω_4 is the frequency of the resulting coherently driven nonlinear-optical polarization density of the form $\mathrm{Re}(\vec{P}_{\omega 4}e^{-i\omega_4 t})$. Energy is conserved in this four-wave mixing process which is implied by the sum of the frequency arguments of $\overset{\leftrightarrow}{\chi}^{(3)}$ being equal to zero. The relationship between nonlinear polarization density and driving fields is given by [25, 26]

$$\vec{P}_{\omega 4} = K(\omega_3, \omega_2, \omega_1)\overset{\leftrightarrow}{\chi}^{(3)}(-\omega_4, \omega_3, \omega_2, \omega_1) : \vec{E}_{\omega 3}\vec{E}_{\omega 2}\vec{E}_{\omega 1} \tag{2}$$

The induced polarization density radiates signal fields of frequency ω_4 which are detected in nonlinear optical experiments. The Maker-Terhune convention [25] which we are using in the following defines the dimensionless factor K such that the zero-frequency limit of the $\chi^{(3)}$ tensor is independent of the method by which it is measured. The definition of K varies between different authors and reported values of $\chi^{(3)}$ can vary by up to an order of magnitude. Consequently, we emphasize that quotations of absolute numbers for the value of $\chi^{(3)}$ without reference standard and without the numerical value for this reference standard are of limited value. Fused silica is an excellent reference standard with a mostly electronic nonlinearity. Our preferred χ_{1111} value for fused silica is 3.95×10^{-15} esu [27, 28]. An often used reference for picosecond degenerate four-wave mixing is CS_2. Its nonlinearity is mostly orientational with a relaxation time on the order of a picosecond. Our preferred value for χ_{1111} of CS_2 is 4.7×10^{-13} esu [27, 29]. These reference values are compatible with the formulas for the nonlinear refractive index, n_2, given below in Eq. (6).

The $\chi^{(3)}$ tensor is a bulk property which is related to the underlying quantum-mechanical molecular properties by [25, 26]

$$\overset{\leftrightarrow}{\chi}^{(3)}(-\omega_4, \omega_3, \omega_2, \omega_1) = F\frac{Ne^4}{\hbar^3}\sum_{a,b,c}^{\text{all states}}$$

$$\times \left(\frac{\langle g|\vec{r}|c\rangle\langle c|\vec{r}|b\rangle\langle b|\vec{r}|a\rangle\langle a|\vec{r}|g\rangle}{[(\omega_{cg} - \omega_3 - \omega_2 - \omega_1) - i\gamma_{cg}][(\omega_{bg} - \omega_3 - \omega_2) - i\gamma_{bg}][(\omega_{ag} - \omega_3) - i\gamma_{ag}]} + \cdots\right)$$

$$\tag{3}$$

where, $e\langle a|\vec{r}|b\rangle$ is a transition dipole moment, ω_{ag} a molecular transition frequency, γ_{ag} a transition line width, and N is the number density of molecules. The factor F summarizes local field factors [26] which are not important for the present discussion. The $\chi^{(3)}$ tensor is an average over all dipole-allowed transition pathways which begin and end in the ground state $|g\rangle$. These transition pathways are represented by the quadruple transition-dipole-moment products in Eq. (3). Note that the sequence of transitions is ground state to one-photon state to two-photon state to one-photon state to ground state. The last excited state in such excitation pathways must be one photon in nature as a single transition-dipole moment connects it to the ground state. Even though these allowed transition pathways do not depend on the input frequencies, their relative weight in the summation of Eq. (3) does. This is due to the resonance-frequency terms in the denominator of Eq. (3). Whenever the sum of some combination of the input frequencies equals an allowed molecular transition frequency a resonance occurs, which emphasizes a particular transition pathway and leads to a strong frequency-dependent variation in $\chi^{(3)}$. The elements of the $\chi^{(3)}$ tensor are complex numbers which translate into a phase for the driven nonlinear polarization density. This phase expresses the relative temporal relationship between the induced nonlinear polarization density and the driving-field term in Eq. (2). When a resonance occurs, the polarization density has a phase near $\pm\pi/2$ whereas far above or below any resonances the phase approaches either π or 0 as can be seen from Eq. (3).

Different nonlinear optical techniques measure $\chi^{(3)}$ for different frequency arguments. In third-harmonic generation, a single input beam at frequency ω generates a signal beam at frequency 3ω. Third-harmonic generation measures

$$\overset{\leftrightarrow}{\chi}^{(3)}(-3\omega, \omega, \omega, \omega) = F\frac{Ne^4}{\hbar^3} \sum_{a,b,c}^{\text{all states}}$$

$$\times \left(\frac{\langle g|\vec{r}|c\rangle\langle c|\vec{r}|b\rangle\langle b|\vec{r}|a\rangle\langle a|\vec{r}|g\rangle}{[(\omega_{cg} - 3\omega) - i\gamma_{cg}][(\omega_{bg} - 2\omega) - i\gamma_{bg}][(\omega_{ag} - \omega) - i\gamma_{ag}]} + \cdots \right) \quad (4a)$$

In degenerate four-wave mixing, three input beams at frequency ω generate signal beams at frequency ω. Degenerate four-wave mixing measures

$$\overset{\leftrightarrow}{\chi}^{(3)}(-\omega, \omega, \omega, -\omega) = F\frac{Ne^4}{\hbar^3} \sum_{a,b,c}^{\text{all states}}$$

$$\times \left(\frac{\langle g|\vec{r}|c\rangle\langle c|\vec{r}|b\rangle\langle b|\vec{r}|a\rangle\langle a|\vec{r}|g\rangle}{[(\omega_{cg} - \omega) - i\gamma_{cg}][(\omega_{bg} - 2\omega) - i\gamma_{bg}][(\omega_{ag} - \omega) - i\gamma_{ag}]} + \cdots \right) \quad (4b)$$

The minus sign on the last frequency argument indicates that one of the input beams receives a photon, whereas the remaining two input beams give one photon each. In electroabsorption, a static or low-frequency field is applied to the sample transversely to the incident probe beam of frequency ω. Electroabsorption measures

$$\text{Im}[\overset{\leftrightarrow}{\chi}^{(3)}(-\omega, \omega, 0, 0)] = F \frac{Ne^4}{\hbar^3} \sum_{a,b,c}^{\text{all states}}$$

$$\times \text{Im}\left(\frac{\langle g|\vec{r}|c\rangle\langle c|\vec{r}|b\rangle\langle b|\vec{r}|a\rangle\langle a|\vec{r}|g\rangle}{[(\omega_{cg} - \omega) - i\gamma_{cg}][(\omega_{bg} - \omega) - i\gamma_{bg}][(\omega_{ag} - \omega) - i\gamma_{ag}]} + \cdots\right) \quad (4c)$$

In general, positive frequency arguments of $\chi^{(3)}$ indicate ''absorbed'' photons, while negative frequency arguments indicate emitted photons. In third-harmonic generation, three photons are absorbed and one photon is emitted, as shown in Figure 1a. The vertical energy axis in this scheme indicates the accumulated photon energy during the nonlinear excitation cycle. (Note that this is *not* an energy scale for the three excited states shown in Figure 1). If the accumulated photon energy does not match the excited state energy at a given point in the excitation cycle, the state is usually referred to as a virtual state. Excited states are revealed through resonances in $\chi^{(3)}$ when the total photon energy deposited up to a given point in the excitation cycle matches the excited state energy at that point. Therefore, following the diagram for THG in Figure 1a, if the energy of the first state in the excitation pathway equals the exciting photon energy, we speak of a one-photon resonance. If the energy of the second state in the excitation pathway equals the energy of two exciting photons, we speak of a two-photon resonance; and if the energy of the third state in the excitation pathway equals the energy of three exciting photons, we speak of a three-photon resonance. Note

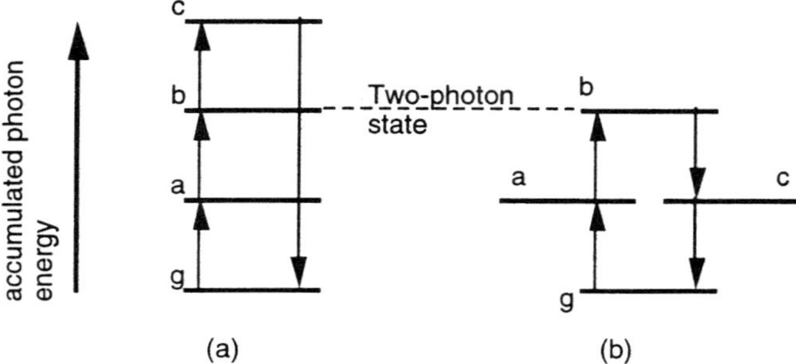

Figure 1 Excitation pathways relevant for two-photon spectroscopy with (a) third-harmonic generation and (b) degenerate-four wave mixing. The state labeling is in accordance with Eq. (3). The two-photon state is labeled b. The excitation energies of states a and c are known from the linear absorption spectrum. The vertical axis indicates the accumulated photon energy during the nonlinear excitation cycle; i.e., it is not related to the excitation energy of the states involved.

that a three-photon resonance reveals an excited-state energy which can be seen in a simple one-photon absorption spectrum when a photon of 3 times the energy of that of the three-photon-resonant photon is used. Three-photon resonances do not, therefore, contain any new spectroscopic information, but rather, may at times obstruct the observation of unknown two-photon states. To illustrate this point let us denote the excitation energy of the lowest one-photon state by E_g. If the excitation energy of a two-photon state is $^2/_3E_g$, a two- and a three-photon resonance in THG would occur at a photon energy of $^1/_3E_g$. In the absence of any other criteria such a resonance would most likely be interpreted as a strong three-photon resonance, while the hidden two-photon resonance would remain undetected. Only when the two-photon state is more than a line width below $^2/_3E_g$ can it be unambiguously identified. Two-photon states below the lowest one-photon allowed state can only be expected in inversion-symmetric molecules such as the fullerenes. Generally two-photon states are at or above E_g. This makes third-harmonic generation spectra often difficult to interpret.

Figure 1b shows the excitation pathway relevant for two-photon spectroscopy with degenerate four-wave mixing. At most, the energy of two photons is absorbed by the molecule before energy is released again. This means that DFWM may have only one- and two-photon resonances. If we assume that a two-photon state has an excitation energy of $2E_g$, we find that a one- and a two-photon resonance will occur simultaneously when the exciting photon energy equals E_g. In this case, the observed resonance might be erroneously interpreted as a strong one-photon resonance. However, when the two-photon-state energy is more than a line width below $2E_g$, the resulting DFWM resonance can be clearly identified as a two-photon resonance. It is, therefore, clear that DFWM can uniquely identify two-photon states up to $2E_g$, whereas THG can do so only for energies up to $^2/_3E_g$. The described behavior concerning one-, two-, and three-photon resonances can be recognized in the denominators of the sample expressions of $\chi^{(3)}$ for DFWM and THG in Eqs. (4a) and (4b).

The behavior of electroabsorption can be discussed in similar terms. Since two of the three absorbed photons have zero frequency, we find that resonances occur whenever the energy of the optical input photon matches the excitation energy of a one- or two-photon state. If a two-photon state is below E_g, it can be unambiguously detected in EA without interference from a one-photon state.

When comparing third-harmonic and degenerate four-wave mixing data, we note that due to the differences in input frequencies the expression of $\chi^{(3)}$ for DFWM has three times as many *distinguishable* resonance terms in the summation over all excitation pathways when compared to that for THG [26]. As a result, two-photon resonances observed with DFWM tend to be roughly 3 times larger when observed with THG relative to the zero-frequency limit of $\chi^{(3)}$.

When a linearly polarized optical beam of intensity I propagates through a medium it experiences a refractive index n and an absorption α, which depend linearly on the intensity. This dependence becomes noticeable at high intensities. It is expressed by

$$n = n_0 + n_2 I$$
$$\alpha = \alpha_0 + \alpha_2 I$$

(5)

where n_2 is referred to as the nonlinear refractive index, and α_2 as the nonlinear absorption. These parameters are related to χ_{1111} measured with degenerate four-wave mixing through the relations

$$n_2 = \frac{48\pi^2}{n_0^2 c} \operatorname{Re}[\chi_{1111}(-\omega, \omega, \omega, -\omega)]$$

$$\alpha_2 = \frac{2\omega}{c} \frac{48\pi^2}{n_0^2 c} \operatorname{Im}[\chi_{1111}(-\omega, \omega, \omega, -\omega)]$$

(6)

where c is the speed of light in vacuum. The above equations are given in esu, i.e., the unit of intensity is erg s^{-1} cm^{-2} which corresponds to 10^{-7} W/cm^2. A χ_{1111} value of 10^{-12} esu together with a n_0 value of 2 results in a value for n_2 of 4.0×10^{-14} cm^2/W. Note that the numerical prefactors in Eq. (6) are specific to the choice of the factor K in Eq. (2), and are consistent with the reference values for fused silica and CS_2 which we quoted above. The nonlinear absorption below the linear optical absorption edge is caused by two-photon absorption. Two-photon absorption spectra can therefore be measured through intensity dependent transmission, if the sample excitation during the measurement can be kept low [11].

3 ELECTRONIC STRUCTURES OF C$_{60}$ AND C$_{70}$

Evaporated solid films of C_{60} or C_{70} molecules may contain many defects and impurities. Nevertheless, the main features in their optical absorption spectra and nonlinear optical response are intrinsic, and remain stable indefinitely at room temperature, in room light and atmosphere. Films are usually amorphous Van der Waals solids with sublimation energies of 1.7 eV for C_{60} and 1.3 eV for C_{70} [30]. C_{60} molecules rotate freely at temperatures above 250 K [31, 32]. Optical [3, 30, 33, 34] and photoemission [35, 36] spectra show that the electronic structures of C_{60} and C_{70} in solid films are very similar to those in the gas phase and liquid solutions.

The symmetries of the excited states of C_{60} are classified in terms of the irreducible representations of its molecular point group, I_h. The ground state of the molecule is an electronic singlet of A_g symmetry and the lowest lying excited electronic state is a triplet 1.55 eV above the ground state [37, 38]. Excitations from the ground state through electric-dipole one-photon transitions are allowed only into singlet states of T_{1u} symmetry [13, 15] (see Figure 2). Above 1.55 eV, the first strongly dipole-allowed one-photon transition seen in gas, liquid solution, and solid films of C$_{60}$ is near 3.6 eV [3, 30, 33, 34] (see Figure 4).

The oscillator strengths of the three lowest T_{1u} states are, according to computations by Braga et al. [14], 0.08, 0.41, and 2.37; i.e., the two lowest dipole allowed transitions are relatively weak. Similar predictions for the oscillator strengths were also made by Negri et al. [15]. Oscillator strengths derived from the absorption spectrum by Leach et al. tend to be a factor of 3 to 5 lower than these predictions [34]. Leach et al. [34] assigned the strongest of the three lowest transitions to a peak near 3.8 eV (0.33 µm) in the absorption spectrum of a liquid solution which corresponds to the transition near 3.6 eV (0.34 µm) seen in films; see Figure 4. The absorption lines observed in liquid solution tend to be shifted by about 0.2 to 0.3 eV toward lower energies when they are observed in thin films [3, 34]. The two lowest T_{1u} states in liquid solution were assigned to a relatively weak but sharp absorption peak near 3.0 eV (0.41 µm), and a weak shoulder near 3.3 eV (0.38 µm) [34]. A weak, but clearly visible peak seen in films near 2.6 eV (0.48 µm) corresponds possibly to these two weak features seen in liquid solution.

Although various theories only agree approximately on the energies of excited levels, they all do agree that there are a half-dozen or more singlet energy levels below the first T_{1u} level having symmetries T_{1g}, T_{2g}, G_g, G_u, T_{2u}, and H_u such that they cannot be excited from the A_g ground level via electric-dipole-allowed one- or two-photon transitions [13, 15–17]. A similar number of triplet states below 3.6 eV is expected, see Figure 3 [13, 15]. (In practice one also sees vibronic transitions and various side bands [34].) Many of these ''optically silent'' levels have been seen nevertheless by very sensitive optical techniques [17, 34, 37–40], as well as by electron scattering and photoemission spectroscopy [35, 41]. In addition, all theoretical calculations agree that below the first T_{1u} level there should lie only one level of H_g symmetry. It is the only level below

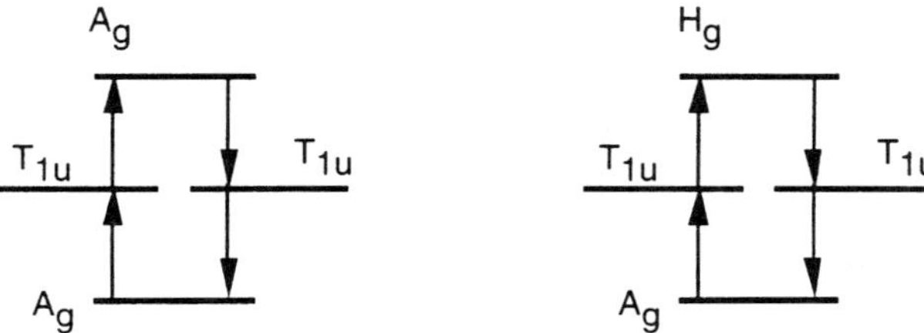

Figure 2 The two types of excitation pathways which contribute to $\chi^{(3)}$ in C$_{60}$. States are labeled with their symmetry symbols.

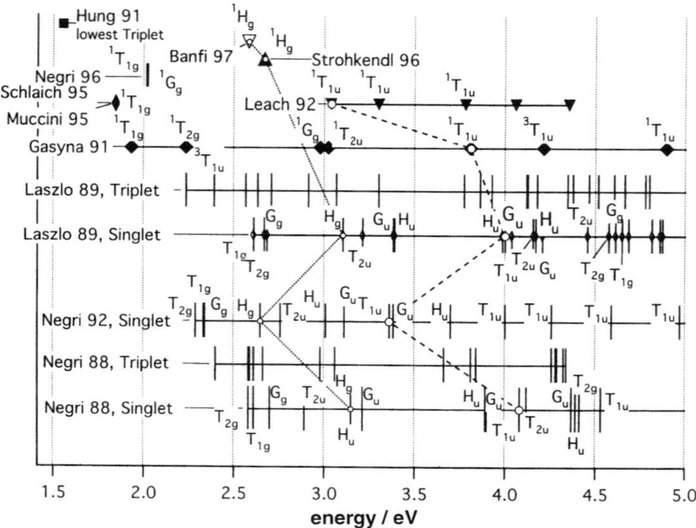

Figure 3　Energy levels in C_{60} according to computations by Laszlo and Udvardi [13] and Negri et al. [15, 16] and experimental level assignments are shown. Predictions and assignments of the lowest T_{1u} singlet level which defines the intrinsic absorption edge and the lowest H_g level which corresponds to the lowest two photon states are connected through dashed lines.

the first T_{1u} state which can be excited by electric-dipole-allowed two-photon absorption (TPA) [13, 15, 16]. This H_g state has fivefold degeneracy. Results from several theoretical computations of the energy level structure as well as experimental assignments are summarized in Figure 3.

All excitation pathways which contribute to $\chi^{(3)}$ must pass through an intermediate T_{1u} state before they can reach a two-photon state of either H_g or A_g symmetry; compare Figure 2. In the expression for $\chi^{(3)}$ in Eq. (3) such excitation pathways would be characterized by a product of transition dipole moments of the form

$$\langle A_g|\vec{r}|T_{1u}\rangle\langle T_{1u}|\vec{r}|H_g\rangle\langle H_g|\vec{r}|T_{1u}\rangle\langle T_{1u}|\vec{r}|A_g\rangle \tag{7a}$$

or

$$\langle A_g|\vec{r}|T_{1u}\rangle\langle T_{1u}|\vec{r}|A_g\rangle\langle A_g|\vec{r}|T_{1u}\rangle\langle T_{1u}|\vec{r}|A_g\rangle \tag{7b}$$

Excited A_g states are very energetic and the only A_g state in the visible and infrared region is the ground state [13]. So far the only observed two-photon resonance arises from an excitation pathway of the form (7a) [8, 9, 11].

The shift in electronic and linear optical properties in the transition from C$_{60}$ to C$_{70}$ can be qualitatively accounted for by the increase in molecular size and reduction of symmetry. While C$_{60}$ belongs to the symmetry group I_h with up to fivefold degeneracies, the highest degeneracy in C$_{70}$ that belongs to the symmetry group D_{5h} is twofold and results in the splitting of energy levels [42]. Since the C$_{60}$ states of symmetry T_{1u} are threefold-degenerate, such line splittings are expected in the linear absorption spectrum of C$_{70}$. Comparison of the thin-film absorption spectra of C$_{60}$ and C$_{70}$ in Figure 4 indicates that the absorption feature at 3.6 eV in C$_{60}$ splits into peaks of energies 3.6 and 3.1 eV. The weak absorption feature in C$_{60}$ near 2.6 eV shifts to 2.5 eV in C$_{70}$ and gains strength and width, where the increase in width may be the result of an unresolved line splitting. In accordance with these small shifts toward lower energies, the refractive index in the zero-frequency limit moves from 1.90 in C$_{60}$ to 1.94 in C$_{70}$ [43].

4 NONLINEAR OPTICAL EXPERIMENTS

Ideally one would like to perform nonlinear optical studies in the gas phase in order to explore molecular properties. However, the low number density under such conditions makes these experiments difficult, if not impossible, to perform. Similarly, one finds the concentrations achievable in liquid solution too low to detect the nonlinear optical response of the solute [44]. Considerations based purely on number density show that the nonlinearities of C$_{60}$ and C$_{70}$ would have to be excessively high to overcome the solvent contributions. Early DFWM measurements on C$_{60}$ solutions with 30 ps pulses which reported such unusually high nonlinearities were found to be erroneous [44]. DFWM performed on solutions by Tang et al. [29] found upper limits for the amplitude of the second hyperpolarizability of C$_{60}$ consistent with the earlier results of Lindle et al. [1, 4] and more recent studies of Strohkendl et al. [8, 9] in solid C$_{60}$ films. Measurements in amorphous thin-film samples of C$_{60}$ and C$_{70}$ are so far the only measurements which have yielded clear upper and lower limits for amplitudes and phases of $\chi^{(3)}$ [1–10, 28]. The first spectra of $\chi^{(3)}$ in C$_{60}$ and C$_{70}$ were obtained by third-harmonic generation at fundamental wavelengths below the linear absorption edge by Meth et al. [2] and Kajzar et al. [3, 5] Figure 4 shows $\chi_{1111}(-3\omega, \omega, \omega, \omega)$ and linear absorption spectra published by Kajzar et al. for C$_{60}$ and C$_{70}$ [50]. Note that we have renormalized the nonlinear data compared to the original publication such that they have the appropriate magnitude relative to our fused silica reference standard of 3.95×10^{-15} esu (originally a fused silica value of 2.8×10^{-14} esu was used by Kajzar et al. [5, 28]). Any peak at wavelength λ in the linear absorption spectrum must, by definition, have a corresponding third-harmonic generation peak at a fundamental wavelength of

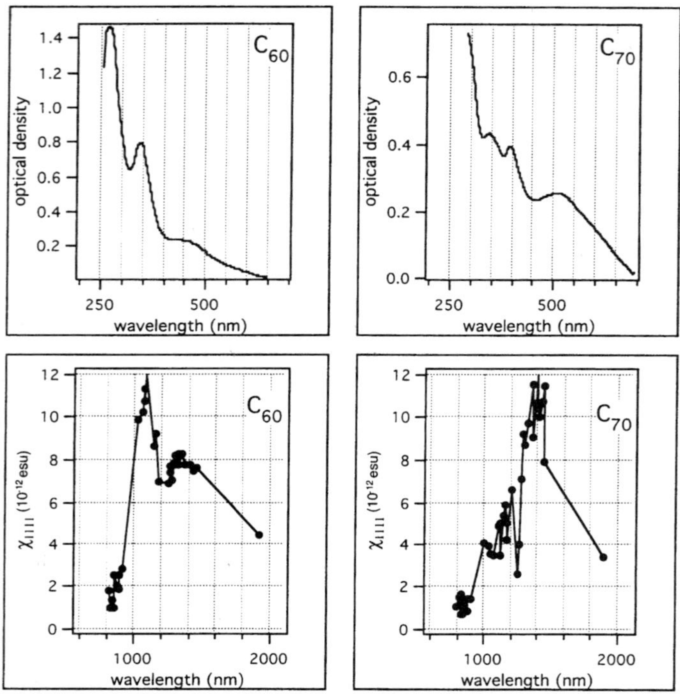

Figure 4 Comparison of linear absorption and third-harmonic generation data in C_{60} and C_{70} films according to Ref. 50. Film thicknesses are 88 nm for C_{60} and 53 nm for C_{70}. Third-harmonic and absorption data are arranged such that vertical lines connect three-photon peaks with their corresponding linear absorption peaks.

3λ as discussed above. We adjusted the data graphs such that any vertical line between absorption and third-harmonic data connects the absorption data point at λ with the corresponding third-harmonic data point at 3λ. When comparing the data, one has to keep in mind that even though the resonance energies are directly translated when going from the linear to the nonlinear optical spectrum, this is not necessarily true for the shape and exact proportions of spectral features. Nevertheless, in both C_{60} and C_{70} we detect striking similarities between third-harmonic and linear absorption spectra in resonance location and spectral proportions, especially when we subtract the linear absorption background arising from strong one-photon resonances above 3.6 eV. Even the line splitting in the absorption spectrum of C_{70}, evident in the peaks near 340 and 400 nm, appears to be recognizable in the third-harmonic spectrum. The resonance peak near 1.3 μm in the nonlinear optical spectrum of C$_{60}$ was convincingly identified as a three-photon resonance by Meth et al. [2] and Kajzar et al. [3]. Figure 5 shows a comparison of nonlinear optical data sets obtained by different groups and techniques in C$_{60}$ films. The third-harmonic resonance near 1.06 μm was also identified by Kajzar et al. as a three photon resonance [3]. This resonance was not seen by Meth et al. [2] due to a relatively large film thickness and the associated difficulty in evaluating the strongly absorbed third-harmonic signal. Research by Schlaich et al. [39] and Muccini et al. [40] revealed an electronic singlet level of symmetry T_{1g} at 1.8 eV above the ground state observed in one- and two-photon-excited luminescence. The excitation of the one- and two-photon forbidden T_{1g} level is made possible through symmetry-breaking molecular vibrations [40] and is, therefore, most likely weak. Nevertheless, based on such evidence Kajzar et al. reassigned the THG resonance at 1.3 μm as a two-photon resonance with the T_{1g} level at 1.8 eV [5]. Based on the presented evidence, we suggest that the broad resonance at 1.3 μm in THG is mostly due to a three-photon resonance. Weak two-photon contributions to this resonance cannot be excluded. Fluctuations in the DFWM spectrum of Strohkendl et al. [8, 9] near 1.3 μm might be the result of such a weak two-photon feature (see Figures 5 and 7). A two-photon feature near 1.3 μm in third-harmonic generation would be observed in

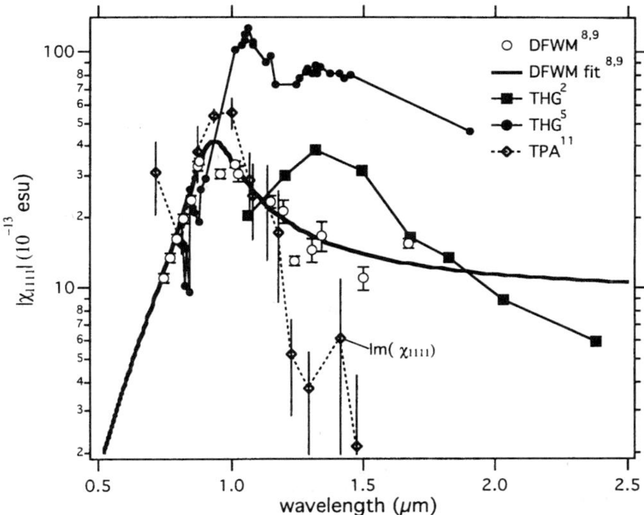

Figure 5   Comparison of degenerate four-wave mixing (DFWM), third-harmonic generation (THG), and two-photon absorption (TPA) data in C$_{60}$ films. $|\chi_{1111}|$ is shown for THG and DFWM, Im(χ_{1111}) for TPA.

electro-absorption at 0.65 μm. [Compare Eq. (4c).] This wavelength is still below the linear absorption edge and any electric-field-induced absorption would give clear evidence of an underlying two-photon state. Neither Jeglinski et al. [45] nor Hess et al. [46] were able to find evidence for an underlying two-photon state again clearly favoring a three-photon-resonance for the interpretation of the THG peak near 1.3 μm.

Degenerate four-wave mixing experiments were traditionally done only at a small number of select wavelengths, at which laser pulses of good temporal and spatial quality were available. Considering the problems involved in measuring the purely electronic nonlinearity which we discussed above, we mention here from the early DFWM experiments only the measurements by Lindle et al. at 1.06 μm with 30 ps pulses which yielded at their time the lowest reported value for χ_{1111} in C_{60} of 7×10^{-12} esu. This value was measured relative to CS_2 where a reference value for χ_{1111} of 4×10^{-13} esu was used. They also measured a value of 7 for the ratio of $|\chi_{1111}/\chi_{1221}|$ in C_{60}.

Strohkendl et al. used a newly developed degenerate four-wave mixing spectrometer, which is based on a tunable 100 fs Ti: Sapphire amplifier and a white-light seeded optical parametric amplifier, to perform the first continuous degenerate four-wave mixing studies on C_{60} and C_{70} films in the wavelength range of 0.74–1.7 μm [8, 9]. This wavelength range covers excited state energies for two photon states of 1.46–3.35 eV.

This wavelength range covers, therefore, essentially the energy interval between the first excited state, which is the triplet at 1.55 eV, and the first strong dipole transition into a T_{1u} state at 3.6 eV. It should therefore reveal the single two-photon resonance of symmetry H_g predicted by theory. During this study it became clear that the thick nonabsorbing substrates (either CaF_2 or fused silica) used made contributions to the observed signals which were comparable to those of the thin films. This led to the development of a new degenerate four-wave mixing technique described in detail in Ref. 28. Typically, degenerate four-wave mixing measurements have yielded only the amplitudes of $\chi^{(3)}$ tensor elements. This new technique also provided a straightforward method to measure the corresponding phases. Knowledge of these phases is important for device applications as well as for the deconvolution of multiline spectra. A brief description of this new technique is given here.

The beam geometry employed is shown in Figure 6. It shows the sample, which is a thin film on a thick substrate, and a signal plane which is parallel to the thin film. Three beams, 1, 2, and 3, intersect inside the sample. They are arranged such that their points of intersection with the signal-plane define three corners of a square. The phase-matched (often also referred to as Bragg-matched) signal beam, which is usually measured in DFWM experiments, marks the fourth corner denoted by A. This phase-matched signal contains contributions from the thin film as well as its thick substrate. Although the substrate has a much smaller

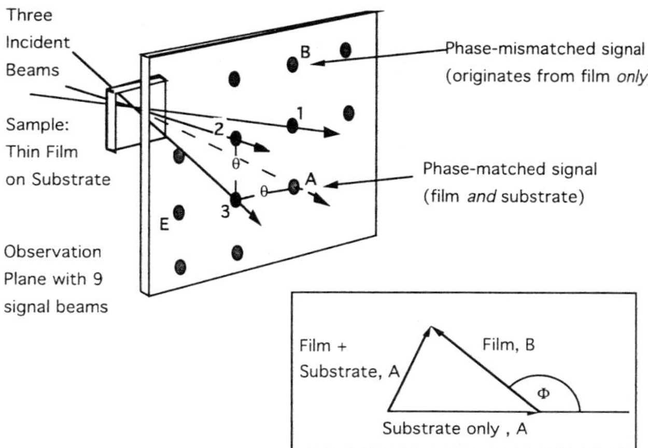

Figure 6 Geometry for phase-mismatched degenerate four-wave mixing. Three input beams generate nine signal beams. The phase-matched signal at point *A* arises from a superposition of the thin-film and the substrate signals. The remaining eight signals are all phase-mismatched and arise essentially from the thin film only. The inset shows the phasor diagram for the addition of the signal-field amplitudes in the phase-matched direction. The phase Φ of the thin film χ_{1111} can be derived from three amplitude measurements, as indicated.

χ^3 than the thin film, its signal contribution is comparable to that of the film because of its large thickness advantage. It is not possible to remove the substrate contribution by a simple background subtraction since the two signal fields have a relative phase ϕ which can have values in the range of 0 to 2π. The signal addition is illustrated by the ''phasor diagram'' in the inset of Figure 6. The thin-film and substrate field amplitudes add vectorially to form the composite signal amplitude. The square of this composite signal amplitude is observed in the DFWM experiment at point *A*. Note that constructive and destructive superposition of film and substrate fields are possible.

Besides the phase-matched signal at point *A*, there are an additional eight DFWM-signal beams emerging from the sample under directions which are phase-mismatched. These phase-mismatched signal beams carry under typical experimental conditions (3° angle between write beams 1 and 2; 10 μm and 1 mm thickness for film and substrate, respectively) the full, unattenuated signal from the thin film with only a small residual contribution from the thick substrate. For the present discussion, we are interested only in the signals at points *A*, *B*, and *E*. When all input-beam polarizations are parallel, the signals in *A*, *B*, and *E* are all generated by the χ_{1111} tensor components of the underlying materials.

By measuring the signals from the thin-film sample in points A and B and the signal from the bare substrate in point A, one is able to form the phasor triangle of Figure 6 and determine the phase of χ_{1111} of the thin film relative to the nonresonant substrate. If the polarization in beam 1 is set orthogonal to that of beams 2 and 3, the signal at A and B is generated by χ_{1122}, and the signal at E by χ_{1221}. Note that the ordering of tensor subscripts is meaningful only in conjunction with a corresponding order of frequency arguments, which we have chosen above as $(-\omega, \omega, \omega, -\omega)$. In general, the tensor component $\chi_{1221}(-\omega, \omega, \omega, -\omega)$ would be the same as $\chi_{1212}(-\omega, \omega, -\omega, \omega)$.

Figure 7 shows the amplitudes of χ_{1111}, χ_{1122}, and χ_{1221}. Figure 8 depicts the phase of χ_{1111} derived from measurements on C_{60} films [6, 8, 9]. The phases derived for χ_{1122} and χ_{1221} measured in the range 0.74–0.88 µm are not shown, since they exhibit essentially the same behavior as that of χ_{1111}. All data sets shown support the single two-photon resonance which is predicted by theory to be of symmetry H_g. A simple Lorentzian line-shape model of a two-photon resonance fits all amplitude and phase data as shown in Figures 7 and 8 [8, 9]. The resonance maximum occurs at 0.93 µm which corresponds to an excitation energy for the two-photon state of 2.67 ± 0.1 eV and a line width of 0.25 eV [8, 9].

The picosecond result for $|\chi_{1111}|$ of C_{60} by Lindle et al. [4] which is quoted as 7×10^{-12} relative to a CS_2 standard of 4×10^{-13} esu is a factor 2.6 greater than the value predicted by the theoretical fit in Figure 7. However, these picosecond data exhibit a temporal signal tail, suggesting the presence of energy-dependent effects in the nonlinear optical signal, which are not related to the electronic nonlinear optical susceptibility. Also, $|\chi_{1111}/\chi_{1221}|$ was found to be 7 whereas the

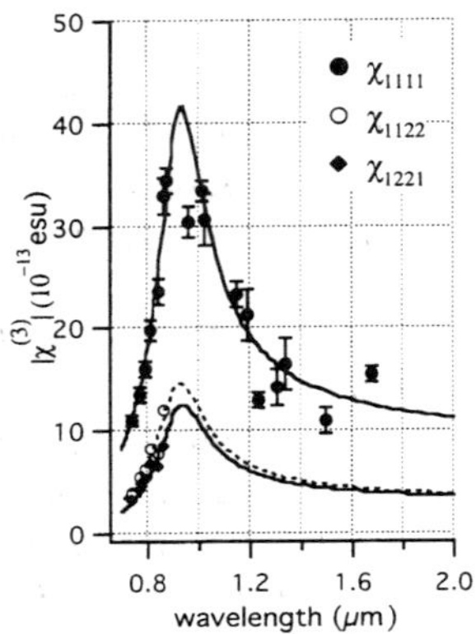

Figure 7 Amplitudes of $\chi^{(3)}$ in C_{60} film derived from degenerate four-wave mixing. (*From Ref. 9.*)

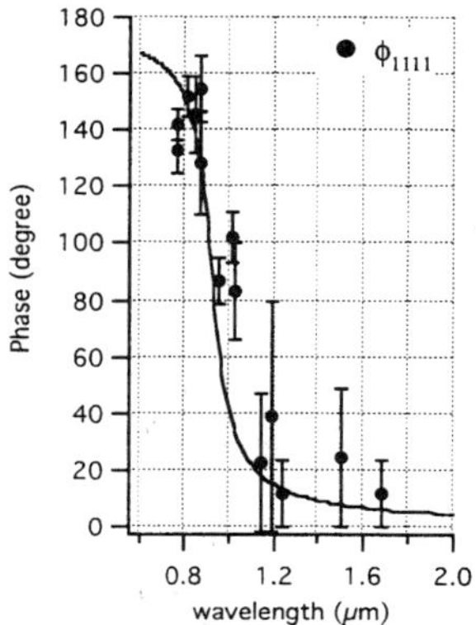

Figure 8 Phase of χ_{1111} in C$_{60}$ film derived from degenerate four-wave mixing. (*From Ref. 9.*)

fit in Figure 7 predicts a ratio of $|\chi_{1111}/\chi_{1221}| = 3.27$ near 1.06 μm. As χ_{1221} is less susceptible to energy-dependent effects, we compare only their value of $|3.27\chi_{1221}|$ to the $|\chi_{1111}|$ value by Strohkendl et al. [8, 9]. We find their data point to be only 40 percent larger, which is in reasonable agreement.

Strong evidence for the validity of the findings by Strohkendl et al. was provided by recent direct two-photon absorption measurements in the range 0.7–1.5 μm by Banfi et al. [11] (see Figure 5). Many of the known difficulties [47] due to energy-dependent effects in measuring two-photon absorption were apparently overcome in these measurements. This was done through the use of short, femtosecond pulses and the use of a frequency doubler as a nonresonant reference standard for two-photon absorption. Banfi et al. identified a single resonance at 2.58 eV with a width of 0.27 eV. A value for the two-photon absorption coefficient of 0.025 MW/cm^2 was found at the resonance peak compared to 0.02 cm/MW by Strohkendl et al. These results agree well within the relative uncertainty of the different reference standards (BBO and fused silica) used in these studies. The strength of the observed two-photon absorption coefficient is essentially the same as that of bulk GaAs, one of the strongest known two-photon absorbers [47]. Overall, the agreement between the results derived from these two different measurement techniques is remarkable and a validation of the employed approaches.

Figure 5 compares the third-harmonic generation data by Meth et al. and Kajzar et al., the degenerate four-wave mixing data by Strohkendl et al., and two-photon absorption data by Banfi et al. The fit of the degenerate four-wave mixing data yields a zero-frequency limit for χ_{1111} of $9 \pm 3 \times 10^{-13}$ esu which corresponds to a nonlinear refractive index n_2 of 3.6×10^{-14} cm^2/W. Qualitative extrapolation of the THG data by Kajzar et al. yields a χ_{1111} value of 3×10^{-12} esu, whereas the THG data of Meth et al. yield a value six times lower.

The absence of the singular two-photon resonance in the third-harmonic spectrum of C_{60}, which is so clearly seen in degenerate four-wave mixing and two-photon absorption, is unexplained. Destructive interference between different excitation pathways in the expansion of $\chi^{(3)}$ may play a role [8, 9] [compare Eq. (3)]. Also, three-photon-resonant contributions originating from the strong transitions in the linear absorption spectrum below 300 nm may have to be taken into account. The absence of the strong absorption line near 270 nm from the third-harmonic spectrum emphasizes the complex interference effects, which have to be considered for the analysis of the third-harmonic generation data. In Figures 9 and 10 we show the amplitude and the phase of χ_{1111} for a C_{70} film derived from degenerate four-wave mixing measurements by Strohkendl et al. [10]. The amplitude spectrum exhibits two peaks at wavelengths 0.940 and 1.03 μm. If these two peaks are interpreted as two two-photon resonances one finds corresponding excited-state energies of 2.64 ± 0.03 and 2.41 ± 0.05 eV. Compared to C_{60} the data scatter for the phase-angles is larger and is probably due to the thinner sample used for C_{70} (3.1 μm compared to 10 μm). Nevertheless, they show the behavior expected for two-photon resonances: phase angles vary from $-120°$ near 0.8 μm to almost $0°$ toward longer wavelengths. The peak near 0.94 μm is relatively sharp. Great care was taken to differentiate this smaller peak from the larger peak centered near 1.03 μm The uncertainty of the averaged peak-value at 0.94 μm derived from 12 measurements taken over the course of a week is shown. Also the dip near 0.96 μm which separates the two peaks was

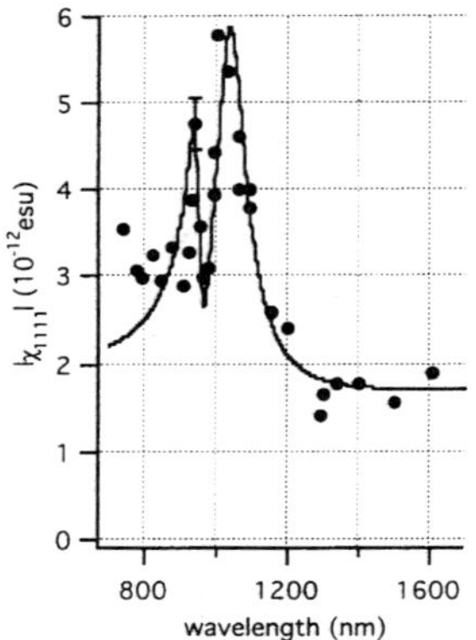

Figure 9 Amplitude of χ_{1111} in C_{70} films measured with degenerate four-wave mixing. The continuous line is a guide for the eye. (*From Ref. 10.*)

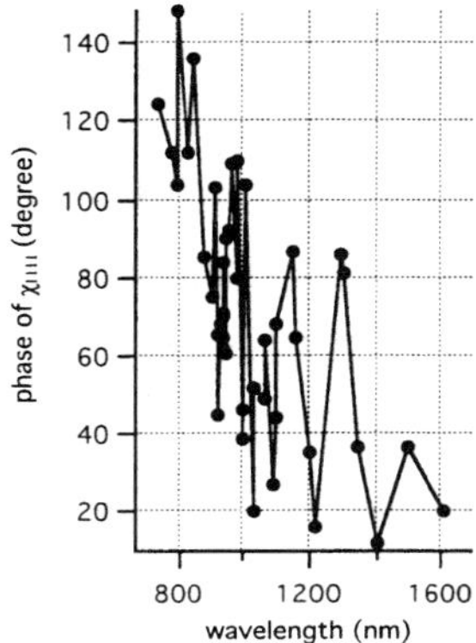

Figure 10 Phase of χ_{1111} in C$_{70}$ films measured with degenerate four-wave mixing. (*From Ref. 10.*)

carefully explored by moving the wavelength back and forth, a measurement for which computer control of the employed optical parametric amplifier was instrumental. The spectral width of the pulses near the dip at 0.98 μm was 14 nm or less.

The two-photon double feature occurs indeed where the third-harmonic generation data show a shoulder which was interpreted by Kajzar et al. [7] as a two-photon resonance. The continuous line shown in Figure 9 is a first attempt to fit the observed two-photon resonances with a model analogous to that used for the C$_{60}$ data. At present, it can only be considered a guide for the eye as the phase data are not properly reflected by this fit. We notice in particular that a phase near $\pi/2$ is observed in-between the two resonance peaks. The absence of any resonance near 1.4 μm from the degenerate four-wave mixing spectrum clearly confirms the resonance at that wavelength in the third-harmonic spectrum as a three-photon resonance.

In the two-photon resonant region of C$_{70}$, the magnitude of χ_{1111} reaches a maximum of 5.9 ± 0.5 × 10^{-12} esu which is about a factor 1.5 higher compared to that of C$_{60}$. Lindle et al. measured at 1.06 μm, with 30 ps degenerate four-wave mixing, a value of 3.5 × 10^{-12} esu for $|3\chi_{1221}|$ which is about 30% below $|\chi_{1111}|$ measured by Strohkendl et al. This can be considered good agreement. The two-photon absorption is similarly large as in C$_{60}$ but less well defined due to the uncertainty in the phase angles.

We extrapolate a zero frequency limit for χ_{1111} in C$_{70}$ of 2 × 10^{-12} esu from the third-harmonic data by Kajzar et al. This is in good agreement with a value of 1.8 ± 0.2 × 10^{-12} esu derived from the degenerate four-wave mixing data by Strohkendl et al. While the third-harmonic data of Kajzar et al. indicate a decrease in the zero-frequency limit of χ_{1111} in the transition from C$_{60}$ to C$_{70}$ by about 30%, one observes the opposite trend in the DFWM data which indicate an increase by a factor 2. Such an increase is consistent with the observed smaller band gap and increased refractive index in C_{70}.

5 DISCUSSION

It is instructive to discuss the third-harmonic spectra computed by Shuai and Brédas [12] in relation to nonlinear optical experimental results as well as to the experimental one-photon spectra, particularly since some of the interpretations of the third-harmonic spectra have made use of these predictions [5, 7]. In C_{60}, Shuai and Brédas find that a two-photon and a three-photon resonance occur essentially at the same fundamental wavelength, i.e., at 1.02 and 1.06 μm, respectively. The prediction of this ''double-resonance'' near 1 μm is in surprisingly good agreement with the experimental results [5, 8, 9]. It is clear that the observed two-photon resonance in C_{60} corresponds to the already mentioned lowest excited state of symmetry H_g with an excited-state energy near 2.7 eV [8, 9, 11, 13, 15, 16]. The T_{1u} state which gives rise to the three-photon resonance corresponds to the absorption peak near 3.6 eV in the film [5]. Experimentally, we find a ratio between the excitation energies of these T_{1u} and H_g states of 1.35 ± 0.05, which is close to the ratios of 1.44 by Shuai and Brédas [12], 1.29 by Negri et al. [15], 1.26 by Laszlo et al. [13], and 1.28 by Negri et al. [16]. In particular, the second calculation by Negri et al. [16] also appears to agree quite well with the observed excitation energies. If one considers the ratio between the energies of the second excited T_{1u} state and the H_g state, one finds ratios of 1.44 by Negri et al. [15], 1.51 by Laszlo et al. [13], and 1.37 by Negri et al. [16]. Also the second excited T_{1u} state may be a good candidate for the observed three-photon resonance near 1.06 μm. Corresponding ratios for the third excited state of T_{1u} symmetry are in the range 1.59–2.16, decidedly higher than the experimental value.

If one assigns indeed the strong absorption feature in the film near 340 nm (3.6 eV) to the second excited T_{1u} state, one can interpret the weaker absorption feature near 450 nm (2.8 eV) as the lowest T_{1u} state. This assignment of the lowest T_{1u} state would agree with that of Leach et al. [34], who found the corresponding absorption feature for C_{60} in liquid solution near 410 nm. Here we note again, that absorption features in liquid solution tend to be shifted toward shorter wavelengths relative to those observed in thin films.

We therefore suggest the energy-level assignments for C_{60} and C_{70} which are shown in Figure 11 and are consistent with all the features seen in the linear absorption spectra as well as in the nonlinear spectra of thin films. The lowest lying T_{1u} state in C_{60} films has an excitation energy of 2.8 eV and gives rise to a three-photon resonance near 1.3 μm in the third-harmonic spectrum [2, 3, 5]. The second excited T_{1u} state is near 3.6 eV and gives rise to a three-photon resonance near 1.06 μm. The H_g state near 2.67 eV [8, 9] (2.58 eV) [11] gives rise

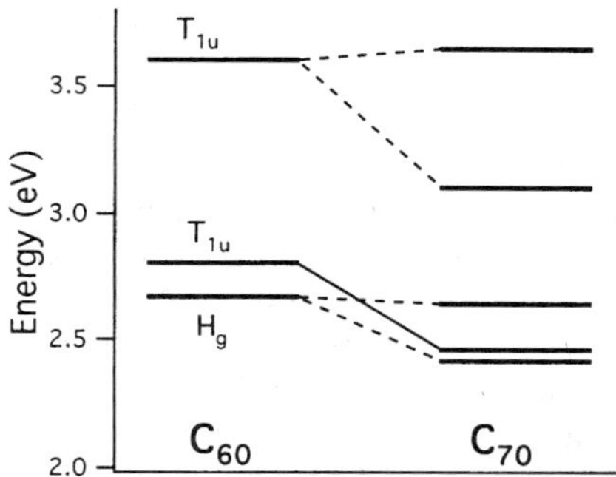

Figure 11   Proposed energy-level model for C$_{60}$ and C$_{70}$ films.

to the two-photon resonance near 0.93 μm (0.95 μm) observed in degenerate four-wave mixing (and two-photon absorption).

In the transition from C$_{60}$ to C$_{70}$ the T_{1u} level at 3.6 eV splits into levels at 3.65 and 3.1 eV and loses relative strength, whereas the 2.8 eV level shifts to 2.46 eV, gains relative strength, and broadens due to an unresolved splitting. The one-photon level in C$_{70}$ at 2.46 eV gives rise to the strong three-photon resonance in THG, which peaks for wavelengths larger than 1.4 μm. The one-photon levels in C$_{70}$ at 3.65 and 3.1 eV participate in the creation of the THG shoulder near 1.0 μm. The levels in C$_{70}$ near 2.64 and 2.39 eV give rise to the observed two-photon peaks in degenerate four-wave mixing. The only remaining question is the absence of the two-photon resonances, which can be seen so clearly in degenerate four-wave mixing and two-photon absorption, from the third-harmonic generation spectra. According to arguments made above, *isolated* two-photon resonances should appear three times stronger in a third-harmonic generation spectrum than in a corresponding degenerate four-wave mixing spectrum. One would therefore predict, from the degenerate four-wave mixing results of Strohkendl et al., that the third-harmonic generation spectra should show two-photon resonances similar in size to the observed strong three-photon resonances. Destructive interference between two-photon and three-photon resonant excitation pathways might play a role. A more detailed analysis of the $\chi^{(3)}$ expansion in terms of the participating states will be needed to resolve this question. The inclusion of higher lying states which are visible in the linear absorption spectrum might be necessary.

The real part of the degenerate four-wave mixing $\chi^{(3)}$ is important for all-optical switching. An intensity induced phase shift of π is necessary to achieve full switching in nonlinear directional couplers [48]. Such a π phase shift can be accumulated by an optical pulse over a propagation distance of 20 mm in C$_{60}$, at a wavelength of 1.5 μm and a nondamaging intensity of 1 GW/cm [2, 9] [compare Eq. (6)]. This propagation distance equals the measured attenuation length [51]; i.e., such devices are feasible. The same π phase is accumulated in C_{70} after a 10 mm propagation distance.

A large number of theoretical computations for the low-frequency limit of $\chi^{(3)}$ in C_{60} has been published. In order to gauge their validity, it is important that such computations also make predictions for the linear refractive index and take screening effects properly into account [21] for the prediction of $\chi^{(3)}$. From a number of eight publications which satisfied these requirements, four were found which predicted the refractive index within a factor 2 [9, 18, 20, 21, 24]. Among these, predictions for the low-frequency limit of $\chi^{(3)}$ were 1 to 2 orders of magnitude below the experimentally observed values [2, 5, 9]. This large discrepancy between theory and experiment remains unexplained [9].

6 CONCLUSION

C_{60} and C_{70} films exhibit strong nonresonant nonlinearities in the telecommunications window near 1.5 μm which are 230 and 460 times larger than that of fused silica. This makes them attractive for nonlinear optical switching devices in this spectral region due to the absence of intrinsic loss mechanisms. This is especially true if the observed extrinsic losses can be reduced through improvements in thin-film quality. Both materials exhibit strong two-photon resonances near 1 μm with two-photon absorption coefficients similar to that of bulk GaAs, one of the strongest known two-photon absorbers. The electronic structure of C_{70} appears to evolve in a quite natural and continuous fashion from that of C_{60}. All of the observed optical effects can be qualitatively explained in terms of lowered excitation energies, due to the increase in the molecular size, and in terms of line-splitting, due to the reduction in symmetry. Based on linear and nonlinear optical data, we have suggested an energy-level model for the low-lying optically active states of C_{60} and C_{70} which can account for the observed linear and nonlinear optical responses. The only exception are the existing third-harmonic generation data which, apparently due to the interference from predictable three-photon resonances, fail to reveal the strong two-photon resonances which can be observed so clearly in the spectra of degenerate four-wave mixing and nonlinear absorption. We believe that this open question can be resolved within the suggested energy-level model through the construction of the corresponding $\chi^{(3)}$ tensor.

ACKNOWLEDGMENTS

The authors would like to acknowledge the contributions of colleagues at the University of Southern California, the Naval Research Laboratory, and other institutions whose work is described in more detail in the accompanying references. F. P. Strohkendl acknowledges support through the National Science Foundation (DMR-9528021) and the Air Force Office of Scientific Research (F49620-95-1-0035, F49620-95-1-0450, F49620-96-1-0035, and F49620-97-1-0307). Z. H. Kafafi thanks the Office of Naval Research for partial financial support.

REFERENCES

1. Z. H. Kafafi, J. R. Lindle, R. G. S. Pong, F. J. Bartoli, L. J. Lingg, and J. Milliken, *Chem. Phys. Lett.* **188**, 492 (1992).
2. J. S. Meth, H. Vanherzeele, and Y. Wang, *Chem. Phys. Lett.* **197**, 26 (1992).
3. F. Kajzar, C. Taliani, R. Zamboni, S. Rossini, and R. Danieli, *Synthetic Metals* **54**, 21 (1993).
4. J. R. Lindle, P. G. S. Pong, F. J. Bartoli, and Z. H. Kafafi, *Phys. Rev. B* **48**, 9447 (1993).
5. F. Kajzar, C. Taliani, R. Danieli, S. Rossini, and R. Zamboni, *Chem. Phys. Lett.* **217**, 418 (1994).
6. F. P. Strohkendl, R. J. Larsen, L. R. Dalton, R. W. Hellwarth, and Z. H. Kafafi, *Proc. SPIE* **2284**, 78 (1994).
7. F. Kajzar, C. Taliani, R. Danieli, S. Rossini, and R. Zamboni, *Phys. Rev. Lett.* **73**, 1617 (1994).
8. F. P. Strohkendl, T. J. Axenson, L. R. Dalton, R. W. Hellwarth, and Z. H. Kafafi, *Proc. SPIE* **2854**, 191 (1996).
9. F. P. Strohkendl, T. J. Axenson, R. J. Larsen, L. R. Dalton, R. W. Hellwarth, and Z. H. Kafafi, *J. Phys. Chem.* **101**, 8802 (1997).
10. F. P. Strohkendl, T. J. Axenson, R. J. Larsen, L. R. Dalton, R. W. Hellwarth, and Z. H. Kafafi, *Proc. SPIE* **3142**, 2 (1997).
11. G. P. Banfi, D. Fortusini, M. Bellini, and P. Milani, *Phys. Rev. B* **56**, 10075 (1997).
12. Z. Shuai, J. L. Brédas, *Phys. Rev. B* **46**, 16135 (1992).
13. I. László and L. Udvardi, *J. Molec. Struct.* **183**, 271 (1989).
14. M. Braga, S. Larsson, A. Rosén, and A. Volosov, *Astron. and Astrophys.* **245**, 232 (1991).
15. F. Negri, G. Orlandi, and F. Zerbetto, *Chem. Phys. Lett.* **144**, 31 (1988).
16. F. Negri, G. Orlandi, and F. Zerbetto, *J. Phys. Chem.* **97**, 6496 (1992).
17. F. Negri, G. Orlandi, and F. Zerbetto, *J. Phys. Chem.* **100**, 10849 (1996).
18. A. A. Quong and M. R. Pederson, *Phys. Rev. B* **46**, 12906 (1992).
19. K. C. Rustagi, L. M. Ramaniah, and S. V. Nair, *Int. J. Mod. Phys. B* **6**, 3941 (1992).
20. F. Williame and L. M. Falicou, *J. Chem. Phys.* **98**, 3941 (1993).
21. Y. Wang, G. F. Bertsch, D. Tomanek, *Z. Phys. D* **25**, 181 (1993).
22. J. Li, J. Feng, and J. Sun, *Chem. Phys. Lett.* **203**, 560 (1993).
23. R. J. Knize, *Opt. Commun.* **106**, 95 (1994).
24. E. Westin, and A. Rosén, *Appl. Phys. A* **60**, 49 (1995).
25. P. D. Maker and R. W. Terhune, *Phys. Rev.* **137**, A801 (1965).

26. See, for example, Robert W. Boyd, *Nonlinear Optics*, Academic Press: San Diego, CA (1992).

27. A. Owyoung, *IEEE J. Quantum. Electron.* **QE-9**, 1064 (1973).

28. F. P. Strohkendl, L. R. Dalton, R. W. Hellwarth, H. W. Sarkas, and Z. H. Kafafi, *J. Opt. Soc. Am. B* **14**, 92 (1997).

29. N. Tang, J. P. Partanen, R. W. Hellwarth, and R. J. Knize, *Phys. Rev. B* **48**, 8404 (1993).

30. H. Kataura, N. Irie, N. Kobayashi, Y. Achiba, K. Kikuchi, T. Hanyu, and S. Yamaguchi, *Jpn. J. Appl. Phys.* **32**, L1667 (1993).

31. D. R. Huffman, *Physics Today*, November 1991, p. 22.

32. P. A. Heiney, J. E. Fischer, A. R. McGhie, W. J. Romanow, A. M. Denenstein, J. P. McCauley Jr., A. B. Smith III, and D. E. Cox, *Phys. Rev. Lett.* **66**, 2911 (1991).

33. B. B. Brady and E. J. Beiting, *J. Chem. Phys.* **97**, 3855 (1992).

34. S. Leach, M. Vervloet, A. Despres, E. Breheret, J. P. Hare, T. J. Dennis, H. W. Kroto, R. Taylor, and D. R. M. Walton, *Chem. Phys.* **160**, 451 (1992).

35. J. H. Weaver, J. L. Martins, T. Komeda, Y. Chen, T. R. Ohno, G. H. Kroll, N. Troullier, R. E. Haufler, and R. E. Smalley, *Phys. Rev. Lett.* **66**, 1741 (1991).

36. J. H. Weaver, *J. Phys. Chem. Solids* **53**, 1433 (1992).

37. R. R. Hung and J. J. Grabowski, *J. Phys. Chem.* **95**, 6073 (1991).

38. Y. Zeng, L. Biczok, and H. Linschitz, *J. Phys. Chem.* **96**, 5237 (1992).

39. H. Schlaich, M. Muccini, J. Feldmann, H. Bässler, E. O. Gobel, R. Zamboni, C. Taliani, J. Erxmeyer, and A. Weidinger, *Chem. Phys. Lett.* **236**, 135 (1995).

40. M. Muccini, R. Danieli, R. Zamboni, C. Taliani, H. Mohn, W. Müller, and H. U. ter Meer, *Chem. Phys. Lett.* **245**, 107 (1995).

41. G. Gensterblum, J. J. Pireaux, P. A. Thiry, R. Caudano, J. P. Vigneron, Ph. Lambin, A. A. Lucas, and W. Krätschmer, *Phys. Rev. Lett.* **67**, 2637 (1991).

42. J. Shumway and S. Sapathy, *Chem. Phys. Lett.* **211**, 595 (1993).

43. S. L. Ren, P. Zhou, Y. Wang, A. M. Rao, M. S. Meier, J. P. Selegue, and P. C. Eklund, *Appl. Phys. Lett.* **61**, 124 (1992).

44. Z. H. Kafafi, F. J. Bartoli, J. R. Lindle, and R. G. S. Pong, *Phys. Rev. Lett.* **68**, 2705 (1992).

45. S. Jeglinski, Z. V. Vardeny, D. Moses, V. I. Srdanov, and F. Wudl, *Synthetic Metals* **50**, 557 (1992).

46. B. C. Hess, D. V. Bowersox, S. H. Mardirosian, and L. D. Unterberger, *Chem. Phys. Lett.* **248**, 141 (1996).

47. E. W. Van Stryland and L. L. Chase, *CRC Laser Handbook of Laser Science and Technology*, suppl. 2 (M. J. Weber, ed.), p. 299 (1995).

48. W. S. Wong, S. Namiki, M. Margalit, H. A. Haus, and E. P. Ippen, *Opt. Lett.* **22**, 1150 (1997).

49. Z. Gasyna, P. N. Schatz, J. P. Hare, T. J. Dennis, H. W. Kroto, R. Taylor, and D.R.M. Walton, *Chem. Phys. Lett.* **183**, 283 (1991).

50. F. Kajzar, C. Taliani, R. Zamboni, S. Rossini, and R. Danieli, *Synthetic Metals* **77**, 257 (1996).

51. M. Jäger, Q. Zhang, G. I. Stegeman, C. D. Merritt, and Z. H. Kafafi, *Proceedings of the American Chemical Society, Division of Polymeric Materials: Science and Engineering*, 222 (1996).

6
Optical Limiting and Excited-State Absorption in Fullerene Solutions and Doped Glasses

R. Kohlman, V. Klimov, L. Smilowitz, and D. McBranch
Los Alamos National Laboratory
Los Alamos, New Mexico

1. INTRODUCTION

The optical properties of the fullerene C_{60} have attracted considerable interest due to its unique molecular symmetry, with π electrons highly delocalized over the nearly spherical surface. Given the truncated isocahedral symmetry (I_h symmetry group), extensive theoretical models have been developed to describe the electronic states [1] reported in linear absorption measurements. Based on the dipole-forbidden nature of the lowest electronic transition in C_{60}, a number of novel effects are observed, including strong vibrational coupling to the lowest transition [2], and optical limiting [3–5]. Optical power limiting (OL) occurs when the optical transmission of a material decreases with increasing laser fluence [6]. OL is desirable for providing passive protection to optical sensors (such as human eyes) from the high fluence output of modern pulsed lasers. The mechanism for obtaining OL in C_{60} is reverse saturable absorption (RSA), which occurs when excited states (excited singlet and triplet states) formed through optical pumping of the ground state have a higher absorption cross section than the ground state [3, 6].

To understand resonant nonlinear optical processes such as RSA in fullerenes, it is essential to characterize the spectral dependence and temporal evolution of excited-state absorption transitions following photoexcitation. Transient absorption (TA) spectroscopies with femtosecond (fs) to nanosecond (ns) timescales provide a means to probe the energies of excited-state transitions and their dynamics. Several wavelength-dependent TA studies have appeared for C_{60} solutions [5, 7–9]. Within 250 fs following photoexcitation [10], broad singlet excited states which extend throughout the visible and near infrared (NIR) are observed which decay with quantum yield approaching unity [9] to a triplet manifold of states. The triplet excited-state transitions are also observed throughout the visible and NIR with lifetimes in the range of μs [9] to ms [8]. The evolution of the singlet and triplet excited-state absorption can be used to predict optimal spectral regions for OL for both picosecond (ps) and ns pulses [4–6].

Due to the broad spectral extent of these excited-state transitions and the weak ground-state absorption, broadband optical limiting is observed throughout the visible and NIR due to RSA [3, 5]. A practical limitation for the use of C_{60} for broadband OL is that its ground-state absorption goes to zero near 650 nm so that increasingly high concentrations of molecules are needed to activate RSA beyond this wavelength. However, it is in this long wavelength region that the triplet excited-state absorption peaks and in which OL should be optimum for ns pulses. Due to its limited solubility and vanishing ground-state absorption, C_{60} cannot be effectively utilized for OL at long wavelengths. Both of these problems can be overcome through selective derivatization across a bond joining two hexagons in the C_{60} molecule (to form a 6,6 fullerene adduct) [11]. These 6,6 derivatized fullerenes have improved solubility in organic solvents relative to neat C_{60} [11, 12] and additional long wavelength ground-state absorption which extends the OL response into the red [5]. 6,6 derivatization also enhances the processability of fullerenes into solid-state hosts such as sol-gel glasses [13–15].

To address the issues related to both understanding the excited-state dynamics, and optimizing the OL response of fullerenes for applications, we review in this chapter recent studies of both TA spectroscopy and wavelength-dependent OL for neat and derivatized C_{60} in solution, films, and sol-gel glasses. Following an experimental section, we first review in Section 3.1 spectral TA studies for neat C_{60} in solution. Our TA studies of neat C_{60} have revealed several noteworthy features. First, we report several new excited-state transitions for the singlet and triplet manifolds, which should serve as useful benchmarks for theoretical models. Second, we present multispectral analysis of the decay dynamics of fullerenes in solution, which allows accurate determination of both the intersystem crossing time (600±100 ps) and the relative strengths of the singlet and triplet excited-state cross sections as a function of wavelength. Due to the improved signal to noise of scanning TA spectroscopy, the ratio of spectra at long delays ($\sim$1 ns) to that at short delays ($\sim$3 ps) can now be used for a quick quantitative prediction of the triplet to singlet excited-state absorption cross-section ratio (σ_T/σ_S), which closely matches σ_T/σ_S obtained from multispectral analysis of the decay dynamics in both C_{60} and 6,6 derivatives. Finally, we have resolved the initial energy relaxation dynamics from higher excited singlet states to the lowest excited state (250 ± 50 fs). Thus, it is now possible to resolve and analyze several excited-state singlet transitions, the initial relaxation dynamics within the singlet manifold, as well as the intersystem crossing.

Section 3.2 reviews results of linear and excited-state absorption studies for 6,6 derivatives of C_{60}. A comparison of the ground-state absorption of 6,6 derivatized and neat C_{60} shows that the derivatization leads to additional absorption in the red and NIR, which can be used to extend the OL of 6,6 derivatized fullerenes into the NIR. The TA spectra for derivatized fullerenes demonstrate that the essential dynamics are similar to those of C_{60}, but with a broadening of the singlet and triplet states. Therefore, broadband OL extended into the NIR is observed also for 6,6 derivatized fullerenes.

In Section 3.3, we present OL data measured at multiple wavelengths for both neat and derivatized fullerenes in solution which confirm the broadband nature of the limiting predicted from the TA spectra. Using a five-level RSA model to fit the OL response, we estimate the ratio of the excited-state to ground-state absorption cross sections (σ^*/σ_0) at multiple wavelengths, and compare these results with predictions from the TA spectra. σ^*/σ_0 increases at long wavelengths, demonstrating optimal OL for fullerenes in the red [16].

In Section 3.4, we investigate the different behavior seen in both TA and OL studies of fullerene thin films and sol-gel glass composites. The excited-state dynamics of fullerenes in films and sol-gel glasses are faster than in solution, indicating some solid-state quenching. In addition, the optical limiting response of neat and derivatized C_{60} incorporated into sol-gel glasses has generally been observed to be weaker than in solution. However, by careful processing of post-doped sol-gel glasses, we demonstrate that the OL response of 6,6 mono-adducts in a sol-gel glass matrix can approach that of the fullerene derivatives in solution, improving the potential of these materials for applications.

2. EXPERIMENTAL

Purified C_{60} (99.99%) was obtained from MER Corp. and introduced into various solvents without further purification. Synthesis of C_{60}-pyrrolidine(3)-5-benzyloxy-1H-pyrrolo[2,3-C]-pyridine (CPBPP) is described in detail elsewhere [16]. The 6,6 derivatives 1-(3-methoxycarbonyl)propyl)-1-phenyl-[6,6]-C_{61} (PCBM) and phenyl-C_{61}-butyric acid cholestryl ester (PCBCR), as well as the 5,6 derivative of PCBCR, were synthesized as published previously [11]. Silica gels containing C_{60}, CPBPP, PCBM, and PCBCR were prepared by mixing into a precursor sol, using either o-dichlorobenzene (DCB), tetrahydrofuran (THF), or toluene as cosolvents, as previously described [14]. These gels are termed "pre-doped" gels. "Post-doped" sol-gel glasses were prepared using porous glass (GELTECH, Inc.) with 75 Å nominal pore size produced by their proprietary process. Treated porous glass samples were placed in a solution containing the fullerene derivatives in various concentrations, and the solvent was dried to trap the fullerenes within the porous matrix.

Time-resolved excited-state absorption spectra were measured using a femtosecond (fs) pump-probe technique. The samples were photoexcited at 400 nm by 100 fs pulses from a frequency-doubled regeneratively amplified mode-locked Ti:sapphire laser (Clark-MXR CPA-1000). The repetition rate of the pulsed output was $f = 1$ kHz, with pump pulse energies up to 5 µJ, corresponding to an excitation density of 10 mJ/cm^2. The pump beam was modulated synchronously at $f/2$ with a mechanical chopper. A fs broadband continuum (450–1000 nm) probe pulse was generated in a 1 mm sapphire plate in the single-filament regime [17]. The probe beam was split into a reference channel (directed through the sample) and a sample channel (focused with reflecting optics through the center of the photo-excited spot on the sample). The pump beam was delayed with respect to the probe beam via a mechanical translation stage (up to 1.5 ns total delay). The reference and the signal channels were sent via a two-leg fiber bundle through a scanning imaging monochromator. The dispersed reference and signal channels were detected with matched large-area p-i-n Si photodiodes coupled to current preamplifiers, sent to a differential amplifier, and then to a digital lock-in amplifier synchronized to $f/2$. This method can be used for recording chirp-free TA spectra by scanning the monochromator and simultaneously adjusting the relative pump-probe delay time according to the calibrated chirp [17]. In addition, time transients at a fixed wavelength can be recorded by scanning the delay line at a fixed wavelength. This technique provides an accuracy of up to 10^{-5} in differential transmission, an improvement of order 100 over similar single-shot CCD camera measurements [17]. For TA measurements, 6,6 mono-adducts and neat C$_{60}$ were prepared as solutions in 2 mm path length cuvettes, with concentration adjusted to yield optical densities near unity at 400 nm.

Broadband optical limiting was measured using the output of an optical parametric oscillator (CASIX OPO BBO-3B with 6 ns pulses from 400–2200 nm) which was pumped by the frequency-tripled output (355 nm) of a Nd:YAG laser (Quanta Ray GCR-3). OL data at 532 nm was obtained directly from the frequency-doubled output of the Nd:YAG laser. Intensity control of the laser beam was obtained by passing the beam through a combination half-waveplate and polarizing beam splitting cube. The beam was split into a reference arm, a sample arm, and an intensity measurement arm. Intensity-dependent transmission was measured by placing the sample at the focus of a 120 mm focal length lens, which focused the beam down to a gaussian spot with beam waist of roughly 47 µm. The transmitted energy in the reference and sample arms was measured using 13 mm^2 Si photodiodes followed by gated integration. Each point represents the average of 20–100 laser shots. Fullerene samples were prepared as solutions in

2 mm path length cuvettes, with concentration adjusted to yield a linear transmission in the range of 50% (when solubility allowed).

3. RESULTS AND DISCUSSION

3.1 Transient Absorption Spectroscopy of C_{60}

The energy-level diagram for C_{60} is shown in Figure 1 [1]. The HOMO-LUMO transition has even parity, and hence is a symmetry-forbidden optical transition to first order in the electric-dipole matrix element. Following Ref. 1, we use the total electron symmetry to label the various energy states. Hence, the HOMO-LUMO forbidden transition is denoted $^1A_g\!-^1T_{1g}$. Excitation into higher states is expected to lead to rapid relaxation to the LUMO band, and the timescale for this initial relaxation has recently been reported [10]. Following the initial population of $^1T_{1g}$ from higher singlet states, additional strong, symmetry-allowed transitions are expected to higher-lying singlet states; these new transitions should

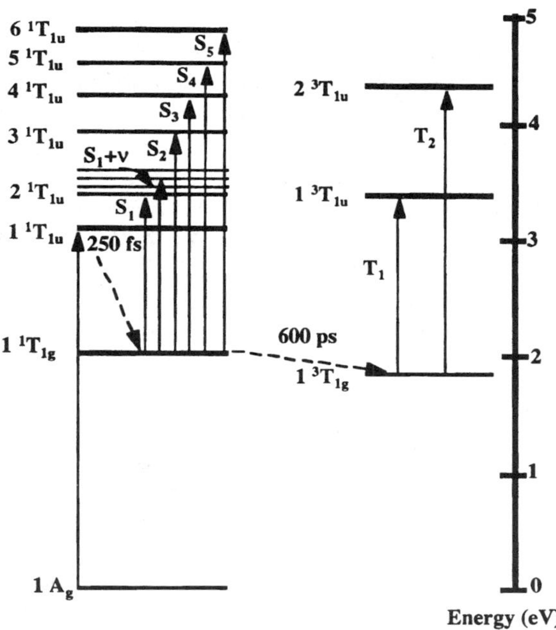

Figure 1 Energy-level diagram for C_{60}. (Adapted from Ref. 1.)

exhibit relaxation dynamics indicative of the singlet excited-state lifetime. Since efficient (~96%) [9] intersystem crossing to the triplet manifold is observed for C_{60} solutions, new spectrally and temporally resolved transitions appear with the intersystem crossing time, indicative of optically allowed transitions to higher excited triplet states.

Early spectrally resolved transient absorption measurements in C_{60} solutions [5, 7, 8, 13] performed in the spectral range 350–1100 nm indicated that at short delay times ($\Delta t < 30$ ps) between pump and probe pulses the excited-state absorption is dominated by a broadband S_1 at 975 nm (1.27 eV) with a weaker band near ~500 nm. The decay of these bands is accompanied by the complementary growth of a new band T_1 around 750 nm (1.65 eV). A weaker band is also seen in the 1 ns spectrum near ~530 nm [8, 14]. These dynamics have been explained in terms of singlet-triplet intersystem crossing with a time constant ranging from 650 ps to 1.2 ns [7, 8, 13]. The bands S_1 and T_1 were assigned to the excited-state absorptions associated with the lowest excited singlet and triplet states, respectively. Due to dramatically improved signal-to-noise ratios obtainable with chirp-free scanning TA spectroscopy, detailed spectral structure can now be resolved over the entire visible range and consequently compared with the energy-level structure shown in Figure 1.

Figure 2a shows TA spectra for C_{60} in toluene measured at $\Delta t = 3$ ps (dashed line), 100 ps (dotted line), and 1 ns (solid line) after excitation. The spectra show the same qualitative features reported earlier [5, 7–9, 18] including singlet absorptions at ~500 nm and ~900 nm at early times, as well as a strong triplet transition near ~750 nm at 1 ns time delay. Spectral analysis of the TA spectrum at 3 ps using a series of Gaussian functions reveals several strong and weak excited singlet bands (Figure 2b). Singlet TA bands are resolved at 964 nm (1.29 eV), 895 nm (1.38 eV), 690 nm (1.80 eV), 568 nm (2.18 eV), 528 nm (2.35 eV), and 475 nm (2.61 eV). Triplet bands can be discerned in the 1 ns TA spectrum at 740 nm (1.68 eV) and 530 nm (2.34 eV) (see Figure 2c) after subtracting the singlet features (assumed to decay with a 600 ps time constant). Since the LUMO in C_{60} is an even-parity state relative to the ground state, the transition energies observed in TA should be closely related to those observed in the ground-state linear absorption spectrum, shifted by the HOMO-LUMO energy difference of approximately 2.0 eV. As detailed in Table 1, this agreement is in fact excellent. Table 1 shows the energies of the allowed ground-state transitions, taken from Ref. 1 and shifted by 2.0 eV, within the energy range of our TA experiment. Also shown in Table 1 are the TA transition energies deduced from Figure 2, the state assignments and symmetries, and a comparison of relative oscillator strengths. Since the LUMO $^1T_{1g}$ state has different symmetry from the ground ^1A_g state, the oscillator strength for the excited-state transitions should differ from the oscillator strength of the peaks seen in linear absorption. Since it is difficult to obtain accurate absolute values of the excited-state oscillator

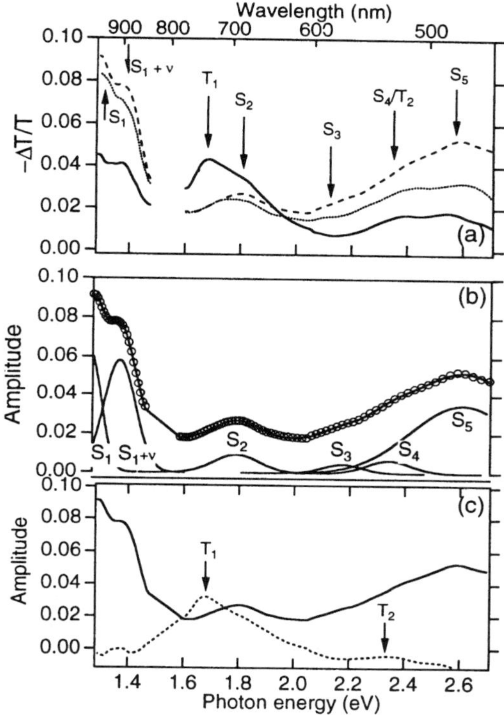

Figure 2 (a) $-\Delta T/T$ for C_{60} in toluene solution taken at three delay times: 3 ps (solid line, singlet excited state absorption) 100 ps (dashed line), and 1 ns (dot-dashed line, triplet excited-state absorption). (b) Nonlinear least squares fit of 3 ps TA spectrum for C_{60} using six gaussians. (c) Spectrum for the triplet at 1 ns obtained by subtracting the time-decayed singlet spectrum.

strengths from TA measurements, we compare in Table 1 the relative oscillator strengths for linear absorption and TA. In particular, the strongest peak in linear absorption which corresponds to a feature in our data is the transition $6^1T_{1u}-^1A_g$, denoted E in Ref. 1. The relative oscillator strengths of the ground-state transitions in Table 1 are normalized to the strongly allowed peak E. By contrast, the relative oscillator strengths for the excited-state transitions are normalized to the area of peak S_5 in Figure 2, corresponding to $6^1T_{1u}-^1T_{1g}$. What we observe from the TA peak analysis is that the relative ordering of oscillator strengths for transitions from $^1T_{1g}$ to the states labeled 3^1T_{1u} through 6^1T_{1u} is very similar to the ordering of oscillator strengths for transitions from ^1A_g to 3^1T_{1u} through 6^1T_{1u}. However, S_1, corresponding to $2^1T_{1u}-1^1T1_g$ is enhanced by a factor of ~40, as

Table 1 Energies and Oscillator Strengths for Transitions from the First Excited Singlet State

Transition	Energy (eV) (Linear Abs)	Energy (eV) $(-\Delta T/T)$	Osc. Strength $(-\Delta T/T)$ Normalized to S_5	Osc. Strength (Lin. Abs.) Normalized to $6^1T_{1u}-{}^1A_g$
$2^1T_{1u} - 1^1T_{1g}\ (S_1)$	1.29	1.28	0.38	0.01
$2^1T_{1u} + v - 1^1T_{1g}\ (S_1 + v)$	1.36	1.38	0.5	0.01
$3^1T_{1u} - 1^1T_{1g}\ (S_2)$	1.78	1.8	0.12	0.16
$4^1T_{1u} - 1^1T_{1g}\ (S_3)$	2.06	2.18	0.05	0.04
$5^1T_{1u} - 1^1T_{1g}\ (S_4)$	2.35	2.34	0.08	0.04
$6^1T_{1u} - 1\ ^1T_{1g}\ (S_5)$	2.82	2.6	1.0	1.0

is its vibronic satellite $S_1 + v$ over the relative values reported for the relatively weak feature observed in linear absorption for $2^1T_{1u}-{}^1A_{1g}$ (labeled B_0 in Ref. 1). This indicates that the state denoted 2^1T_{1u} in Ref. 1 likely has different symmetry from the rest of the series $3^1T_{1u}-6^1T_{1u}$. Calculations for the oscillator strength of the excited-state transitions S_1-S_5 are still needed for comparison, and to our knowledge have not been reported.

Though it is not possible to test the triplet excited-state transitions against the ground-state absorption in the same manner as for the singlet transitions, the analysis of two triplet transitions should provide accurate new transition values for comparison to theories. The relative spacing of two triplet excited-state transitions from the lowest excited triplet 1^3T_{1g} is qualitatively reproduced in available theoretical work, but the absolute position of this state in calculations is off by as much as 40%, relative to the experimentally established range of 1.4–1.8 eV [9]. We have put this state in Figure 1 at 1.7 eV [1] relative to the ground-state energy, which implies additional allowed triplet states (m^3T_{1u}) at 3.38 and 4.04 eV.

In conjunction with the spectral studies, we performed measurements of the TA dynamics in the first 1.0 ns for several wavelengths in the spectral range from 450–960 nm (Figure 3) to study the intersystem crossing dynamics in more detail [10]. Depending on the spectral energy, the TA curves show either growth or decay, corresponding to a dominant contribution from either the triplet or singlet excited state, respectively. In particular, we observed a decaying TA signal in two separated spectral regions (450–550 nm and 820–960 nm), consistent with early-time TA spectra exhibiting two bands attributed to singlet excited-

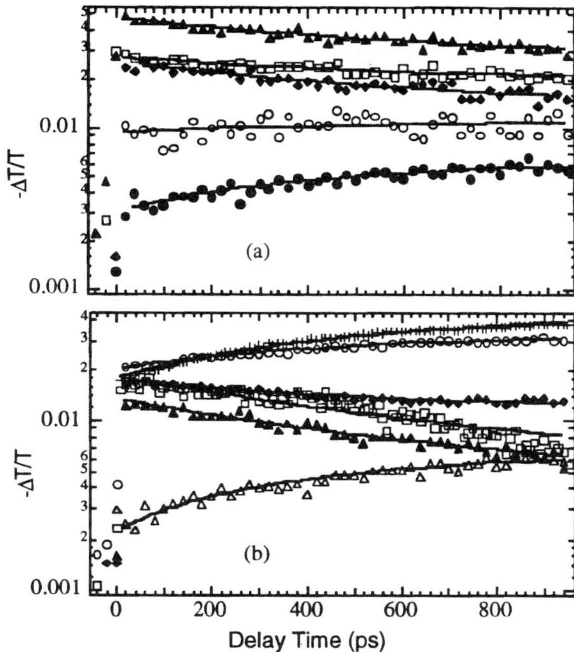

Figure 3 $-\Delta T/T$ versus delay time up to 1 ns for C_{60} in toluene solution, taken at 11 wavelengths, showing transition from singlet decay (450 nm $< \lambda <$ 650 nm), to triplet growth (650 nm $< \lambda <$ 800 nm), and back to singlet decay ($\lambda >$ 800 nm). (a) 450 nm (open squares), 500 nm (solid triangles), 550 nm (solid diamonds), 650 nm (open circles), and 700 nm (solid circles); (b) 725 nm (open triangles), 750 nm (crosses), 780 nm (open circles), 820 nm (solid diamonds), 890 nm (solid triangles), and 920 nm (open triangles). Fits are plotted as solid lines (see text).

state absorption. In analyzing the decays at multiple wavelengths, it is important to note that there is not a spectral region where only singlet or triplet features exist; rather, both triplet and singlet excited-state features overlap across the whole spectrum from 450 to 960 nm.

Previous studies showed that in isolated C_{60} molecules, intersystem crossing occurs with quantum yield 0.96 [9]. Therefore, the dominant decay pathway for the singlet is intersystem crossing to the triplet manifold. Under these conditions, the TA dynamics on ps timescales can be accounted for with one time constant τ which describes both the decay of the singlet and the growth of the triplet features over the whole spectral range. Introducing the wavelength-dependent excited-state absorption cross sections $\sigma_S(\lambda)$ (singlet) and $\sigma_T(\lambda)$ (triplet), we can model the dynamics at multiple wavelengths as follows:

$$\frac{-\Delta T}{T(\lambda, t)} = n_{ex}[\sigma_S(\lambda)e^{-t/\tau} + \sigma_T(\lambda)(1 - e^{-t/\tau})]$$

$$= n_{ex}\sigma_T(\lambda)\left(1 - \frac{\sigma_T(\lambda) - \sigma_S(\lambda)}{\sigma_T(\lambda)}e^{-t/\tau}\right)$$

(1)

where n_{ex} is the concentration of fullerene molecules in the excited state. It is noted that Eq. (1) assumes an intersystem crossing quantum yield of 1. The TA dynamics fit well to Eq. (1) using a single time constant $\tau = 600 \pm 100$ ps for 11 separate wavelengths measured (Figure 3). This intersystem crossing time agrees with single-wavelength measurements in Ref. 7. Others have reported values for τ as high as 1.2 ns [8].

To demonstrate the importance of multispectral analysis over single-wavelength analysis to determine the intersystem crossing time, we compare in Figure 4 the TA dynamics for C_{60} at 725, 750, and 780 nm (wavelengths at which triplet growth dominates the TA spectrum) with the optimal single-wavelength fits of Eq. (1) at each individual wavelength. The intersystem crossing times τ obtained at these wavelengths are, respectively, 1.2, 540, and 660 ps. Even within this narrow band of wavelengths, a τ which spans the range reported in the literature is obtained. This result demonstrates the problem with obtaining τ from single wavelength fits: the spectrally broad overlapping singlet and triplet bands contribute differently at each wavelength. However, the singlet and triplet decay or rise time is identical regardless of the wavelength. Therefore, a self-consistent modeling of the dynamics for multiple wavelengths is necessary to determine a unique

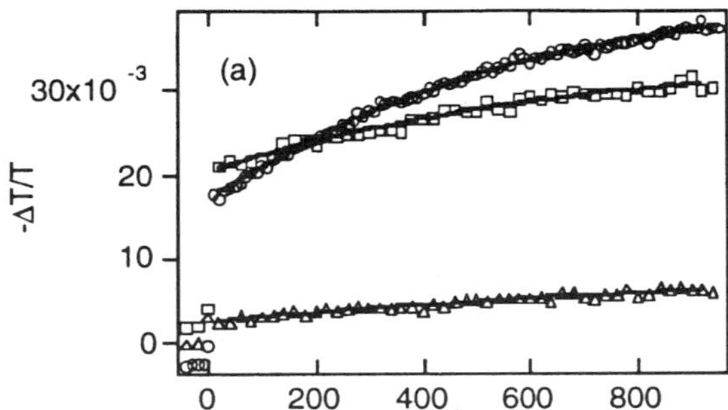

Figure 4 $-\Delta T/T$ versus delay time up to 1 ns for C_{60} in toluene solution at 725 nm (triangles), 750 nm (circles), and 780 nm (squares) along with single-wavelength fits using Eq. (1).

intersystem crossing time for the triplet. We deduce the value $\tau = 600 \pm 100$ ps based on fits of Eq. (1) to dynamics at 11 separate wavelengths (covering ranges of both predominant singlet growth and triplet decay) (see Figure 3). These results therefore establish a firm (and unique) number for $\tau \approx 600$ ps, which until now has been in some dispute. Also, this analysis clearly demonstrates the difficulty in obtaining a unique value for τ using single wavelength fits.

Given the discrepancy in reported values for τ, it is important also to comment on the effect of extrinsic factors such as competing channels for intersystem crossing on the measured dynamics and the resulting determination of τ. If a decay channel is present in the system which competes with the crossing from the singlet to the triplet manifold, with rate constant $1/\tau_{\text{ex}}$, then the equation describing the TA dynamics (assuming the competing state does not contribute directly to TA) is given by

$$\frac{-\Delta T}{T(\lambda,\, t)} = n_{\text{ex}}[\sigma_S(\lambda)e^{-t/\tau_1} + n_T\sigma_T(\lambda)(1 - e^{-t/\tau_1})] \tag{2}$$

where $1/\tau_1 \,(= 1/\tau + 1/\tau_{\text{ext}})$ is the decay time of the singlet, τ is the intersystem crossing time, and $n_T = n_{\text{ex}}\,\tau_1/\tau$ (the quantum yield is no longer unity). Such a channel may result, e.g., from clustering of C_{60} molecules [14, 19, 20]. From inspection of Eq. (2), additional decay channels would shorten the intersystem crossing time determined from fitting. The level of detail in existing publications [8] reporting 1.2 ns for the intersystem crossing time is insufficient to determine the fitting procedure used. From the above, however, two possible resolutions to the discrepancy are: (1) Studies reporting 1.2 ns may have used single-wavelength fits which can be misleading (see Figure 4); (2) our data may reflect additional decay channels which compete with intersystem crossing, leading to an effectively shorter τ.

In addition to the intersystem crossing time, the multispectral analysis using Eq. (1) yields the spectral distribution of the ratio σ_T/σ_S, which is plotted in Figure 5. In this way, detailed analysis of the spectral dependence of the relaxation dynamics allows separation of the singlet and triplet excited-state absorption bands, a separation which cannot be achieved by any other means. In addition to the relative values of σ_S and σ_T shown in Figure 5, the absolute values could also be obtained from this data, which requires detailed knowledge of the laser beam properties, or a comparison with a sample with calibrated excited-state cross section at a similar intensity. Interestingly, we also compare in Figure 5 a simple approximate spectral measurement of σ_T/σ_S which was obtained by dividing the 1 ns $\Delta T/T$ spectrum (Figure 2) by the spectrum obtained at 3 ps delay. The solid line in Figure 5 shows the result of this ratio, which represents a considerably simpler method for obtaining σ_T/σ_S over a wide spectral range, and which serves as an independent check that the detailed temporal fitting procedure gives

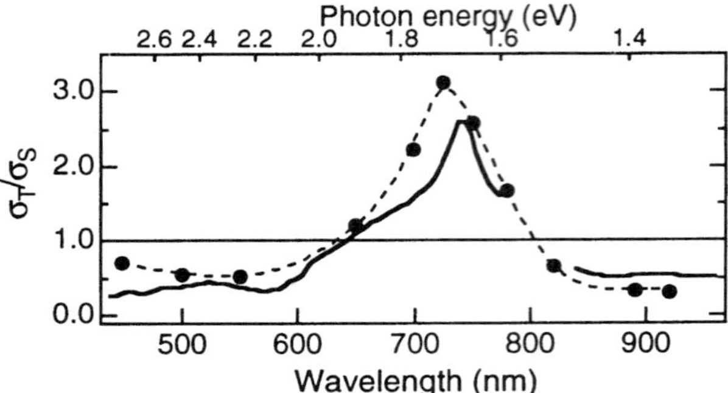

Figure 5 Comparison of σ_T/σ_S obtained from ratio of fs TA spectra at 1 ns and 3 ps (solid line) with ratio obtained from fitting time transients (solid circles). (From Refs. 10, 14).

sensible results. Of course, very accurate, low-noise TA spectra are needed in order to obtain meaningful results in a ratio, and this result highlights the dramatic improvements which can be realized using high-sensitivity $\Delta T/T$ scanning methods over previous methods based on CCD camera detection [17]. Figure 5 demonstrates that in the wavelength range from 620 to 810 nm, σ_T is larger than σ_S. In this wavelength range, OL is predicted to be more effective for ns pulses than for ps pulses, due to the long lifetime and large σ_T of the triplet state. In other parts of the spectrum OL should be stronger for ps pulses.

In addition to these long-scan measurements, TA dynamics were monitored on much shorter timescales. Figure 6 plots the short-time dynamics, along with the pump-probe cross-correlation function recorded using the instantaneous two-photon absorption in ZnS. Except for weak features seen at $\Delta t = 0$ (pump and probe pulses overlap inside the sample), the transients for all spectral energies in the range 450–920 nm have almost the same build-up dynamics. Fitting to an exponential growth yields time constants of 200–300 fs. The fast bleaching seen at 450 and 550 nm at $\Delta t = 0$ is attributed to pump-induced bleaching of the ground-state absorption, which is quickly overwhelmed by the increased absorption associated with the lowest excited singlet state. The resolution-limited fast increased absorption seen at 750 and 820 nm can be attributed to a two photon absorption process which involves one photon from the pump pulse and the other photon from the probe pulse. This fast response serves to calibrate the absolute arrival time of the probe pulse relative to the pump, and shows unequivocally that the subsequent growth dynamics involves relaxation from the initially excited state to the lowest excited singlet state.

The TA build-up dynamics are determined by the nature of the initial photo-

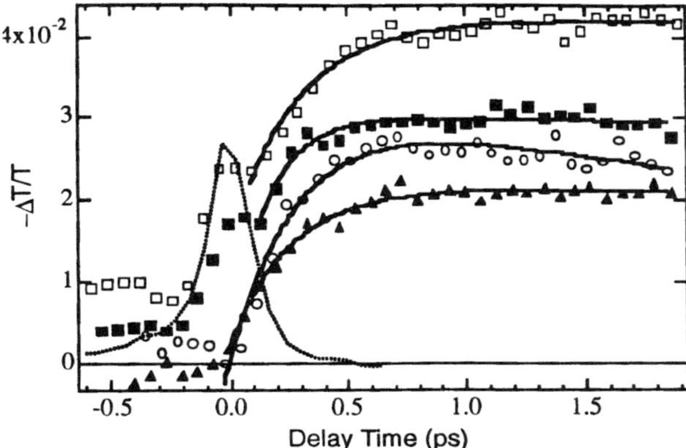

Figure 6 $-\Delta T/T$ versus delay time up to 2 ps for C_{60} in toluene solution, for four wavelengths: 450 nm (open circles), 550 nm (solid diamonds), 750 nm (solid squares), and 820 nm (open squares). Pump/probe cross correlation is included for reference (dotted line).

excitations. The first allowed optical transition for C_{60} in hexane solution occurs at 3.04 eV ($1\,^1T_{1u}$) [1], very close to our pump photon energy of 3.1 eV. The energy-relaxation $1\,^1T_{1u}-1\,^1T_{1g}$ is manifested in the build-up of increased absorption to the series of allowed transitions, as detailed in Table 1. The TA build-up time is determined by rates of 1u to 1g energy conversion and subsequent energy redistribution within the 1g vibronic manifold. The fact that we don't see any TA features associated with transitions involving the initial 1u state suggests that the conversion time is extremely short (<100 fs) and unresolved in our experiment. Therefore, 200 fs measured for the TA build-up should be attributed to energy redistribution within the $^1T_{1g}$ vibrational manifold.

3.2 Linear and Transient Absorption of 6,6 Substituted Fullerenes

Figure 7 shows σ_0, the ground-state absorption cross section, for C_{60} in toluene; the absorption drops toward zero beyond 650 nm. Since there is little or no ground-state absorption to initiate population of the excited states in the range where the triplet excited states have their maximum absorption, the long-lived triplet states are inefficiently activated for OL beyond ~650 nm. It is also not possible to increase the absorption in the NIR by preparing more concentrated samples, due to the limited solubility of C_{60}. Figure 7 compares σ_0 of the derivatized fullerenes PCBM, CPBPP, and 5,6 as well as 6,6 derivatized PCBCR with

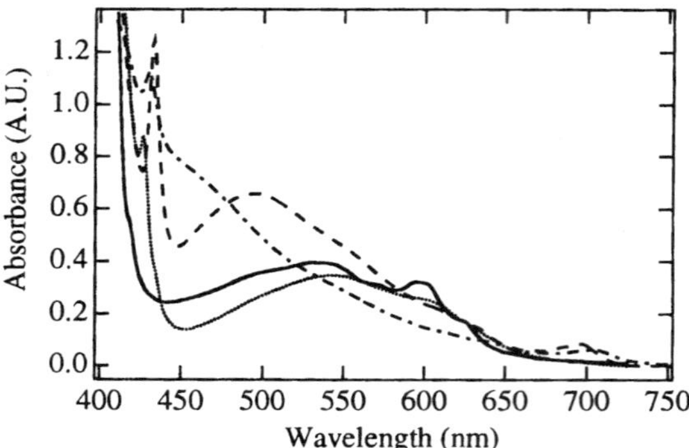

Figure 7 Ground-state absorption for solutions of C_{60} (solid line), PCBM [6,6 PCBCR] (dashed line), 5,6 PCBCR (dotted line), and CPBPP (dot-dash line).

C_{60}. It is noted that the σ_0 for PCBM and 6,6 PCBCR are nearly identical so that only σ_0 for PCBM is shown in Figure 7. For each of the 6,6 mono-adducts, the derivatization not only enhances the solubility of the fullerene but also activates new vibrationally coupled optical transitions in the wavelength range 650–740 nm [5]. If the excited-state response is not drastically affected by the derivatization, the presence of additional ground-state absorption in the NIR can be used to enhance the wavelength range of OL since the OL process now has a weak NIR ground-state absorption. For 5,6 PCBCR, the ground-state absorption is not drastically changed from that of C_{60}.

Figure 8a shows the TA spectra measured for CPBPP. The spectral dynamics of CPBPP are very similar to those of C_{60}. There is a region of dominant singlet decay near ~500 and ~900 nm. Around 700 nm, the induced absorption has slow dynamics and no clear triplet peak growth, attributed to broadened overlapping singlets and triplets in this wavelength range [16]. These results are similar to the spectral dynamics observed in PCBM and other 6,6 derivatives [5, 21]. A multispectral analysis of the decay transients using Eq. (1) is shown in Figure 8b for PCBM. For this 6,6 mono-adduct, $\tau \approx 600$ ps, unaltered from the result for neat C_{60}. This indicates that the addition of a 6,6 mono-adduct and the resulting reduction in symmetry and π electrons does not significantly affect the intersystem crossing time. $\sigma_T/\sigma_S(\lambda)$ obtained from the ratio of the 1 ns to 3 ps TA spectra for PCBM and CPBPP are compared with that of C_{60} in Figure 8c. $\sigma_T/\sigma_S(\lambda)$ obtained from fitting time transients for PCBM is shown as well to point out the generality of the agreement between these two techniques for obtaining $\sigma_T/\sigma_S(\lambda)$. For both PCBM and CPBPP, there is not a wavelength region where

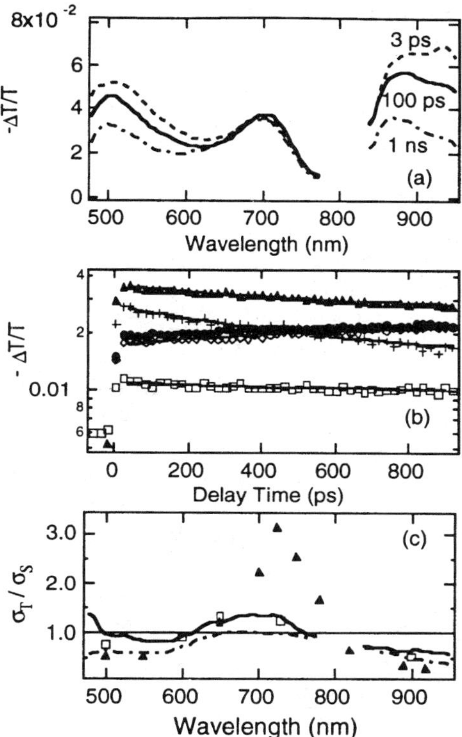

Figure 8 (a) Fs TA spectra for CPBPP at 3 ps (dashed line), 100 ps (solid line), and 1 ns (dot-dashed line). (b) σ_T/σ_S from TA for PCBM (solid line) and CPBPP (dot-dashed line line). (b) $-\Delta T/T$ versus delay time up to 1 ns for C_{60} in toluene solution, taken at 11 wavelengths, showing transition from singlet decay (450 nm $< \lambda <$ 650 nm), to triplet growth (650 nm $< \lambda <$ 800 nm), and back to singlet decay ($\lambda >$ 800 nm). 500 nm (solid triangles), 600 nm (open squares), 650 nm (open diamonds), 730 nm (solid circles), and 900 nm (crosses). (c) σ_T/σ_S from time transient modeling for C_{60} (solid triangles) and PCBM (solid circles) are also shown for comparison.

σ_T is much greater than σ_S, indicating that the RSA response should be more uniform in time at different wavelengths than for C_{60}.

3.3 Optical Limiting with Fullerene Solutions

After the initial reports of optical limiting in C_{60} several mechanisms were suggested to account for the OL behavior including reverse saturable absorption

(RSA) [3], nonlinear scattering [22], thermal blooming [4, 23], and multiphonon absorption [4]. Transient absorption spectra for the derivatized and neat C_{60}, as shown earlier in Sections 3.1 and 3.2, demonstrate that RSA is the dominant mechanism, though other processes become important at high input fluences [4, 22, 23]. RSA has been modeled in C_{60} using a simple five-level model [6] which includes the ground state, two excited singlet states, and two excited triplet states. For a strong OL response, the excited-state absorption cross section for the singlet (σ_S) or the triplet (σ_T) must exceed that of the ground state (σ_0) at the same wavelength [6]. These simple models may require additional levels at very high fluences where multiple excited singlet or triplet states may be accessed during a pulse, or in contrast require only three levels (ground and singlet states) when the input pulse has a time duration much shorter than the intersystem crossing time. The OL response also depends very strongly on whether the excited singlet or triplet states decay away too rapidly to obtain a significant population during the pulse. As shown above, fullerene singlet and triplet excited-state absorptions that extend throughout the visible and near infrared are induced upon ground-state pumping, in regions where σ_0 is small [5, 7–9, 13]. Therefore, derivatized and neat C_{60} are expected to provide OL with a potentially broadband response [3–5, 23, 24]. For ns pulses the OL response is predicted to be strongest near the triplet peak at $\sim$750 nm. The intersystem crossing time is short enough ($\sim$600 ps) compared with the 6 ns pulse to obtain a saturated population of the triplet states which do not decay away for $\sim\mu$s-ms. In spectral regions where σ_S is larger than σ_T (such as 532 nm), the picosecond OL response is expected to be stronger than the ns OL, as has been shown experimentally [4].

When the OL of C_{60} was first reported [3], C_{60} in toluene possessed the strongest OL response at 532 nm among other known OL materials. Subsequent research into RSA materials has shown that heavy metal substituted ph-thalocyanines and porphyrins have a stronger OL response at 532 nm attributed to strong triplet absorption [25–27]. However, these materials have two drawbacks: The OL response is not very broadband since the ground-state absorption determines a window of order $\sim$100 nm within which σ_0 is small and σ_T is large; and the phthalocyanines and porphyrins are very strongly colored by their ground-state absorption features so that a neutral color which is desirable for applications such as aircraft pilot eye protection is difficult to obtain. In contrast, fullerenes are expected to have a more broadband response and in addition the ground-state absorption is much more uniform in the visible range.

We demonstrate broadband optical limiting for C_{60} in Figure 9 [5]. This same figure illustrates that OL comparable to C_{60} can be obtained with 6,6 derivatives of C_{60}. In Figure 9, the OL response for 6,6 PCBCR in toluene is compared with C_{60} in 1,2 o-dichlorobenzene (ODCB) at 532 nm and 700 nm. The use of ODCB (a much stronger solvent) for C_{60} was necessary in order to achieve a concentration high enough to obtain 20% absorbance at 700 nm. For both materi-

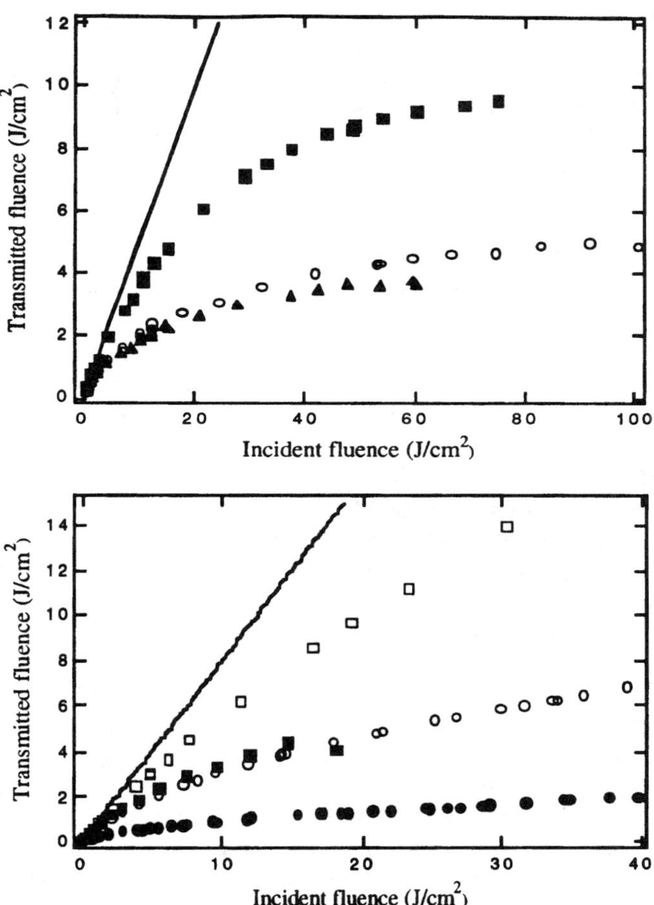

Figure 9 (a) Intensity-dependent transmission at 532 nm for C_{60} (solid triangles), 6,6 PCBCR (open circles), and 5,6 PCBCR (solid squares). (b) Intensity-dependent transmission for C_{60} at 532 nm (open squares, low-intensity transmission $T = 80\%$) and 700 nm (filled squares, $T = 80\%$) and for PCBCR at 700 nm (open circles, $T = 80\%$; filled circles, $T = 50\%$). (From Ref. 5.)

als at 532 nm (Figure 9a), the initial transmission is adjusted to $\sim$50%. The OL response is comparable for C_{60} and 6,6 PCBCR, indicating that the symmetry breaking 6,6 adduct which broadens the ground-state absorption in PCBCR does not hinder the intersystem crossing from the singlet to the triplet, as was shown for PCBM in Section 3.2. The excited-state absorption cross section values are also not significantly reduced [5]. The advantage of using 6,6 mono-adducts to C_{60} for OL is demonstrated in Figure 9b. The optical limiting of C_{60} and 6,6 PCBCR is compared at 700 nm, with each sample having a linear transmission of $\sim$80%. The OL for C_{60} at 532 nm (with initial transmission of 80%) is shown for comparison. From the fact that the ground-state absorption decreases at long wavelength while the excited-state absorption remains high, the OL response for C_{60} would be predicted to be stronger at long wavelength, consistent with the experimental results. Comparing C_{60} with 6,6 PCBCR, the OL response is comparable at 700 nm, except that the damage threshold (where there is a sharp drop in transmission) for C_{60} is significantly lower ($\sim$0.7 mJ), in comparison with 2 mJ for 6,6 PCBCR. This difference in damage threshold may be attributed to the much higher concentration of C_{60} compared with 6,6, PCBCR needed to obtain an initial transmission of 80%. Damage at the cell wall results from solid C_{60} which precipitates from the saturated solution to coat the surface of the cell. Also shown in Figure 9b is the OL response of 6,6 PCBCR with an initial transmission of 50% at 700 nm. The OL response is clearly stronger than the others shown. This improvement is due to the increased solubility of the 6,6 PCBCR and the enhanced ground-state absorption (to initiate OL) which occurs upon derivatization. It is noted that for C_{60}, the lowest initial transmission that could be obtained at 700 nm was 80% due to the limited solubility.

Figure 9a also shows the OL response of a 5,6 substituted C_{60}, 5,6 PCBCR [5]. Since the 5,6 derivatization of C_{60} does not alter the number of π electrons, this derivatization should leave the optical properties most similar to neat C_{60}. The OL for 5,6 PCBCR is significantly weaker than both C_{60} and 6,6 PCBCR. This is a surprising result since the geometry of 5,6 PCBCR most resembles C_{60} while the symmetry has been changed (and number of π electrons reduced) in the 6,6-adduct. This result is consistent with the weaker relative excited state absorption measured for 5,6 PCBCR compared with 6,6 PCBCR and C_{60} [5]. Calculations of the strength of ground and excited state absorption cross sections for additions to C_{60} at different sites are still necessary for comparisons with our experimental results.

For 6,6-adducts, strong excited-state absorption and weak ground-state absorption extend throughout the visible and near-infrared, indicating that these materials should be broadband optical limiters. Figure 10 shows the wavelength-dependent optical limiting measured with 6 ns pulses for solutions of PCBM (a) and CPBPP (b), adjusted to have a linear transmission of 50%. Both compounds show strongly increasing OL with increasing wavelength between 532 nm and

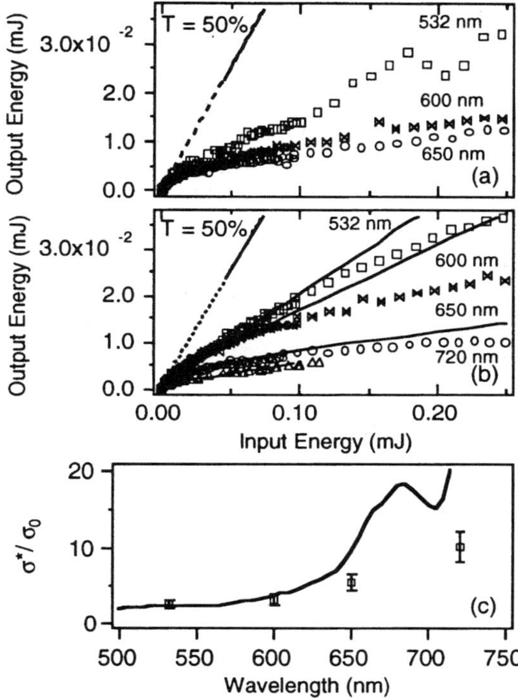

Figure 10 Optical limiting curves for PCBM (a) and CPBPP (b) as a function of wavelength. (c) Comparison of σ^*/σ_0 for CPBPP obtained from five-level fits to OL in (b) (open squares) and from TA spectra at 1 ns and ground-state absorption (solid line).

650 nm, similar to the response shown for neat C_{60} and 6,6 PCBCR [5], confirming that all 6,6 monoadducts exhibit similar broadband optical limiting. In addition, CPBPP at 720 nm (where the initial transmission was 80%) also shows strong OL. The clamping level decreases monotonically with increasing wavelength, consistent with the broad singlet and triplet induced absorption and the decreasing ground state absorption at long wavelengths. The solid lines in Figure 10b are five-level fitting results for each of the wavelengths. In solution, the deviation from the fits for high-input energies is due to additional limiting due to thermal mechanisms such as thermal lensing [22, 23].

The ratio σ^*/σ_0 ($\sigma^* = \nu_S\sigma_S + \nu_T\sigma_T$, where ν_x and σ_x are the population fraction and absorption cross section, respectively, for the excited singlet and triplet) deduced from fitting the OL data for CPBPP is shown in Figure 10c. For comparison, the ratio σ^*/σ_0 evaluated from the TA spectra at 1 ns and the ground-state absorption is shown also in Figure 10(c). $\sigma^*/\sigma_0 \sim 3$ is obtained at 532 nm, in agreement with other reports [22]. At longer wavelengths, σ^*/σ_0 increases monotonically up to ~ 10 at 720 nm. This behavior is in qualitative agreement with the predicted σ^*/σ_0 from TA spectra. This figure demonstrates that fullerenes are broadband optical limiters which are optimal for red and near IR applications. The σ^*/σ_0 obtained from five-level model fits do not grow as quickly as predicted from the TA and ground-state absorption spectra. This may be due to particular aspects of the modeling: The beam is propagated as a top hat beam profile that saturates the excited states more rapidly than typical of a more realistic gaussian beam profile; and in practice saturation fluence is always harder to attain than in simple models [27]. Therefore, we should take the σ^*/σ_0 obtained from fits to a five-level model as a *lower limit* to the actual values.

3.4 TA Dynamics and Optical Limiting of Fullerene Films and Sol-Gel Glasses

To take advantage of the broadband optical limiting of fullerenes for passive eye or sensor protection, a solid-state material is most convenient. However, there are clearly solid-state effects which can drastically reduce the optical limiting response of fullerenes. Figure 11a compares the dynamics of the transient absorption at 750 nm (where the triplet component is dominant in solution) for C_{60} in solution, thin films, and sol-gel glasses. The clear growth of the triplet signal is seen with time for C_{60} in solution. In contrast, the singlet in thin films of C_{60} is quenched more rapidly (on ps or sub-ps time scales) than the intersystem crossing time, so that there is no significant triplet population or optical limiting [13]. Some authors have suggested the importance of exciton-exciton annihilation in C_{60} thin films [19, 20]. This provides one explanation for the strong and rapid deactivation pathway for the singlet which occurs when there is aggregation of the C_{60}. For C_{60} pre-doped sol-gel glasses, the dynamics are much slower than thin films, indicating molecular dispersions of the C_{60} [13, 28]. However, the dynamics of the C_{60} pre-doped sol-gel glasses indicate some quenching or competing pathways to the triplet, as there is no clear growth of the triplet state as seen for C_{60} in solution [14, 18]. These studies indicate that effective solid-state optical limiters using fullerenes must take the form of dispersed solid solutions in a host matrix. Similar results are shown for PCBM at 600 nm (where there is an isosbestic point between singlet decay and triplet growth) in pre-doped sol-gel glassess compared with solutions in Figure 11b. The dynamics are clearly more rapid in the pre-doped PCBM sol-gel glass than in the solvents. The DCB solutions also show clearly faster relaxation kinetics than those in toluene, indicative perhaps of aggregation or solvent interactions, which can lead to new relaxation pathways which compete with the triplet state. Since PCBM gels were most commonly prepared using DCB as a co-solvent to take advantage of the rapid

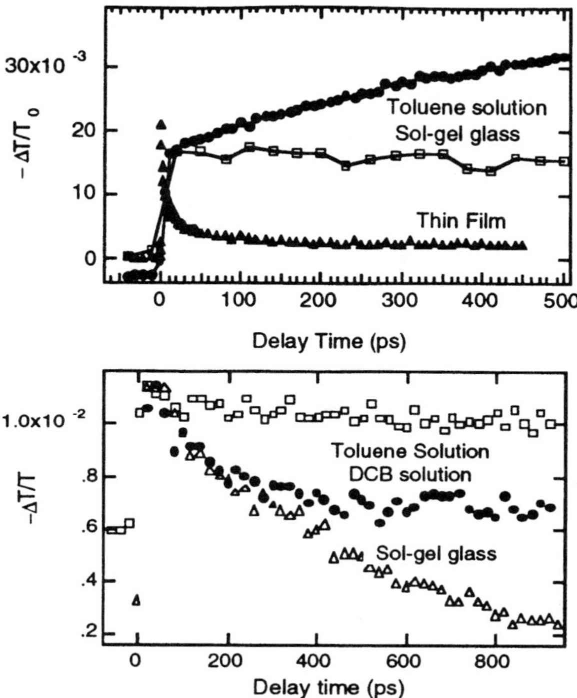

Figure 11 (Top) Dynamics at 750 nm of C_{60} in toluene compared with thin film and pre-doped sol-gel glass. (Bottom) Dynamics at 600 nm of PCBM in toluene and DCB compared with pre-doped sol-gel.

gelation times, this may also explain some of the observed more rapid kinetics in the pre-doped gels.

The incorporation of fullerenes into sol-gel glasses has resulted in both advantages and drawbacks. The sol-gel glass itself is highly transparent and has a very high resistance to optical damage. Here we mean by optical damage that the transmission of the sample shows an abrupt drop and irreproducibility upon cycling the input light fluence. For PCBM incorporated into a sol-gel matrix, the damage threshold occurs when the input optical energy exceeds 1.5 ± 0.5 mJ, compared with a damage threshold of 0.1 ± 0.05 mJ for PCBM in toluene. The mechanism for optical damage in solutions and glasses is also different. In solutions (as described in Section 3.3), the damage occurs due to precipitation at the cell wall, and the damage threshold decreases with increasing concentration. In the glasses, damage occurs due to thermal features of the sol-gel glass host, and the damage threshold increases with increasing concentration due to reduction of the fluence at the focus due to activation of molecules prior to the focus. Therefore, incorporation of fullerenes into a sol-gel glass extends the input fluence level that the optical limiter can sustain before damage, increasing the dynamic range of the optical limiting for applications. A drawback of incorporation of fullerenes into sol-gel glasses is that, in side-by-side comparisons, PCBM pre-doped sol-gel glasses demonstrate less effective optical limiting than solutions [14]. In Figure 12, we show the optical limiting at 650 nm of PCBM in toluene (2 mm path length), in a pre-doped sol-gel glass (1.4 mm thick), and in a post-doped sol-gel glass (2 mm thick), all prepared to have a linear absorption $\sim$50%. The transmission (OL) for high-input energy of PCBM in toluene decreases four times that of PCBM in a typical pre-doped sol-gel glass. Based on values deduced from five-level model fits to the OL response, the effective excited-state absorption cross section σ^* in the glass (2.0×10^{-18} cm^2) is less than half that of PCBM in toluene (5.5×10^{-18} cm^2). Comparing the behavior of a post-doped PCBM sol-gel glass in Figure 12, the OL response is improved such that $\sigma^* \sim 4.0 \times 10^{-18}$ cm^2 for the post-doped sol-gel glass. Other post-doped PCBM sol-gel glasses showed values as high as $\sigma^* \sim 4.8 \times 10^{-18}$ cm^2, indicating that $\sim$90% of the OL response of PCBM in toluene can be observed in post-doped sol-gel glasses. Since the solution optical limiting also includes thermal self-defocusing mechanisms which are lower in solid-state samples, it is thought that the solution results represent an upper limit for the performance to be expected in solid samples. Future improvements in processing may enhance this OL response of both pre-doped and post-doped sol-gel glasses to more closely approximate solution limiters, though with a higher damage threshold. At this point, solid-state PCBM

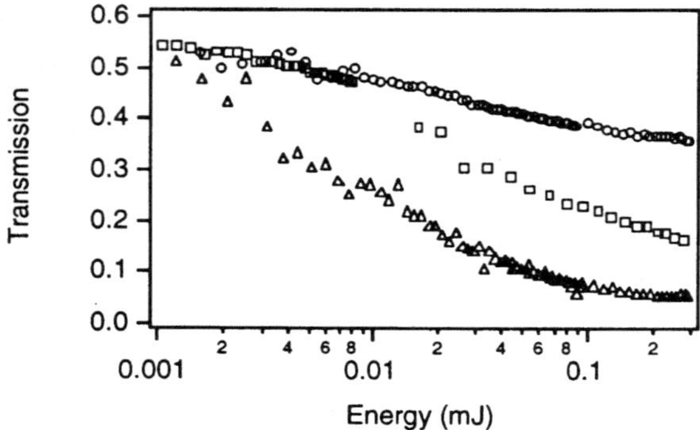

Figure 12 Optical limiting at 650 nm for PCBM in toluene (open triangles), pre-doped sol-gel (open circles), and post-doped sol-gel (open squares).

optical limiters with a performance comparable to the fullerene in solution can be prepared. Due to enhanced solubility, the concentration of 6,6 fullerene derivatives such as PCBM in pre-doped sol-gels can be varied to achieve a wide range of linear transmissions for a variety of applications [13–15].

4. CONCLUSIONS

In this chapter, we have presented transient absorption data for neat and 6,6 derivatives of C_{60} that characterize the essential singlet and triplet excited states which form upon photoexcitation. Using the improved sensitivity of scanning TA spectroscopy, we have made a detailed comparison of excited singlet absorption spectra with energy-level diagrams established from linear absorption for C_{60}. We report excellent quantitative agreement between the energy-level structure, assuming excited-state absorption from the LUMO band with the same parity as the ground state. Qualitative agreement with the predicted triplet specta was also found. Multispectral analysis of decay dynamics using a simple model of singlet decay and complementary triplet growth with a single time constant provided an intersystem crossing time of $\tau = 600 \pm 100$ ps for both C_{60} and the 6,6 derivatized fullerene PCBM in solution. Since τ does not change significantly upon 6,6 derivatization of C_{60}, this indicates that the loss in symmetry and π-electrons due to derivatization does not significantly alter the singlet-triplet dynamics. Also, this multispectral analysis provides a more solid number for the intersystem crossing time than has been reported previously for C_{60} using analysis of decay at a single wavelength. In addition, multispectral analysis provides σ_T/σ_S as a function of wavelength, a useful tool for predicting wavelength ranges where effective ps or ns optical limiting occur. σ_T/σ_S can also be obtained directly from the ratio of TA spectra at 1 ns and 3 ps, showing very good agreement with the ratio obtained from multispectral analysis. For 6,6 fullerene derivatives, which exhibit an infrared-extended ground-state absorption compared with C_{60}, there is no spectral region where the σ_T dominates σ_S as strongly as is observed in C_{60}, indicating that the OL should be more uniform for different length pulses across the wavelength spectrum.

We used the transient absorption spectra to predict broadband optical limiting for neat and 6,6 derivatized fullerenes. Wavelength-dependent studies show that the ns OL response improves monotonically at longer wave-lengths for all 6,6 mono-adducts and for neat C_{60}, demonstrating broadband limiting with $\sigma^*/\sigma_0 \gtrsim 10$ at 720 nm, and confirming predictions from the TA spectra. To take advantage of this broadband limiting, we have studied the dynamics and optical limiting of fullerenes in films and sol-gel glasses. Forming a solid-state optical limiter from fullerenes provides certain challenges. In thin films where neat or derivatized C_{60} can aggregate, pathways for rapid quenching of the singlet state dominate the dynamics. In sol-gel glasses, the fullerenes form a molecularly dispersed phase which demonstrates optical limiting and a damage threshold much higher than the same fullerene in solution. However, the OL response is typically weaker than the same fullerene in solution. New approaches to processing sol-gel encapsulated fullerenes demonstrate that the OL performance of solid-state dispersed fullerenes can be engineered to approach the response of solution limiters, making them more attractive for applications.

ACKNOWLEDGMENTS

We thank J. M. Robinson, A. Koskelo, B. R. Mattes, and F. Wudl for assistance and useful discussion. D. Guldi also provided important comments. 6,6 monoadducts were synthesized by X. Shi (LANL) and H. Wang (UCSB). Pre-doped sol-gel glasses were provided by M. Grigorova (LANL). Post-doped sol-gel glasses were provided by W. Moreshead (GELTECH, Inc.). This work was supported in part by Los Alamos National Laboratory Directed Research and Development funds and Collaborative University of California/Los Alamos Research funds, under the auspices of the U.S. Department of Energy. Additional funding was provided by the U.S. Army Small Business Technology Transfer program, under contract DAAH04-96-C-0077.

REFERENCES

1. S. Leach, M. Vervloet, A. Després, E. Bréheret, J. P. Hare, T. J. Dennis, H. W. Kroto, R. Taylor, and D. R. M. Walton, "Electronic spectra and transitions of the fullerene C_{60}," *Chem. Phys.* **160**, 451, 1992.
2. Y. Wang, J. Holden, A. Rao, P. Eklund, U. Venkateswaran, D. Eastwood, R. Lidberg, G. Dresselhaus, and M. Dresselhaus, "Optical absorption and photoluminescence in pristine and photopolymerized C_{60} solid films," *Phys. Rev. B* **51**, 4547–4556, 1995.
3. L. Tutt and A. Kost, "Optical limiting performance of C_{60} and C_{70} solutions," *Nature* **356**, 225–226, 1992.
4. D. McLean, R. Sutherland, M. Brant, D. Brandelik, P. Fleitz, and T. Pottenger, "Nonlinear absorption study of a C_{60}-toluene solution," *Opt. Lett.* **18**, pp. 858–860, 1993.
5. L. Smilowitz, D. McBranch, V. Klimov, J. Robinson, A. Koskelo, M. Grigorova, B. Mattes, H. Wang, and F. Wudl, "Enhanced optical limiting in derivatized fullerenes," *Opt. Lett.* **21**, p. 922, 1996.
6. L. Tutt and T. Boggess, "A review of optical limiting mechanisms and devices using organics, fullerenes, semiconductors, and other materials," *Prog. Quant. Elect.* **17**, 299–338, 1993.

7. R. J. Sension, C. M. Phillips, A. Z. Szarka, W. J. Romanov, A. R. McGhie, J. P. M. Jr., A. B. S. III, and R. M. Hochstrasser, "Transient absorption studies of C_{60} in solution," *J. Phys. Chem.* **95**, 6075, 1991.

8. T. W. Ebbesen, K. Tanigaki, and S. Kuroshima, "Excited-state properties of C_{60}," *Chem. Phys. Lett.* **181**, 501, 1991.

9. J. Arbogast, A. Darmanyan, C. Foote, Y. Rubin, F. Diederich, M. Alvarez, S. Anz, and R. Whetten, "Photophysical properties of C_{60}," *J. Phys. Chem.* **95**, 11–12, 1991.

10. V. Klimov, R. Kohlman, and D. McBranch, "Broadband femtosecond transient absorption in C_{60}: Dynamics of excited singlet and triplet states," in preparation.

11. J. C. Hummelen, B. W. Knight, F. Lepec, F. Wudl, J. Yao, and C. L. Wilkins, "Preparation and characterization of fulleroid and methanofullerene derivatives," *J. Org. Chem.* **60**, 532, 1995.

12. M. Prato, "Fullerene chemistry for materials science applications," *J. Mater. Chem.*, in press, 1997.

13. D. McBranch, L. Smilowitz, V. Klimov, J. Robinson, B. Mattes, A. Koskelo, J. Hummelen, F. Wudl, N. Borrelli, and J. Withers, "Optical limiting in fullerene solutions and doped glasses," *SPIE Proceedings, Fullerenes and Photonics II* **2530**, 195, 1995.

14. D. McBranch, V. Klimov, L. Smilowitz, M. Grigorova, B. R. Mattes, J. Robinson, A. Koskelo, H. Wang, and F. Wudl, "Femtosecond excited-state absorption dynamics and optical limiting in fullerene solutions, sol-gel glasses, and thin films," *SPIE Proceedings, Fullerenes and Photonics III* **2854**, 140, 1996.

15. M. Maggini, G. Scorrano, M. Prato, G. Brusatin, P. Innocenzi, M. Guglielmi, A. Renier, R. Signorini, M. Meneghetti, and R. Bozio, "C_{60} derivatives in sol-gel glasses," *R. Adv. Mater.* **7**, 404–406, 1995.

16. R. Kohlman, V. Klimov, M. Grigorova, X. Shi, B. R. Mattes, D. McBranch, H. Wang, F. Wudl, J.-P. Nogues, and W. Moreshead, "Ultrafast and nonlinear optical characterization of optical limiting processes in fullerenes," *SPIE Proceedings, Fullerenes and Photonics IV*, in press, 1997.

17. V. Klimov and D. McBranch, "Femtosecond high-sensitivity, chirp-free transient absorption spectroscopy," *Opt. Lett.*, submitted.

18. V. Klimov, D. McBranch, L. Smilowitz, J. Robinson, B. R. Mattes, A. Koskelo, H. Wang, and F. Wudl, "Femtosecond to nanosecond dynamics of C_{60}: implications for excited-state nonlinearities," *Res. Chem. Intermed.*, special issue, in press, 1997.

19. S. L. Dexheimer, W. A. Varecka, C. V. Shank, D. Mittelman, and A. Zettl, "Nonexponential relaxation in solid C_{60} via time-dependent singlet exciton annihilation," *Chem. Phys. Lett.* **235**, 552, 1995.

20. S. R. Flom, F. J. Bartoli, H. W. Sarkas, C. D. Merritt, and Z. H. Kafafi, "Resonant nonlinear optical properties and excited-state dynamics of pristine, oxygen-doped, and photopolymerized C_{60} in the solid state," *Phys. Rev. B* **51**, 11376–11381, 1995.

21. D. Guldi and K.-D. Asmus, "Photophysical properties of mono- and multiply-functionalized fullerene derivatives," *J. Phys. Chem.* **101**, 1472–1481, 1997.

22. S. Mishra, H. Rawat, M. P. Joshi, and S. Mehendale, "The role of nonlinear scattering in optical limiting in C_{60} solutions," *J. Phys. B* **27**, pp. L157–L163, 1994.

23. B. Justus, Z. Kafafi, and A. Huston, "Excited-state absorption-enhanced thermal optical limiting in C_{60}," *Opt. Lett.*, **18**, *19*, 1603–1605, 1993.

24. M. Joshi, S. Mishra, H. Rawat, S. Mehendale, and K. Rustagi, "Investigation of optical limiting in C_{60} solution," *Appl. Phys. Lett.* **62**, *15*, 1763, 1993.

25. J. Perry, K. Mansour, S. Marder, K. Perry, J. D. Alvarez, and I. Choong, "Enhanced reverse saturable absorption and optical limiting in heavy-atom substituted phthalocyanines," *Opt. Lett.* **19**, 625–627, 1994.

26. J. Shirk, R. Pong, F. Bartoli, and A. Snow, "Optical limiter using a lead phthalocyanine," *Appl. Phys. Lett.* **63**, 1880–1882, 1993.

27. T. Xia, D. J. Hagan, A. Dogariu, A. A. Said, and E. W. V. Stryland, "Optimization of optical limiting devices based on excited-state absorption," *Appl. Opt.* **36**, *18*, 4110–4122, 1997.

28. D. McBranch, B. R. Mattes, A. Koskelo, J. M. Robinson, and S. P. Love, "C_{60}-doped silicon dioxide sonogels for optical limiting," *SPIE Proceedings, Fullerenes and Photonics I* **2284**, 15–20, 1994.

7
Magnetic Resonance Studies of Photoexcited Fullerenes

P. A. Lane
University of Sheffield
Sheffield, England

Zeev Valy Vardeny
University of Utah
Salt Lake City, Utah

Joseph Shinar
Ames Laboratory, U.S. Department of Energy, and Iowa State University
Ames, Iowa

1. INTRODUCTION

Electron spin resonance (ESR) and optically detected magnetic resonance (ODMR) have yielded valuable information about photoexcitation dynamics and the electronic properties of fullerenes. Both of these techniques rely upon the paramagnetic interaction of excitations with an applied magnetic field. For an isolated particle with nonzero spin placed in a magnetic field, the interaction can be described by the following Hamiltonian:

$$\mathcal{H} = \mu_B \vec{S} \cdot \overleftrightarrow{g} \cdot \vec{H} \tag{1}$$

where $\vec{S}$ is the spin of the particle, $\overleftrightarrow{g}$ is a tensor defining the interaction of the particle with the field, and $\vec{H}$ is the applied magnetic field. The g tensor is diagonal in a basis set corresponding to the symmetry of the molecule and possesses three principal values. Although C_{60} is isotropic, ESR measurements [1] have found g values for C_{60}^- to be slightly anisotropic ($g_\perp = 1.9937$ and $g_\parallel = 1.9987$ with respect to the external magnetic field). This anisotropy has been attributed to a Jahn-Teller distortion of the molecule, arising from lifting of the degeneracy of the ground state [2, 3]. C_{60}^- consequently deviates from icosohedral symmetry.

For ESR measurements, the sample is placed in a resonant microwave cavity between the pole pieces of an electromagnet. The magnetic field is gradually increased, which induces a Zeeman splitting of the excitation energy levels. Resonant absorption or emission of microwaves occurs when the splitting of the energy levels matches the energy of the microwave photons ($h\nu = g\mu_B H$). High sensitivity is achieved by modulating the magnetic field with a pair of coils built into the cavity and using lock-in detection of microwave absorption at the modulation frequency. This technique results in a signal proportional to the derivative of microwave absorption. Neutral C_{60} molecules have no unpaired spins and therefore exhibit no ESR signal. Since C_{60} is highly electronegative and can therefore be easily doped, a number of ESR studies have been performed on C_{60} ions. C_{60}^{-} has one unpaired electron and exhibits a narrow ESR spectrum, shown in Figure 1a [4]. The g value is 1.999, slightly below that of the free electron, and the derivative peak-to-peak linewidth $\Delta H_{PP} = 2.5G$ at room temperature. C_{60}^{2-} is a triplet state with $S = 1$ and its ESR spectrum, shown in Figure 1b [4], is consequently broader than that of C_{60}^{-}. ESR measurements of C_{60} films show that a paramagnetic impurity is present in films evaporated from solution, but not those sublimed under vacuum [5].

Interest in the photoexcited triplet state of fullerenes was spurred by the report of an intense light-induced ESR (LESR) signal in C_{60} and C_{70} solutions

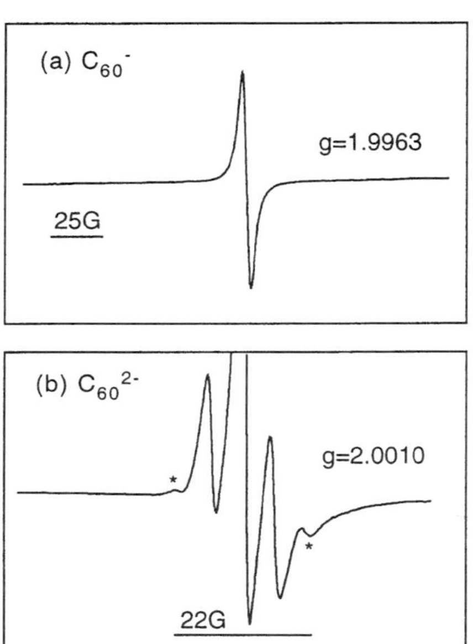

Figure 1   The ESR spectra of (a) C_{60}^{-} and (b) C_{60}^{2-}. (*From Ref. 4; used with permission.*)

by Wasielewski et al. [6]. The spin Hamiltonian for triplet ($S = 1$) states is written

$$\mathcal{H}_{S-S} = g\mu_B \vec{H} \cdot \vec{S} + D\left(S_z^2 - \frac{1}{3}S^2\right) + E\left(S_x^2 - S_y^2\right) \tag{2}$$

where D and E are the zero field splitting (ZFS) parameters of the triplet exciton. These are due to magnetic dipole interactions between the two unpaired spins composing the triplet and are given by

$$D = \frac{3g^2\mu_B^2}{4}\left\langle\frac{r^2 - 3z^2}{r^5}\right\rangle \tag{3}$$

$$E = \frac{3g^2\mu_B^2}{4}\left\langle\frac{y^2 - x^2}{r^5}\right\rangle \tag{4}$$

where x, y, z, and r are the reduced coordinates of the spins in the triplet state. The energy levels of a triplet state with nonzero D and E are shown in Figure 2a. As with a spin-$\frac{1}{2}$ excitation, resonant absorption of microwaves occurs when

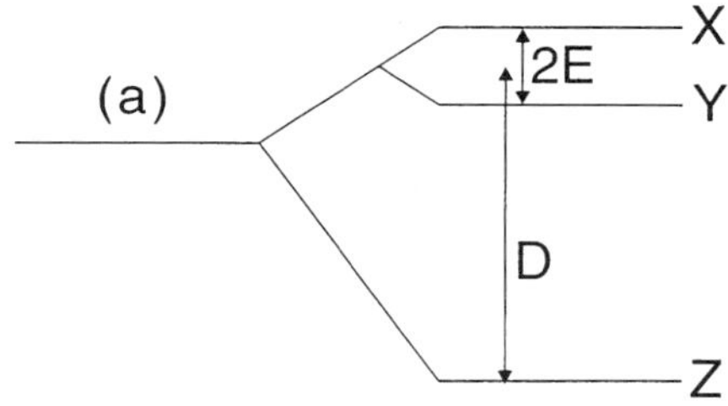

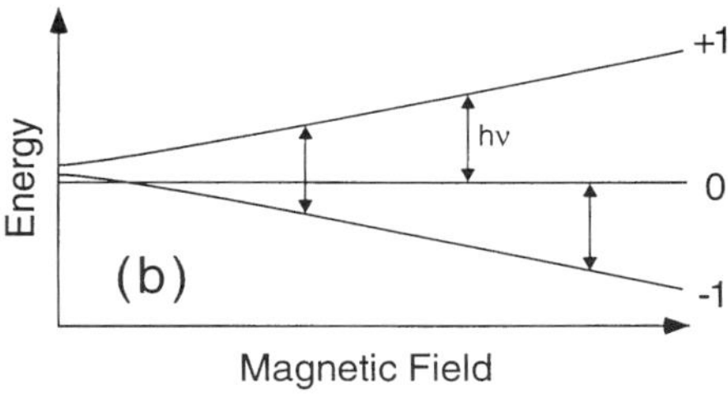

Figure 2 (a) Splitting of triplet sublevels in zero field. (b) Zeeman splitting of a single triplet with the magnetic field aligned parallel to one of the principal axes.

the splitting between two spin sublevels matches the microwave photon energy. The eigenvalues of the spin Hamiltonian depend upon both the magnitude of the applied field and its direction relative to the principal axes of the molecule. The orientation of the molecule relative to $\vec{H}$ determines the energies of the triplet sublevels and, hence, the resonant fields. Figure 2b shows a diagram of energy vs. applied field intensity for the applied field aligned along the z principal axis. The sublevel with its axis parallel to the applied field is unaffected by the field and effectively becomes the $m_S = 0$ sublevel. The other two sublevels are mixed, with one split upward ($m_S = +1$) and the other split downward ($m_S = -1$).

Magnetic resonance studies of photoexcited fullerenes have included fullerenes isolated in translucent glasses (most common is a solution in toluene with a glassing agent such as polystyrene), polycrystalline films, and single crystals. A single triplet has three resonant fields, two due to $\Delta m_S = \pm 1$ transitions and one due to $\Delta m_S = \pm 2$ transitions. For amorphous or polycrystalline samples, a powder pattern is formed due to the random orientation of the various triplets with respect to the applied field. While analysis is somewhat more complex than for a single crystal, it is possible to simulate the powder pattern and extract the ZFS parameters. The "full-field" powder pattern due to $\Delta m_S = \pm 1$ transitions has the following critical points:

$$\text{Singularities at: } H = H_0 \pm \frac{D - 3E}{2g\beta} \tag{5}$$

$$\text{Shoulders at: } H = H_0 \pm \frac{D + 3E}{2g\beta} \tag{6}$$

$$\text{Steps at: } H = H_0 \pm \frac{D}{g\beta} \tag{7}$$

whereas the "half-field" powder pattern due to $\Delta m_S = \pm 2$ transitions has the following critical points:

$$\text{Singularity at: } H = \sqrt{\left(\frac{H_0}{2}\right)^2 - \frac{D^2 + 3E^2}{g^2\beta^2}} \tag{8}$$

$$\text{Shoulder at: } H = \left(\frac{H_0}{2}\right)\left[1 - \frac{1}{2}\left(\frac{D - E}{H_0}\right)^2\right] \tag{9}$$

Here β is the Bohr magneton (9.274078×10^{-28} J/G) and $H_0 = h\nu/g\beta$.

Magnetic resonance also has an effect on photoexcitation dynamics as the absorption of microwaves will equalize the populations of the two coupled spin sublevels. Consider two triplet sublevels X and Y with sublevel populations n_i, generation rates G_i, and recombination rates R_i. Assuming monomolecular decay and excluding spin-lattice relaxation,

$$\frac{dn_i}{dt} = G_i - n_i R_i \tag{10}$$

Under steady-state conditions $dn_i/dt = 0$ and thus

$$n_i = \frac{G_i}{R_i} \tag{11}$$

At the resonant field, the two sublevels are coupled and have generation and recombination rates $\tilde{G} = (G_X + G_Y)/2$ and $\tilde{R} = (R_X + R_Y)/2$, and populations $\tilde{n} = \tilde{G}/\tilde{R}$. The difference in the total triplet population δn at resonance and under saturation by microwaves is

$$\delta n = n_X + n_Y - 2\tilde{n} = (n_Y - n_X)\frac{R_X - R_Y}{R_X + R_Y} \tag{12}$$

Population redistribution among magnetic sublevels produces a change in either the absorbed or emitted light associated with the excitation. The change in light is thus used as a detector of magnetic resonance, replacing direct observation of microwave absorption by the paramagnetic species. This scaling up from the microwave to the optical range makes ODMR extremely sensitive: up to 10^5 times more sensitive than conventional ESR.

In this chapter, we show several simulated triplet ODMR spectra; the details of the model are summarized here. The ODMR spectrum is proportional to the transition probability between two spin sublevels and to the difference in level population (spin polarization) in the absence of magnetic resonance. The eigenfunctions of the spin Hamiltonian can be expressed in terms of the zero-field sublevels $|i\rangle = c_{ix}|x\rangle + c_{iy}|y\rangle + c_{iz}|z\rangle$, where the coefficients c_{ij} are functions of $\vec{H}$ and triplet orientation relative to the field [7]. At resonance and under microwave saturation, the populations of two spin sublevels coupled by magnetic resonance are equalized. In order to model a powder pattern, the spin Hamiltonian must be solved and the spin polarization calculated. The generation and recombination rates for the high-field triplet sublevels are

$$G_i = |c_{ix}|^2 G_x + |c_{iy}|^2 G_y + |c_{iz}|^2 G_z \tag{13}$$

$$R_i = |c_{ix}|^2 R_x + |c_{iy}|^2 R_y + |c_{iz}|^2 R_z \tag{14}$$

where $G_{x,\,y,\,z}$ and $R_{x,\,y,\,z}$ are the respective generation and recombination rates of the zero-field triplet sublevels. The possibility of transitions between triplet sublevels (spin-lattice relaxation) is also included in the model. Taking all these factors into account, the rate equation for each triplet sublevel is

$$\frac{dn_i}{dt} = G_i - n_i R_i - S(n_i - n_{iB}) \tag{15}$$

where n_{iB} is the thermalized (Boltzmann) population of each sublevel and S is the spin-lattice relaxation rate. Under steady-state conditions, $dn_i/dt = 0$, leading to a set of three coupled equations, the solution of which gives the steady-state sublevel populations.

A schematic diagram of an ODMR spectrometer is shown in Figure 3. The ODMR spectrometer resembles a conventional photoluminescence (PL) or photoinduced absorption (PA) spectrometer, with the sample placed in a microwave cavity between the pole pieces of an electromagnet. The sample is constantly illuminated by a pump beam; amplitude-modulated microwaves are coupled into the cavity through a waveguide. Changes δI in PL or δT in probe transmission are detected by lockin-amplification of the photodiode signal. We describe the two different kinds of measurements as PL- and PA-detected magnetic resonance (PLDMR and PADMR, respectively). Two kinds of PADMR

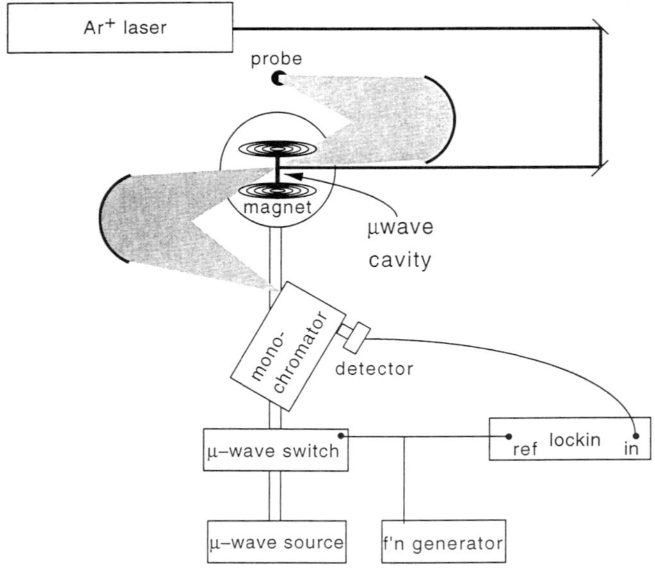

Figure 3 A schematic diagram of a high-field ODMR spectrometer. The pump light is left off for PLDMR measurements and turned on for PADMR measurements.

spectra can be measured: H-PADMR spectra in which δT is measured at a fixed probe wavelength λ while sweeping the magnetic field H, and λ-PADMR where δT is measured at fixed H on resonance, while λ-probe is varied. PADMR spectroscopy is especially valuable in deconstructing a complex PA spectrum, such as that seen for C_{60} films.

Interpreting the ODMR spectra of polycrystalline or amorphous samples can be difficult due to the complexity of triplet powder patterns, especially where more than one triplet species is present. As will be discussed below, various magnetic resonance studies of C_{60} and C_{70} have revealed contributions from multiple triplet excitons. Due to the energy splitting of the triplet sublevels, it is possible to perform an ODMR measurement without a magnetic field. In conventional magnetic resonance, the sample is radiated with fixed frequency microwaves and the magnetic field is swept to generate a spectrum. In a zero-field ODMR measurement, the microwave frequency is swept by using an LC resonant circuit with a variable capacitor that permits tuning of the resonant frequency [8]. The main advantage of zero-field ODMR is the lack of any angular dependence of the triplet-state sublevels. Only three resonant lines are observed, which occur at frequencies corresponding to ZFS parameters $h\nu_1 = |D + E|$, $h\nu_2 = |D - E|$, and $h\nu_3 = |2E|$. D and E are defined such that $|D| > |3E| \Rightarrow \nu_1 > \nu_2 > \nu_3$. Zero field ODMR cannot be used to study spin-$\frac{1}{2}$ excitations, due to the degeneracy of energy levels in zero field. High-field and zero-field ODMR thus complement one another and both kinds of measurements are reviewed in this chapter.

2. STUDIES OF ISOLATED FULLERENES

We first discuss photoexcitation dynamics of isolated fullerenes. All liquid samples studied consist of C_{60} or C_{70} dissolved in degassed organic solvent. The optical properties of C_{60} and C_{70} have varied only weakly upon the solvent used, indicating that solution spectra are due to isolated fullerenes. A glassing agent such as polystyrene was added to solutions, so that light could easily penetrate frozen samples. For some measurements, the solvent was evaporated, producing a suspension of the C_{60} or C_{70} in polystyrene.

2.1 C_{60}

The photophysics of C_{60} molecules are fairly well understood. Upon excitation, a free-singlet exciton is generated which self-localizes to a polaronic singlet exciton within 15 ps [9, 10]. Optical transitions from the lowest excited singlet state to the ground state are dipole forbidden. As a consequence of this and rapid intersystem crossing (ISC) to the triplet manifold, the integrated PL yield of C_{60} is only 7×10^{-4} whereas the triplet yield approaches unity. The fluorescence and optical absorption spectra of C_{60} in a 6.6×10^{-6} M methylcyclohexane solution is shown in Figure 4a. The two spectra are mirror images of one another, permitting correlation of the various features. These are labeled M_i in the fluorescence spectrum and M_i' in the absorption spectrum. A detailed analysis of the two spectra, including the dominant role of X traps, is given elsewhere [11, 12].

Given the high triplet yield, it is rather surprising that phosphorescence cannot be readily observed from C_{60}. Zeng et al. [13] exploited the external heavy atom effect to induce phosphorescence from a C_{60} solution in a methylcyclohexane/2-methyltetrahydrofuran/ethyl-iodide glass. Later, van den Heuvel et al. [14] demonstrated that phosphorescence from C_{60} molecules is masked by

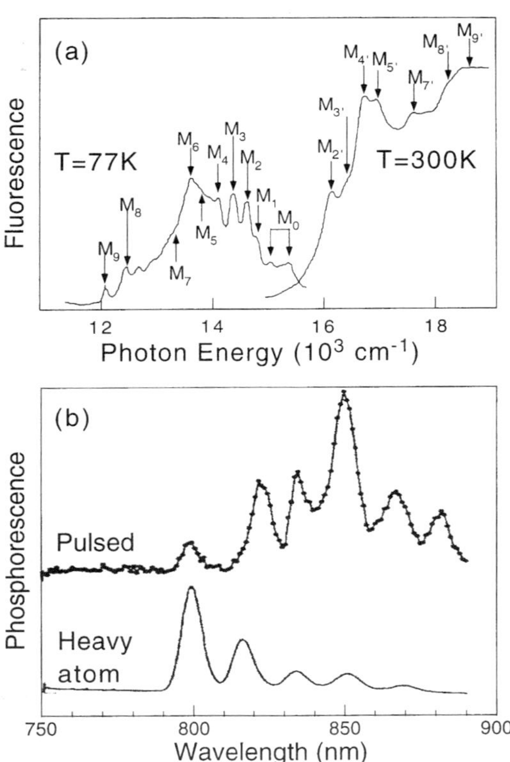

Figure 4   (a) The absorption and luminescence spectra of C_{60} in solution. (*From Ref. 11; used with permission.*) (b) The phosphorescence spectrum of C_{60}, using either pulsed excitation or the external heavy atom effect. (*From Ref. 14; used with permission.*)

the much more intense fluorescence. The phosphorescence spectrum of C_{60}, measured by both methods, is shown in Figure 4b [14]. Both phosphorescence spectra contain a number of sharp bands between 795 and 890 nm. The 0-0 transition is located at 798.1 ± 0.5 nm, giving a triplet-state energy $E_T = 1.553 ± .001$ eV (12,530 ± 8 cm^{-1}). This is in quite good agreement with photoacoustic calorimetry estimates ($E_T = 1.56 ± .03$ eV) [15]. The absorption spectrum of $^3C_{60}$ has been measured by steady-state photoinduced absorption (PA) [16]. The spectrum contains a strong band at 1.65 eV and a shoulder at 1.8 eV. Intensity-dependent measurements show that photoexcitation decay is monomolecular, as one would expect for isolated molecules [17].

The triplet state of photoexcited fullerenes was first detected by Wasielewski et al. [6] who reported the observation of an intense LESR signal in matrix-isolated C_{60} and C_{70} glasses. Figure 5a shows the LESR spectrum of C_{60} in toluene, measured at 5 K and photoexcited by a filtered xenon lamp [6]. The shape of the spectrum indicates resonant absorption below 3.3 kG ($g \approx 2$) followed by resonant emission above 3.3 kG and then changing to absorption. The outer

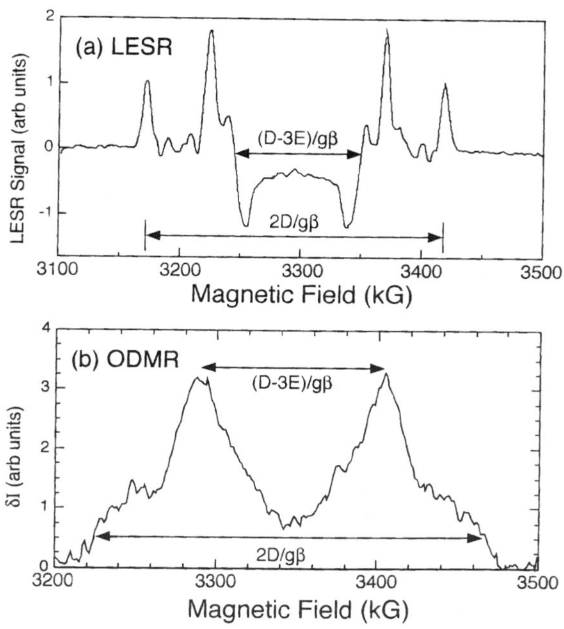

Figure 5 (a) The LESR spectrum of C_{60}:T/PS. (*From Ref. 6; used with permission.*) (b) The PLDMR spectrum of C_{60}:T/PS. (*From Ref. 18.*)

pair of maxima, at 3174 and 3418 G, are due to the steps of the full-field powder pattern whereas the inner zero crossings, at 3246 and 3350 G, are due to the singularities. Using Eqs. (5) and (7), the calculated ZFS parameters are $|D_{C_{60}}| = .0114$ cm^{-1} (122 G) and $|E_{C_{60}}| = .00056$ cm^{-1} (6.0 G). The X-Band ($v = 9.35$ GHz) PLDMR spectrum of C_{60} in toluene with 10% polystyrene (C_{60}:T/PS) is shown in Figure 5b [18]. The sample was excited at 488 nm by an argon ion laser and measured at 5K; the LESR and ODMR spectra are slightly offset due to the different microwave frequencies. The spectrum contains peaks at 3286 and 3403 G, which correspond to the singularities of the triplet powder pattern, and steps at 3221 and 3464 G. From Eqs. (5) and (7), we calculate the ZFS parameters $|D| = .0113$ cm^{-1} (121 G) and $|E| = .0002$ cm^{-1} (2 G). Similar ZFS parameters have been reported by several groups [19–21]; in all cases the observed spectrum has been attributed to the lowest-lying triplet state of the C_{60} molecule ($^3C_{60}$). The half-field powder pattern due to $\Delta m_s = 2$ transitions could not be detected in X-band LESR or PLDMR measurements. This is a consequence of a high-field spin polarization where $|n_0 - n_{\pm 1}| \gg |n_{+1} - n_{-1}|$.

The observation of a triplet state with nonzero D and E requires that $^3C_{60}$ not have icosohedral symmetry. The high symmetry of the C_{60} molecule makes the electronic excited states readily subject to Jahn-Teller (JT) effects due to vibronic coupling. Quantum chemical calculations [22–24] have suggested that the symmetry of the lowest excited state of C_{60} is of D_{5d} symmetry, with six degenerate distortions along the directions of the six pairs of opposite pentagonal faces on the C_{60} molecule. Quantum fluctuations among the six degenerate JT distortions have been postulated to lead to dynamic JT effects [24], such as pseudo-rotation associated with the Berry phase [25] and modifications of the Raman and infrared multiphonon spectra [26]. Magnetic resonance spectroscopy is well suited to studying these effects since it is quite sensitive to the local environment experienced by the triplet states. Several studies have reported evidence for two close-lying triplet states, which can be decomposed by careful analysis of the LESR or ODMR spectra.

The S-Band ($v = 2.998$ GHz) PADMR spectrum, measured at a probe energy of 1.65 eV, is shown in Figure 6. At 2 K, the PADMR spectrum shows a 250 G wide triplet powder pattern centered about $g \approx 2$ (1071 G) with $\delta n/n = -.006$. As with the PLDMR spectrum, the PADMR spectrum contains four singularities: two steps separated by 225 G and two peaks separated by $\Delta H = 122$ G. Even at 2K, the peaks in the magnetic resonance spectrum are quite broad; the full width at half maximum $2\Gamma \cong 28$ G. As the temperature increases, the width of the triplet powder pattern collapses and an inner pair of shoulders split by ~ 50 G develops. These shoulders are due to a second triplet state, with re-duced ZFS parameters compared to the triplet seen at 2K. For ease of notation, we refer to the two triplets as T_A and T_B. With increasing temperature, T_B gains

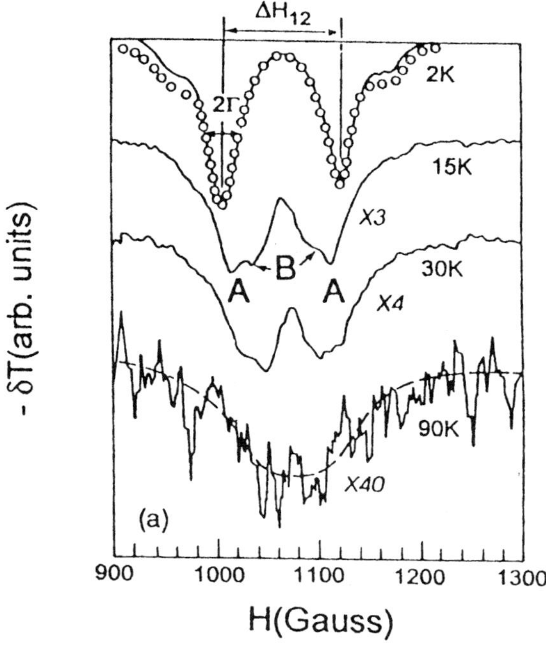

Figure 6 The PADMR spectra of C_{60}:PS, measured at various temperatures. *A* and *B* stand for triplets T_A and T_B, respectively. A fit to the powder pattern of T_A at 2 K is shown (open circles). (*From Ref. 21.*)

strength relative to T_A. There is a similar narrowing in the X-Band ($v = 9.35$ GHz) PLDMR spectrum of C_{60} (Figure 7), measured between 5 and 50 K [27].

The existence of multiple triplet states in C_{60}:T/PS glasses is proven by zero-field PADMR measurements [20]. Figure 8 shows the zero-field PADMR spectra of C_{60}:T/PS observed at 6, 20, and 40 K with a probe energy of 1.61 eV [20]. Two strong overlapping bands are visible at 330 and 357 MHz. They appear to be asymmetric with FWHM of 63 MHz. This is equivalent to $2\Gamma = 22.5$ G, in qualitative agreement with the high-field PADMR spectrum shown in Figure 6. The low temperature spectra also show a broad band centered at 100 MHz with a FWHM of 94 MHz. At 40K the two sharp lines at 330 and 357 MHz disappear leaving a broad unstructured band centered around 300 MHz. No PADMR signals were observed between 400 MHz and 1 GHz. The λ-PADMR spectrum of C_{60}, measured with the microwave frequency set to 354 MHZ, is shown in Figure 9 [20]. The peak of this spectrum is red-shifted by about 20 nm

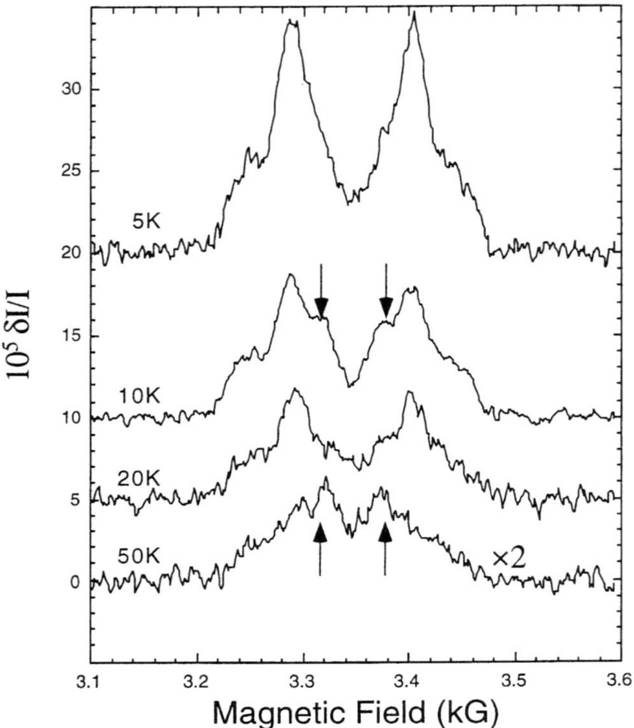

Figure 7 The PLDMR spectra of C_{60}:T/PS, measured at various temperatures. (*From Ref. 27.*)

with respect to the PA spectrum shown in Figure 4c, which may be due to some site selectivity of the triplets detected by PADMR. The λ-PADMR of the 103 MHz band is identical to the spectrum shown in Figure 9.

Double resonance measurements were undertaken to correlate the observed bands. In this measurement, the fixed (hole burning) frequency is set to one of the peaks and a second frequency is scanned. The lineshape of the holes is in principle a complex function of the ZFS, but may be approximated by a Lorentzian line [28]. Figure 10 shows the hole burning spectra with the fixed rf frequency set at 103, 330, and 354 MHz. The spectra at 330 and 354 MHz clearly belong to the same state. In both cases, there is a $|2E|$ signal at 28.6 MHz. The three signals observed are the $|D| + |E|$, $|D| - |E|$, and $|2E|$ resonances of triplet

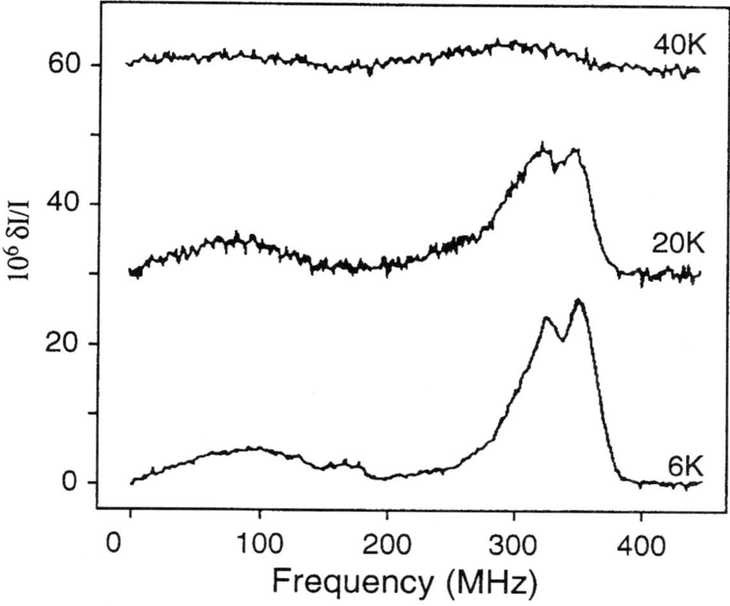

Figure 8 The zero-field PADMR spectrum of C_{60}:T/PS. (*From Ref. 20; used with permission.*)

A, which yield ZFS parameters $|D_A|$ = .01146 cm^{-1} and $|E_A|$ = .00048 cm^{-1}. The double resonance spectrum of the 100 MHz band shows weak signals which appear as broad bands at 255 and 340 MHz. If the low frequency band is taken as the $|2E|$ transition of T_B and the high bands as $|D| + |E|$ and $|D| - |E|$, respectively the ZFS parameters of T_B are $|D_B| \approx .0099$ cm^{-1} and $|E_B| \approx .0017 \pm .0010$ cm^{-1}. The relatively low signal-to-noise ratio would make exact determination of $|D_B|$ and $|E_B|$ difficult. However, given that the $|2E|$ band is extremely broad, there is likely to be a distribution of ZFS parameters associated with this triplet. The above ZFS parameters predict that the singularities of T_B occur at 1045 and 1097 G, in quite good agreement with the PADMR spectra shown in Figure 6.

The variation of the various ODMR spectra with temperature suggest that a dynamic process is involved, which may be thermally activated. The apparent correlation between ΔH and Γ with increasing temperature can be directly plotted

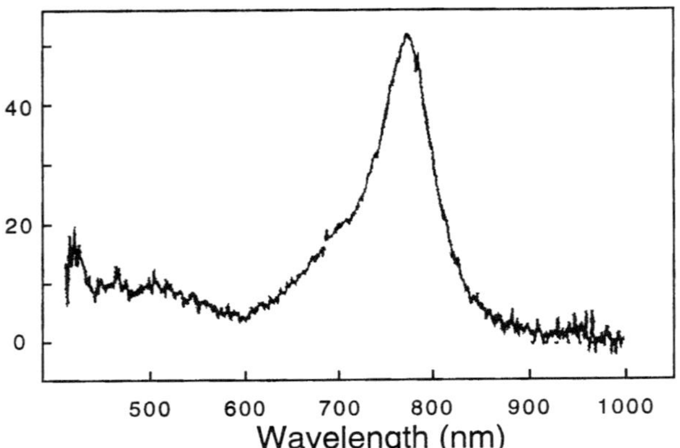

Figure 9   The zero-field λ-PADMR spectrum of C_{60}:T/PS. (*From Ref. 20; used with permission.*)

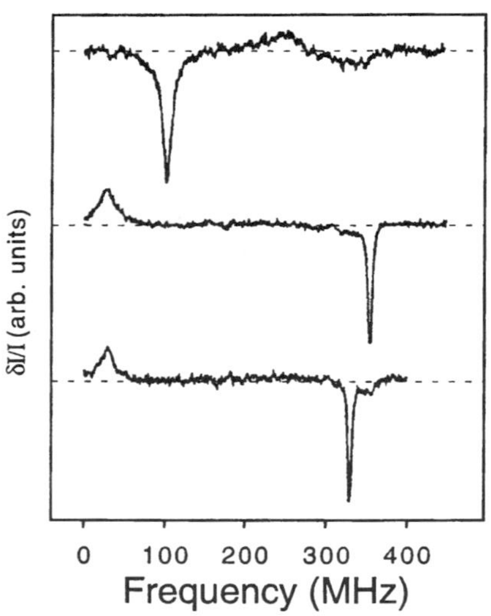

Figure 10   The zero-field double resonance (hole burning) spectra of C_{60}:T/PS. (*From Ref. 28; used with permission.*)

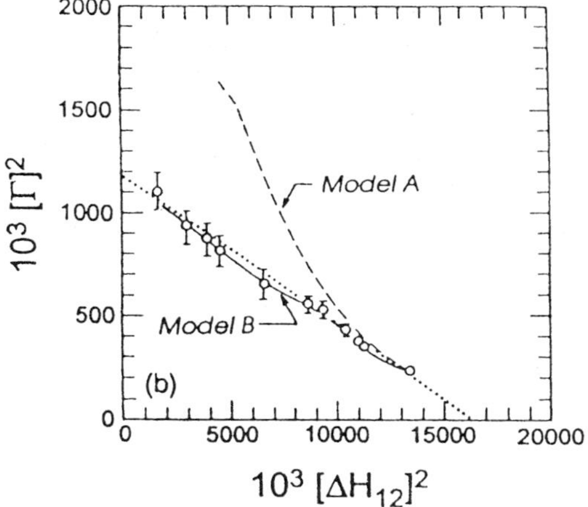

Figure 11 The correlation between Γ and ΔH_{12} in the powder patterns of T_A at various temperatures. The dotted line is a straight line fit; the broken line is calculated using the rotational diffusion model [30]. The full line was calculated using the DJTE model [21].

to show that a spin dynamic process is dominant at all temperatures (Figure 11). The data can be fit to the relationship:

$$13.7\Gamma^2 = (125)^2 - (\Delta H)^2 \tag{16}$$

which is shown as the dotted line in Figure 11. In the Anderson spin-exchange narrowing model [29], the system can resonate at field H_1 or H_2, where $\Delta H_0 = H_2 - H_1$, and the relationship between Γ and ΔH is given by

$$8(\Gamma - \Gamma_0)^2 = (\Delta H_0)^2 - (\Delta H)^2 \tag{17}$$

where Γ_0 is the intrinsic linewidth. Γ is larger than Γ_0 due to exchange narrowing and the difference is proportional to the exchange frequency ν, which usually increases with T. While the forms of Eqs. (16) and (17) are similar, the 13.7 coefficient in Eq. (16) indicates that the powder pattern collapses towards H_0 faster than predicted by the Anderson model, showing that more complicated spin dynamics exist in $^3C_{60}$.

To understand the spin dynamics in $^3C_{60}$, we have considered the possibility that the triplet spin undergoes a continuous rotational diffusion on the fullerene surface [30, 31]. The prediction of this model is shown as a broken line in Figure 11, but does not fit the experimental data. A large hopping angle ($>30°$) between the triplet axes involved in the reorientation dynamics was needed to correctly describe the relationship between Γ and ΔH. Theory predicts that there are six equivalent $^3C_{60}$ Jahn-Teller deformations along the directions of the six pairs of pentagonal faces on the fullerene surface [23, 24]. It was assumed that the principal axes of the triplet ZFS tensors are diagonal along each of the uniaxial Jahn-Teller deformations, but can change direction with respect to $\vec{H}$. This hopping is due to tunneling among the equivalent orientations, giving rise to a dynamic Jahn-Teller effect among the six degenerate Jahn-Teller minima. The density-matrix formalism [32] was used to calculate the triplet powder pattern of T_A with various hopping rates v. In this method:

$$S(H) = \sum_{k=1}^{6} \sum_{l=1}^{3} \int \text{Im}[\rho_k^l (\Omega, H)] \, d\Omega \tag{18}$$

where ρ_k^l are the ρ's of the six degenerate Jahn-Teller distortions, with three microwave transitions each, and Ω is the angle between the triplet z axis and $\vec{H}$. ρ_k^l are subject to the coupled equations:

$$\frac{d\rho_k^l}{dt} (\Omega, H) = \rho_k^l (\Omega, H) \left\{ \frac{ig\beta[H - H_k^l (\Omega)]}{h} - \frac{1}{T_2} - 5v \right\}$$
$$+ v \sum_{m \neq k} \rho_k^l (\Omega, H) + is_k^l (\Omega, H) \tag{19}$$

where T_2 is the spin-spin relaxation time, $H_k^l(\Omega)$ is the static resonant field value, and $s_k^l(\Omega)$ is the steady-state transition rate. After solving Eq. (19) in steady state ($d\rho/dt = 0$), the triplet powder pattern was calculated using Eq. (18).

A simulation of the PADMR spectrum of T_A at 2 K is shown by the open circles in Figure 6. The fit was obtained with ZFS parameters $|D_A| = .0115$ cm^{-1} and $|E_A| = 0$ cm^{-1}, hopping rate $v = 2$ MHz, and $1/T_2 = 6$ G. By fitting the PADMR spectra of T_A at various temperatures, it was found that v depends linearly on T (Figure 12a). This result is in agreement with a one-phonon process. The solid line in Figure 11 shows the relationship between Γ^2 and ΔH^2 that was obtained by fitting the triplet powder pattern of T_A at each temperature using the dynamic Jahn-Teller model. In contrast to the continuous rotational diffusion model (dashed line), the dynamic Jahn-Teller model provides an excellent fit to the data.

The PADMR spectrum of $^3C_{60}$ and its decomposition into triplets A and B are shown in Figure 13. The triplet powder pattern of T_B was originally simulated with ZFS parameters $|D_B| = .0060$ cm^{-1} and $|E_B| = 0$ cm^{-1} [16]. However, a similar spectrum can be generated using the ZFS parameters mentioned above [20]. We note a pulsed LESR study by Bennati et al. [19] also reported two triplet states with ZFS $|D_A| = .0117$ cm^{-1}, $|E_A| = .0014$ cm^{-1}, $|D_B| = .0047$ cm^{-1}, and $|E_B| = 0$ cm^{-1}. Given that the magnetic resonance spectrum of $^3C_{60}$ contains contributions from two triplet species with overlapping spectra, determining a unique

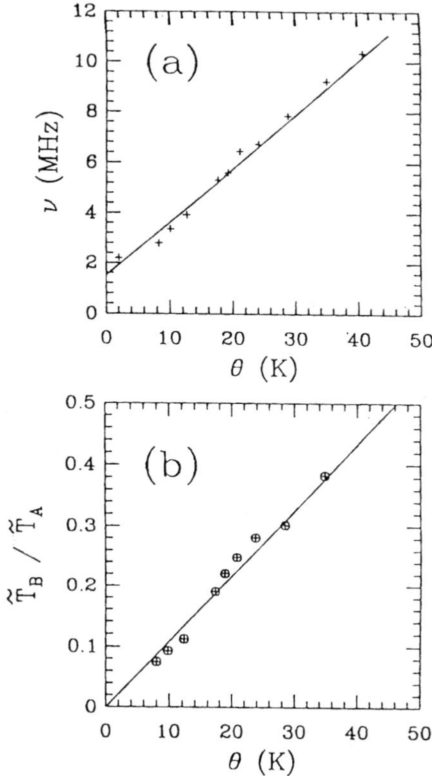

Figure 12   (a) The hopping frequency obtained from the fits of the DJTE model to the powder pattern of T_A, as a function of temperature. (b) The population ratios of T_B to T_A as a function of temperature. (*From Ref. 21.*)

set of ZFS parameters is a nontrivial task. One way to determine the relative weights of the two triplet species is to subtract the model spectrum of T_A from the PADMR spectrum. The relative populations of the two triplets can then be obtained by dividing their integrated PADMR spectra:

$$\frac{n_B}{n_A} \propto \int \delta T_B \, dH \div \int \delta T_A \, dH \tag{20}$$

As shown in Figure 12b, n_B/n_A increases linearly with temperature. Comparison of the 90 K spectrum in Figure 6 to the 50K spectrum in Figure 7 shows that triplet B is also subject to a dynamic Jahn-Teller effect. The two singularities of

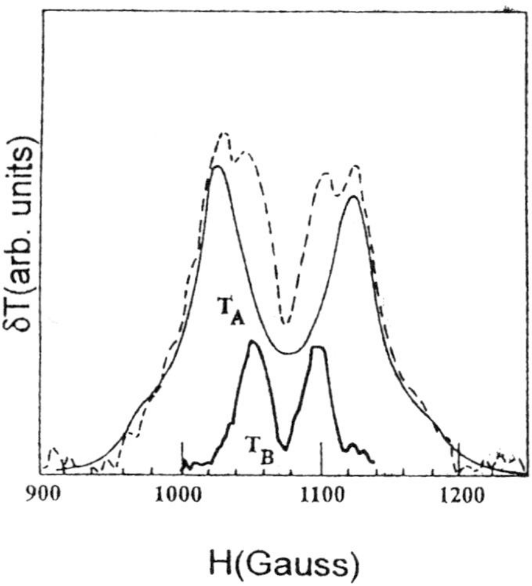

Figure 13 H-PADMR spectrum of C_{60}:PS at 17 K (dashed line) and its decomposition into two triplet powder patterns, T_A and T_B. (*From Ref. 21.*)

the powder pattern of triplet B also collapse toward H_0, though at a higher temperature than for triplet A.

Determining the origin of triplet B is a difficult task. This triplet cannot originate from impurities, as its λ-PADMR spectrum is identical to that of triplet A [16, 33]. Angerhofer et al. [20] suggested that this triplet may be due to residual aggregates of undissolved C_{60} in solution. To check this assumption, the H-PADMR spectra of a variety of C_{60}:PS glasses with varying concentrations were measured. In all cases, two triplet powder patterns were found. The temperature dependence of the relative T_B population was similar, though not exactly the same as shown in Figure 12b. Surprisingly, even for different measurements of the same sample under similar conditions, the relative contributions of triplets A and B may vary. This may indicate strain in the glassy matrix, which would depend upon the sample history. The various ZFS discussed above are summarized in Table 1.

2.1 C_{70}

The excited-state dynamics of C_{70} differ from C_{60}: its triplet state is significantly longer-lived and the PL of C_{70} is primarily due to phosphorescence. Figure 14 shows the fluorescence and phosphorescence spectra of C_{70}:PS, measured at 77 K [34]. The integrated phosphorescence was about 12 times as intense as the

Table 1 Zero-Field Splitting Parameters of $^3C_{60}$ Molecules

Reference	Technique	$\|D\|(10^{-4}\ cm^{-1})$	$\|E\|(10^{-4}\ cm^{-1})$
Wasielewski et al. [6]	LESR	114	5.6
Bennati et al. [19]	Pulsed LESR	114	1.4
		47	0
Lane et al. [18]	PLDMR	113	2
Wei et al. [21]	PADMR	115	0
		60	0
Angerhofer et al. [20]	Zero-field PADMR	114.6	4.8
		99	17

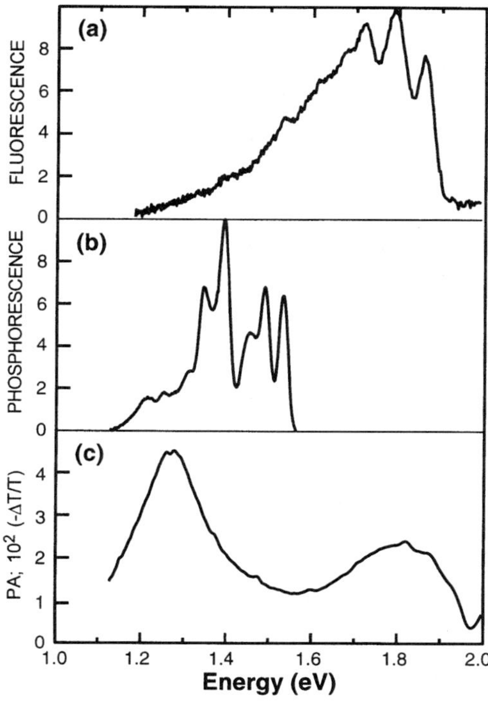

Figure 14 The (a) fluorescence, (b) phosphorescence, and (c) photoinduced absorption spectrum of C_{70}:PS. (*From Ref. 34.*)

fluorescence. Peaks in the fluorescence spectrum are observed at 1.86, 1.79, and 1.72 eV. While this could be due to a 70 meV vibronic progression, it is more likely that the spacing of the peaks is coincidental. Fluorescence from C_{60} has been shown to be defect-related [35] and it is reasonable to make a similar assumption for C_{70}. We therefore assign the peaks to several independent false origins. The phosphorescence spectrum is structured, with strong peaks visible at 1.533, 1.490, 1.455, 1.394, and 1.346 eV, respectively, and weaker peaks at 1.31, 1.25, and 1.21 eV. We take the highest-energy transitions as the 0–0 transitions of the respective states and calculate a singlet-triplet gap of 330 meV. Reports of the lifetime of $^3C_{70}$ have varied considerbly, ranging from 41.7 μs to 51 ms. Fraelich and Weisman [36] measured the effects of oxygen-induced quench-

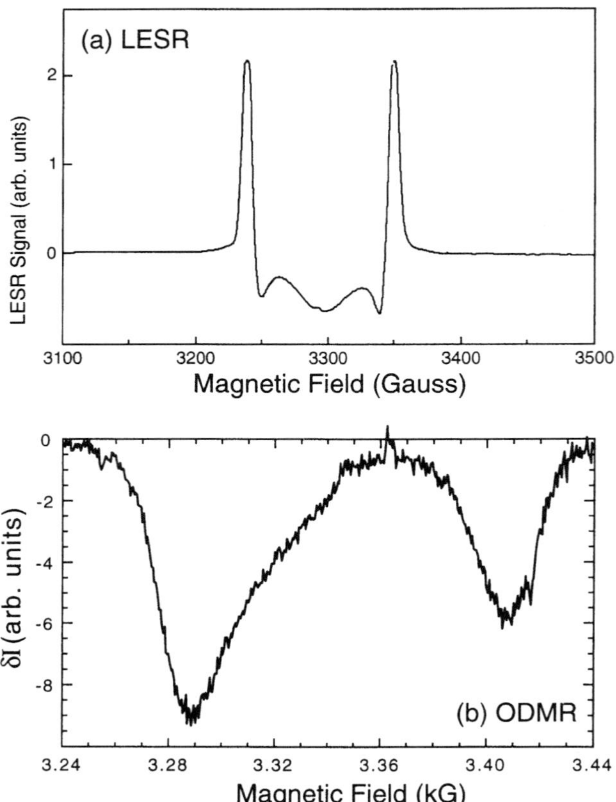

Figure 15 The (a) LESR and (b) PLDMR spectra of C_{70}:T/PS. [*Part (a) from Ref. 6; used with permission. Part (b) from Ref. 34.*]

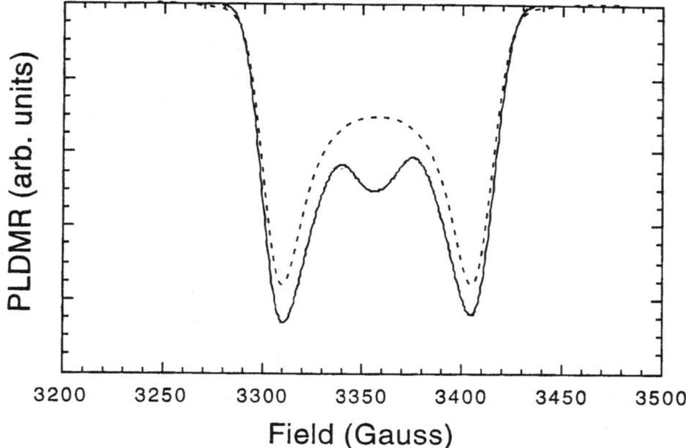

Figure 16 Calculated PLDMR spectra of C_{70} using two different sets of ZFS parameters. Solid line: $D = .0056$ cm^{-1}, $E = .00063$ cm^{-1}; dashed line: $D = .0089$ cm^{-1}, $E = 0$ cm^{-1}.

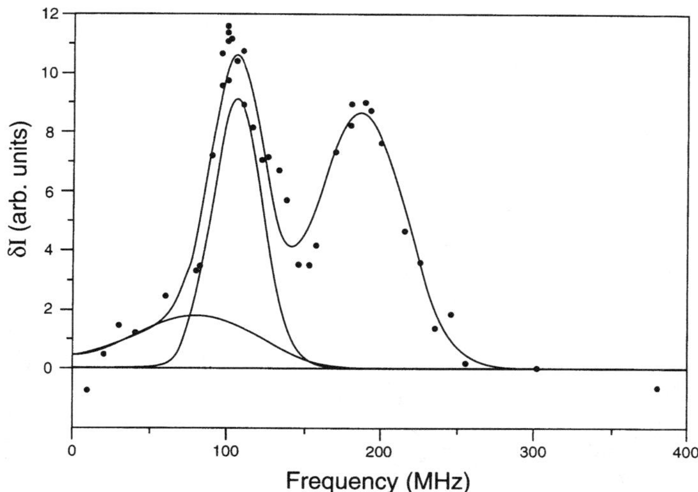

Figure 17 The zero field PLDMR spectrum of C_{70}:T/PS. (*From Ref. 38; used with permission.*)

ing, self-quenching, and triplet-triplet annhilation on the triplet lifetime. When these effects are excluded, they report a lifetime $\tau_T = 12$ ms at 295 K [37]; Wasielewski et al. [6] reported $\tau_T = 51$ ms at 9 K. The PA spectrum of C_{70}:PS, shown in Figure 14c, contains two intense bands: T_1 at 1.3 eV and T_2 at 1.8 eV. The intensity of both bands increases linearly with the excitation power. Hence, the recombination kinetics are monomolecular and the fullerene molecules are isolated.

Figure 15 shows the (a) LESR [6] and (b) PLDMR spectra of C_{70}, measured at 5 K utilizing X-band microwaves. Both spectra are significantly narrower than those of C_{60}. The LESR spectrum has an outer pair of lines separated by 112 G, but no other sharp features. The PLDMR spectrum contains two peaks, separated by 95 G, but not the steps associated with $^3C_{60}$. The absent features in the respective magnetic resonance spectra make determination of ZFS parameters ambiguous. Wasielewski et al. took the two outer features as the steps of the triplet powder pattern and accordingly calculated $|D| = .0056$ cm^{-1}. If, however, we treat the peaks of the PLDMR spectrum (and the corresponding zero level crossings of the LESR spectrum) as the singularities of the triplet powder pattern, then we calculate $|D| = .0089$ cm^{-1}. Most groups studying C_{70} by LESR have taken the former approach, though acceptable simulation spectra can be generated using either set of parameters. Simulated PLDMR spectra with $|D| = .0056$ cm^{-1} or $|D| = .0089$ cm^{-1} are shown in Figure 16.

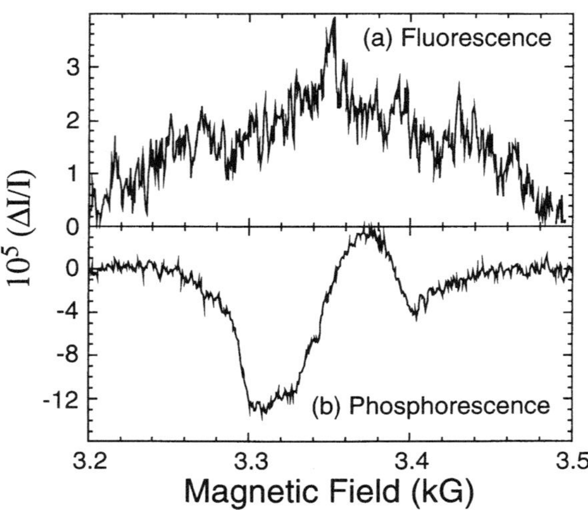

Figure 18   PLDMR spectra of C_{70}:T/PS, measured within the (a) fluorescence and (b) phosphorescence bands. (*From Refs. 27, 34.*)

As with C_{60}, we turn to zero-field ODMR to resolve the ambiguity. The result of pulsed zero-field ODMR experiments at 1.2 K are shown in Figure 17 [38]. The spectrum exhibits lines at 106 and 187 MHz. Taking these as the $|D| + |E|$ and $|D| - |E|$ transitions, this spectrum is due to a triplet exciton with ZFS parameters $|D| = .0049$ cm^{-1} and $|E| = .0014$ cm^{-1}. These ZFS predict a $|2E|$ peak at 82 MHz, which is not observed. Given that this peak was quite weak in comparable measurements of C_{60}, the absence of this peak is not surprising. Rather, these results indicate that the spin-polarization in fullerenes is strongest between the z sublevel ($|n_z - n_{x,y}| \gg |n_x - n_y|$). A hole burning measurement such as that performed by Angerhofer et al. [20] should detect this transition.

As C_{70} exhibits both fluorescence and phosphorescence, comparison of the PLDMR spectra of the two bands is interesting. Figure 18 shows the PLDMR spectrum of C_{70}:T/PS, measured at 20 K within the (a) fluorescence bands and (b) phosphorescence. The fluorescence ODMR is PL-enhancing and symmetric about $g \approx 2$, with shoulders separated by 220 G and a relatively narrow peak at $g \approx 2$. This triplet, which we label triplet B, has ZFS parameters $|D_B| = .0103$ cm^{-1} and $|E_B| = .0034$ cm^{-1}. The signal from triplet B is a factor of 20 times weaker than that of triplet A; the fact that it is not present in the phosphorescence

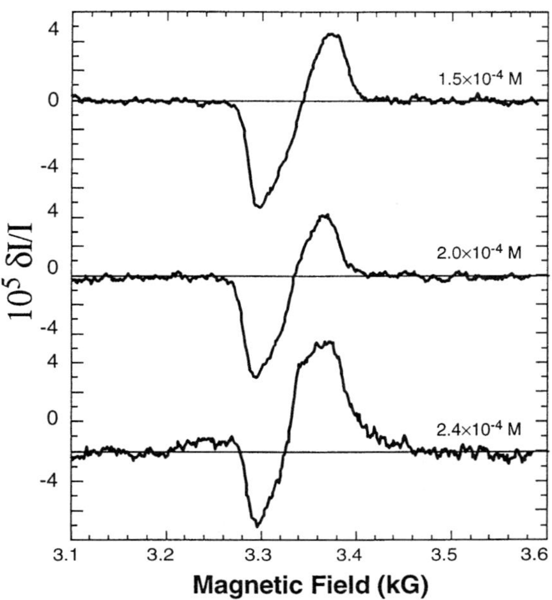

Figure 19 PLDMR spectrum of C_{70}:T/PS, measured at various concentrations.

ODMR indicates that emission from this triplet is forbidden. The complete departure from axial symmetry coupled with the absence of phosphorescence suggests that triplet B resides on distorted molecules, possibly due to strain in the T/PS matrix or interactions with nearby C_{70} molecules. Angerhofer et al. [20] suggested that residual clusters may play a significant role in the excited-state properties of fullerenes in solution. Figure 19 displays the PLDMR at 20 K of C_{70}:T/PS samples with increasing concentration. The phosphorescence yield decreases with increasing C_{70} concentration. The PLDMR of the most concentrated sample clearly shows the effects of coupling between C_{70} molecules.

The phosphorescence ODMR of C_{70} is quite different from the fluorescence ODMR of either C_{60} or C_{70}. Its lineshape is asymmetric and is dependent on both the temperature and microwave modulation frequency ν_C. Figures 20 and 21 show the dependence of the triplet PLDMR of C_{70}:T/PS on these parameters. Depending upon the temperature and ν_C, one of two asymmetric powder patterns is seen. The first type consists of two PL-quenching peaks split by $\sim$110 G; the low-field peak is much more intense than the high-field peak. The second type consists of a low-field quenching peak and a high-field enhancing peak. The enhancing peak occurs at a lower field (3380 G) than the high-field quenching

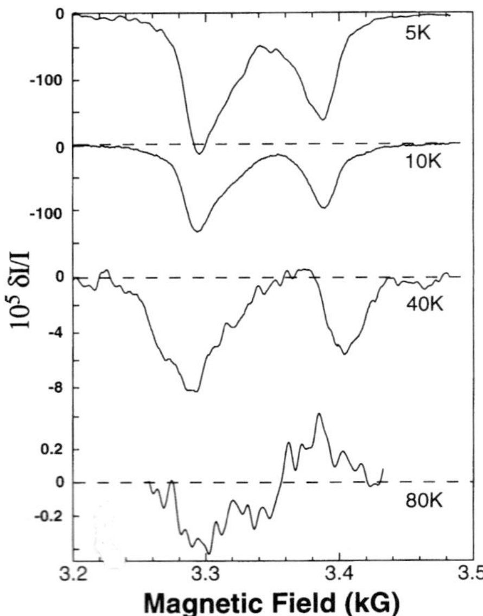

Figure 20   Dependence of the PLDMR spectrum of C_{70}:T/PS on temperature. (*From Ref. 34.*)

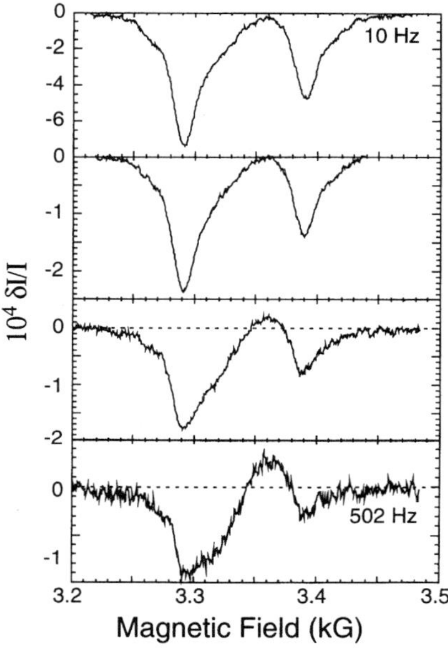

Figure 21 Dependence of the PLDMR spectrum of C_{70}:T/PS on the microwave modulation frequency. (*From Ref. 27.*)

peak (3410 G). Careful examination of the low-field portion of the spectrum ($H <$ 3340 G) reveals a shoulder on the low-field side of the spectrum, at $\sim$3310 G. If we assume the outer features are due to the steps of the powder pattern and the inner features are related to the singularities, we calculate $|D| = .0054$ cm^{-1} and $|E| = .0004$ cm^{-1}. The misleading nature of the phosphorescence PLDMR spectrum is due to the fact that it also depends upon the radiative yield of each sublevel. Consequently, the singularities are masked. The variations of the triplet lineshape as ν_C is varied are due to the interplay between the triplet sublevel

Table 2 Zero-Field Splitting Parameters of $^3C_{70}$ Molecules

Reference	Technique	$|D|(10^{-4}$ cm$^{-1})$	$|E|(10^{-4}$ cm$^{-1})$
Wasielewski et al. [6]	LESR	56	6.3
Lane et al. [34]	PLDMR	89	0
		103	34
Saal et al. [38]	Zero-field PADMR	49	14

lifetimes and varying spin-lattice relaxation rates. A detailed treatment is given elsewhere [34]. The ZFS of triplet excited states of C_{70} are summarized in Table 2.

3. STUDIES OF FULLERENE SINGLE CRYSTALS AND THIN FILMS

Crystalline C_{60} is a van der Waals solid with a lattice constant of 14.17 Å and a nearest-neighbor distance of 10.02 Å [39]. Below 261 K, the lattice structure of C_{60} is simple cubic and above 261 K, the lattice structure is face centered cubic [40, 41]. Magnetic resonance studies have been performed on C_{60} single crystals and thin films of C_{60} or C_{70} produced by vacuum sublimation or evaporated from degassed solutions. We start our survey with ESR and ODMR measurements of C_{60} single crystals and review studies of fullerene thin films.

3.1 Single Crystals of C_{60}

We first review electron spin echo detected ESR (ESE-ESR) measurements of a C_{60} single crystal mounted in a 95 GHz spectrometer at 1.2 K [42]. The details of single-crystal growth have been described elsewhere [43]. The use of a high-frequency spectrometer permitted measurements on a single crystal of submillimeter dimensions. The sample was irradiated at 532 nm by the second harmonic of a Nd:YAG laser; two-pulse echo experiments were performed using microwave pulses of 50 and 100 ns in duration. Whereas conventional ESR uses field modulation and produces a derivative signal, ESE-ESR generates signals proportional to microwave absorption or emission. The spectrum reveals the existence of two distinct triplet species with similar, but not identical, ZFS parameters and coinciding principal axes. Each triplet species occurs in six equivalent sites which can be distinguished by rotating the crystal with respect to the magnetic field. We adopt the notation of Groenen et al. [42]: the triplets are labeled α and β and the sites 1 and 2.

Figure 22 shows the ESE-ESR spectrum with $\vec{H}$ along the [001] or [110] axes of the single crystal [42]. All z axes lie along fourfold cube axes; if z is parallel to the [001] direction, then x and y lie parallel to the [110] and [$\bar{1}$10] axes, or vice versa. For example, the direction of the x principal axis of triplet α in site 1 would be referred to as x_1^α. For the spectrum shown in Figure 22a, the magnetic field is simultaneously parallel to x_1^α, y_2^α, x_1^β, and y_2^β. For the spectrum in Figure 22b, the magnetic field is simultaneously parallel to z_1^α, z_1^β, z_2^α, and z_2^β. The $\Delta m_s = 1$ signals in the ESE spectrum are labeled according to the axis which is parallel to $\vec{H}$. Upon rotating $\vec{H}$ away from [110], the corresponding ESE signal splits into two peaks of equal intensity. The ZFS parameters are $|D_\alpha| = .0070$ cm^{-1}, $|E_\alpha| = .0021$ cm^{-1}, $|D_\beta| = .0086$ cm^{-1}, $|E_\beta| = .0016$ cm^{-1}. The angle

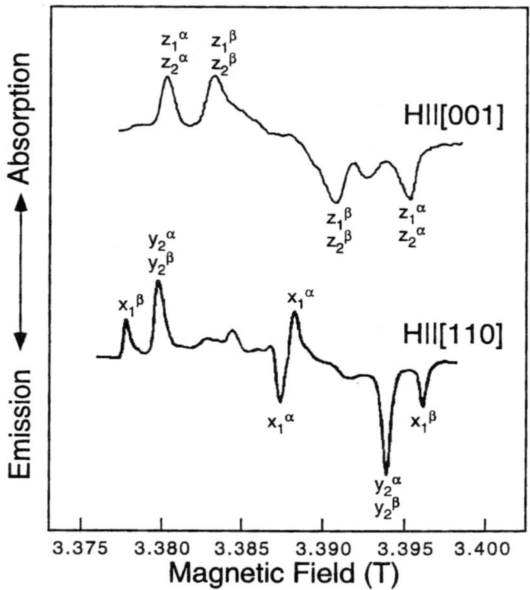

Figure 22 Transient ESR spectrum of a C_{60} single crystal. (*From Ref. 41; used with permission.*)

between the x axes of triplet α in the two different sites is $31° \pm 2°$ and the corresponding angle for triplet β is $65° \pm 2°$. The ZFS parameters of α and β are smaller than those for the isolated fullerene molecules. Based upon this observation and the fact that the triplet principal axes are oriented along crystal axes of the cubic lattice led Groenen et al. [42] to conclude that the triplet excitons are delocalized across dimers or chains of C_{60} molecules.

Zero-field PLDMR measurements performed on C_{60} single crystals [44] detected some of the expected transitions of α and β. The PLDMR spectrum of C_{60} crystals in zero field at 1.2 K upon laser irradiation at 514 nm is shown in Figure 23. Scanning the microwave frequency reveals four transitions at 122, 202, 270 and 301 Mhz, each about 8 MHz (2.85 G) wide. The ratio of the intensities of the 202 and 270 MHz transitions varies from sample to sample and even depends upon the spot irradiated. On the other hand, the ratio of the intensities of the 202 and 301 MHz transitions remains fixed. Hence, these two transitions are corre- lated and are due to the same triplet exciton. The calculated ZFS parameters are $|D| = .00839$ cm^{-1} and $|E| = .00165$ cm^{-1}, identifying this triplet as β. These ZFS predict a third transition at ~ 100 MHz; this low-frequency transition was

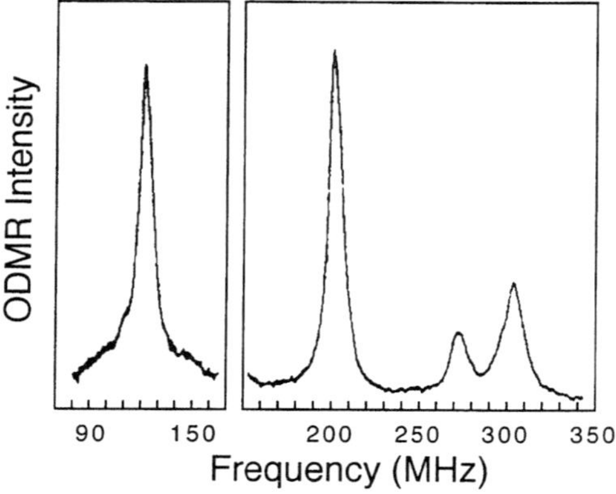

Figure 23   Zero field PLDMR spectrum of a C_{60} single crystal. (*From Ref. 43; used with permission.*)

relatively weak in isolated C_{60} molecules {Angerhofer} and is not clearly resolved. However, the 122 MHz transition is notably asymmetric, with slightly more weight on the low-field side. This asymmetry could be due to a weak peak originating from |2E| transition of triplet β.

The intensity of the 122 MHz transition cannot be directly compared to the other transitions as it was measured with a different coil in the microwave circuit. If we consider the ZFS parameters of triplet α detected in high-field ESE measurements, we expect to find transitions at 126, 147, and 273 MHz. We therefore assign the transitions at 122 and 270 MHz to triplet α and calculate ZFS $|D_\alpha| = .0070$ cm^{-1} and $|E_\alpha| = .0020$ cm^{-1}. Triplet α should have a third transition at 148 MHz (|D − E|/h), which is not seen. This resonance is due to transitions between the Z sublevel and the X or Y sublevel, depending upon the relative signs of D and E. Its absence indicates that two of the triplet sublevels, one of which must be the Z sublevel, have similar steady-state populations ($n_z \cong n_x$ or $n_z \cong n_y$). No transitions were observed at 321 or 363 MHz, where zero-field transitions of $^3C_{60}$ (identified above) should occur. It was also observed that as the excitation wavelength is increased from 514 to 685 nm, triplet β gains intensity relative to triplet α. The extinction coefficient at 685 is 20 times smaller than that at 514 nm, resulting in much greater penetration depth at 685 nm. This implies that triplet α occurs at the surface of the crystal whereas triplet β is found in the bulk. Comparison of C_{60} powder to single crystals would verify this hypothesis.

3.2 Fullerene Films Evaporated from Solution

C_{60}

Figure 24 shows the full-field X-band PLDMR spectrum of a C_{60} film prepared by evaporation from a toluene solution and sealed in vacuum. The film was photoexcited with 50 mW at 488 nm and the spectrum was measured at 20 K with 800 mW of microwave power. The PLDMR spectrum exhibits three distinct features: There is a narrow peak with FWHM 5 G and centered at 3339.8 G ($g = 2.0017 \pm .0005$), which coincides with a similar feature in the LESR of such films [18]. We assign this resonance to spin-$\frac{1}{2}$ excitations and comment on its nature below. The strongest feature in the full-field PLDMR spectrum is a triplet powder pattern which displays shoulders separated by 37, 90, and 235 G. The third feature is a 640 G wide triplet powder pattern, which was faintly observed in the LESR spectrum of this film, but not in measurements of isolated fullerenes [18]. This feature is shown on a magnified scale on the lower half of Figure 24. The half-field powder pattern of the C_{60} film, shown in Figure 25, clearly shows contributions from two distinct types of triplet excitons.

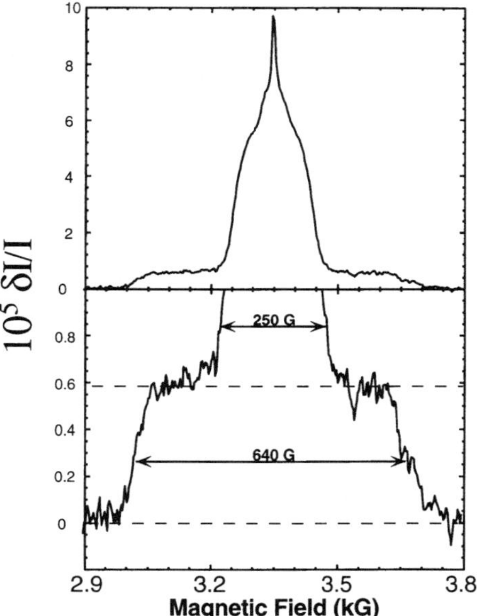

Figure 24 Full-field PLDMR spectrum of a C_{60} film evaporated from solution. The lower portion shows the broad triplet powder pattern on a magnified scale. (*From Ref. 27.*)

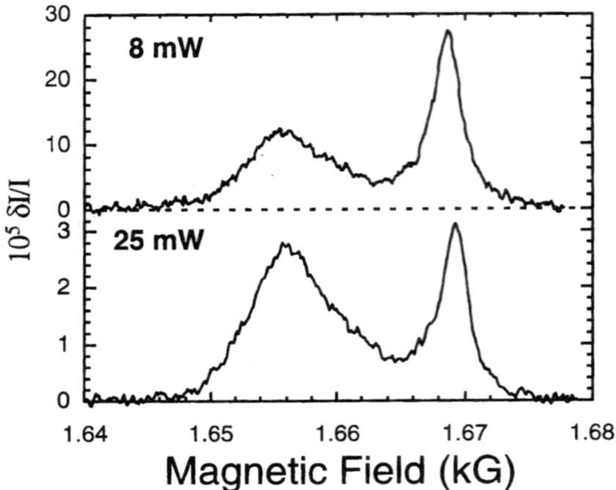

Figure 25 Half-field PLDMR spectrum of C_{60} film, excited with 8 and 25 mW of laser power at 488 nm. (*From Ref. 27.*)

Figure 26 shows the narrow resonance on a reduced field scan with 50, 200, and 800 mW of microwave power. The linewidth of the resonance is not appreciably affected and its magnitude follows the expected square-root dependence on microwave power [7]. The peak is clearly due to spin-½ excitations and was originally assigned to enhanced recombination of photogenerated polar-

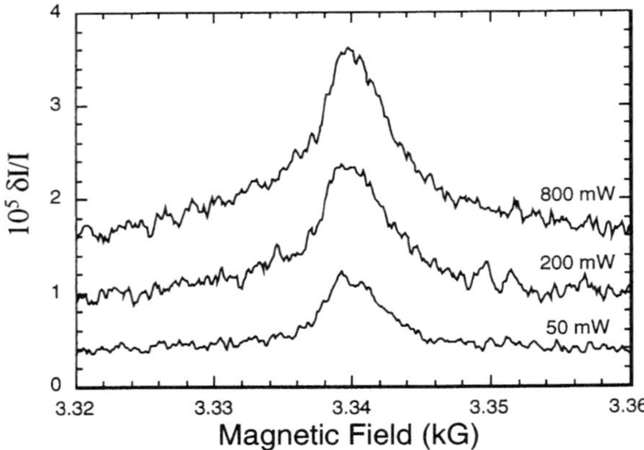

Figure 26 Spin-1/2 resonance of C_{60} film, measured with 50, 200, and 800 mW of microwave power at 9.35 Ghz. (*From Ref. 27.*)

ons [18, 27]. ESR measurements have shown that C_{60} powder generally contains traces of a spin-$\frac{1}{2}$ magnetic impurity, with $g \approx 2.0023$ and width 2.5 G [45]. The similarity of these two features led Steren et al. [46] to suggest that they are related. In order to confirm this hypothesis, a C_{60} film was deposited onto a silica gel surface by wetting the gel with a solution of C_{60} and drying in a vacuum oven at 120°C. The procedure was repeated at least ten times to produce C_{60} aggregates dispersed over the substrate surface. Conventional ESR measurements of this film show that a paramagnetic impurity is present (Figure 27a). This impurity is present in the pristine C_{60} powder, though its nature is not clear. The ESR signal is due to resonant absorption, not emission, of microwaves. This is consistent with a Boltzmann energy distribution where the population of the upper level $n_{+1/2} = n_{-1/2}e^{-g\beta H/k_B T}$. The ESE-ESR spectrum, shown in Figure 27b, has a transient signal which is always in emission and decays within a few microseconds after the laser pulse. The ESE-ESR spectrum is proportional to microwave absorption, whereas the conventional ESR spectrum is proportional to the derivative of microwave absorption. Taking this into account, the two signals have the same lineshape.

The time evolution of the transient ESR signal following pulsed laser excitation of C_{60} at 300 K is shown in Figure 28 [46]. After photoexcitation the signal first turns into emission and then relaxes toward the dark signal intensity. This observation establishes that the transient ESR signal is due to a photoinduced change in the spin polarization of paramagnetic centers. Steren et al. [46] attrib-

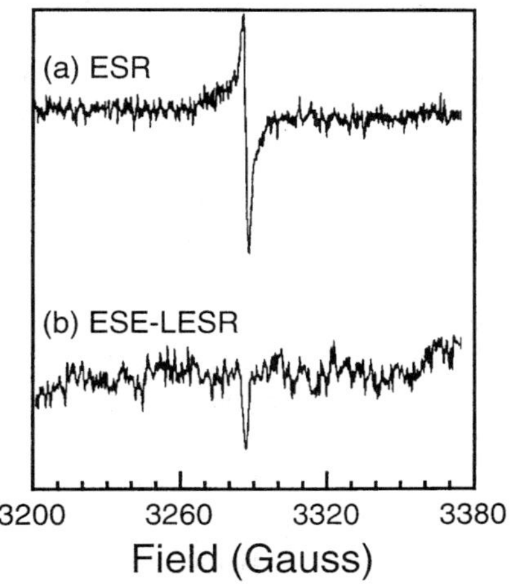

Figure 27 (a) ESR spectrum of C_{60} film. (b) Transient LESR spectrum of C_{60} film. (*From Ref. 46; used with permission.*)

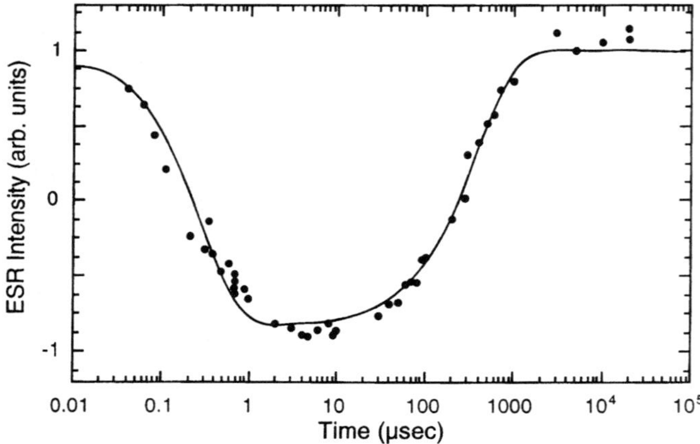

Figure 28   Time evolution of the transient LESR spectrum of C_{60} film. (*From Ref. 46; used with permission.*)

uted the change in the spin polarization to spin-selective quenching of triplet excitons by paramagnetic centers. The radical triplet pairs, in the region of non-zero exchange interaction, can form doublet (L − S) and quartet ($L + S$) spin states. Encounters producing doublets instantaneously quench the triplets. Encounters producing quartets cannot, due to spin-conservation rules. Zero-field splitting can admix quartet and doublet states so that quartets acquire some doublet state character. As the formation rate of quartets and doublets is spin-dependent, radical/triplet interactions will result in a change of the spin polarization of the system [47, 48].

This hypothesis was independently verified by PA and PADMR measurements of C_{60} evaporated from toluene onto a sapphire substrate. The S-Band PLDMR and H-PADMR (at 1.2 eV) spectrum of this film are shown in Figure 29a [49]. Both spectra contain full- and half-field triplet powder patterns and a narrow resonance at $g \approx 2$. The triplet and spin-$\frac{1}{2}$ signals were correlated to one another by H-PADMR measurements taken with the probe energy at 0.9, 1.7, and 2.3 eV (Figure 29b). In each case, the full-field triplet powder pattern is accompanied by a narrow spin-$\frac{1}{2}$ resonance which suggests that the two features are correlated. Indeed, the PA spectra of triplet excitons and photogenerated carriers in vacuum-sublimed C_{60} films do not resemble each other (see below). Hence, in agreement with Steren et al. [46], we attribute the spin-$\frac{1}{2}$ feature in C_{60} films evaporated solution to radical/triplet interactions.

We next turn to the triplet resonances. The half-field peak at 1669 G and the narrow, full-field triplet powder pattern are due to the same species. The ZFS

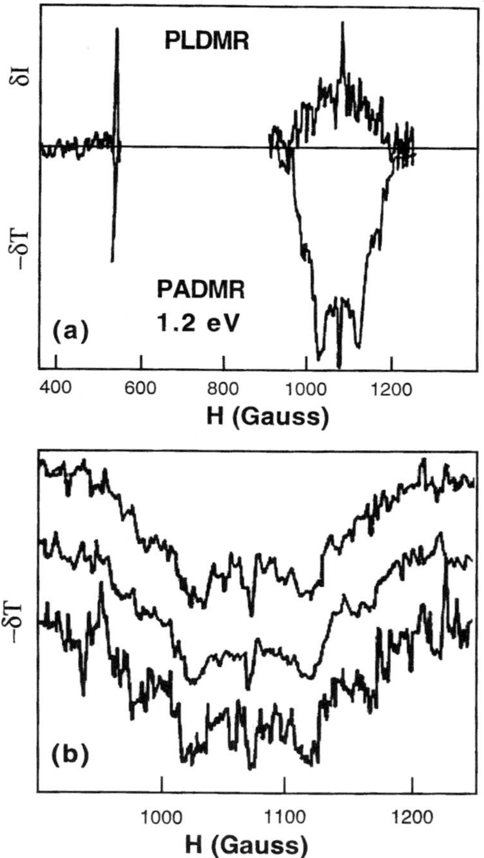

Figure 29 (a) PLDMR and (b) PADMR spectra of C_{60} film. The probe energy was set to 1.2 eV. (b) PADMR spectra of C_{60} film with probe energy at (top to bottom) 0.9, 1.7, and 2.3 eV, respectively.

of triplets α and β predict half-field peaks at 1669.1 and 1668.9 G, respectively, in good agreement with experiment. As both the full-field and half-field powder patterns of these two triplets overlap one another, it is not possible to determine their relative contributions to the PLDMR spectrum. The half-field peak at 1665.4 G and the broad, full-field powder pattern are likewise correlated. Using Eqs. (7) and (8), we calculate ZFS parameters of $|D| = .031$ cm^{-1} and $|E| = .010$ cm^{-1}. Both these ZFS are much larger than those for both matrix isolated fullerenes and single crystal C_{60}. The nonzero $|E|$ requires that the local environment of this triplet is distorted not only from icosohedral symmetry but also from cylindrical symmetry. The much larger principal ZFS parameter, $|D|$, may mean that this triplet is localized, possibly pinned to sites distorted by interactions between neighboring fullerenes or defect sites in the crystal. A more detailed approach can be followed by relating the ZFS parameters to the reduced coordinates of the two spins composing the triplet. The three principal solutions of the spin Hamiltonian, Eq. (2), are:

$$X = \frac{1}{2} g^2 \mu_B^2 \left\langle \frac{r^2 - 3x^2}{r^5} \right\rangle \tag{21}$$

$$Y = \frac{1}{2} g^2 \mu_B^2 \left\langle \frac{r^2 - 3y^2}{r^5} \right\rangle \tag{22}$$

$$Z = \frac{1}{2} g^2 \mu_B^2 \left\langle \frac{r^2 - 3z^2}{r^5} \right\rangle \tag{23}$$

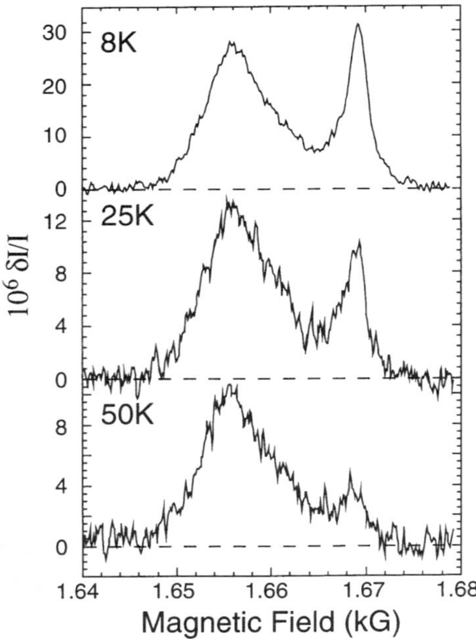

Figure 30   The half-field PLDMR spectrum of C_{60} film, measured at various temperatures. (*From Refs. 18, 27.*)

A simple relationship exists between the two sets of zero-field parameters: $D = -3Z/2$ and $E = (X - Y)/2$. A given set of ZFS parameters may not generate a unique set of reduced triplet coordinates, but it can permit a rough spatial estimate of the triplet extent. Furthermore, a high-field measurement on a polycrystalline sample cannot determine the signs of D and E. For $|D| = .031$ cm^{-1} and $|E| = .010$ cm^{-1}, the reduced triplet coordinates $\sqrt{\langle x^2 \rangle} = 0$ Å, $\sqrt{\langle y^2 \rangle} = 2.0$ Å, and $\sqrt{\langle z^2 \rangle} = 2.9$ Å yield the correct set of ZFS parameters. This gives a triplet which is planar and has a rough extent $\sqrt{\langle r^2 \rangle} = 3.5$ Å, the size of one of the five- or six-membered rings on the fullerene molecule. It was not possible to generate an acceptable set of ZFS parameters with a more delocalized triplet. We therefore conclude that this triplet exciton is localized. The localized triplet is much more prominent in the half-field powder patterns. This is expected, since the $\Delta m_s = 2$ transition scales as $(D/h\nu)^2$ [50]. The delocalized triplet powder pattern could be easily saturated by increasing the photoexcitation intensity. Its amplitude decreased sharply at laser powers above 8 mW (Figure 25). This excitation power

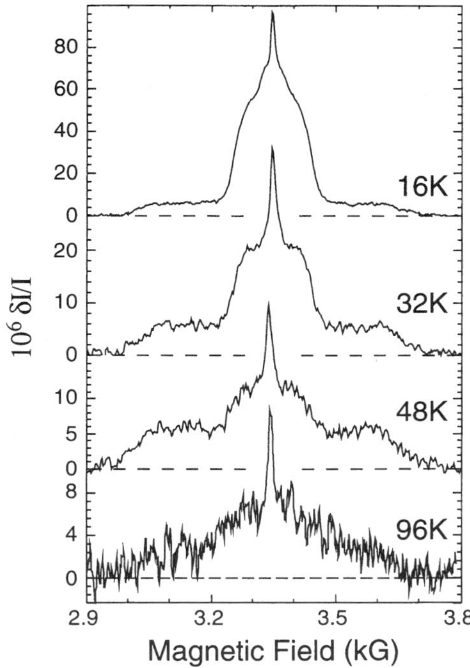

Figure 31 The full-field PLDMR spectrum of C_{60} film, measured at various temperatures. (*From Refs. 18, 27.*)

corresponds to the absorption of roughly 10^3 photons per molecule per second. The two triplet species also have quite different temperature dependencies. Figure 30 displays the half-field PLDMR of C_{60} between 8 and 50 K and Figure 31 the full-field PLDMR of C_{60} films between 16 and 96 K. The full-field spectrum at 96K contains only the spin-$\frac{1}{2}$ peak and the triplet powder pattern of the localized triplet exciton. The rapid decrease of the intensity of the delocalized exciton PLDMR is not surprising, in view of the rapidly increasing rotational motion of the molecule. This decrease may also be related to higher triplet mobilities as T increases.

Agostini et al. observed a resonance similar to that of the localized triplet exciton in the LESR and zero field PLDMR spectra of functionalized derivatives of C_{60} embedded in PMMA [51]. When the sample is dissolved in a solution containing PMMA and evaporated to form a solid glassy matrix (PMMA/E), the LESR spectrum resembles that of C_{60}:T/PS. If the sample is prepared by in situ polymerization of the methyl methacrylate monomer solution containing fullerene (PMMA/T), a much broader spectrum is observed. The zero field PLDMR spectrum of PMMA/T has transitions at 180, 659, and 844 MHz (Figure 32). These transitions yield ZFS $|D_{loc}| = .0251$ cm^{-1} and $|E_{loc}| = .0030$ cm^{-1}. A second derivative was measured with similar results ($|D_{loc}| = .0248$ cm^{-1}, $|E_{loc}| = .0033$ cm^{-1}). This triplet was attributed to a multiple adduct arising from cross-linking between the polymer chains and fullerene derivatives. The ZFS parameters of the triplet states detected in C_{60} single crystals, films evaporated from solution, and functionalized derivatives are summarized in Table 3.

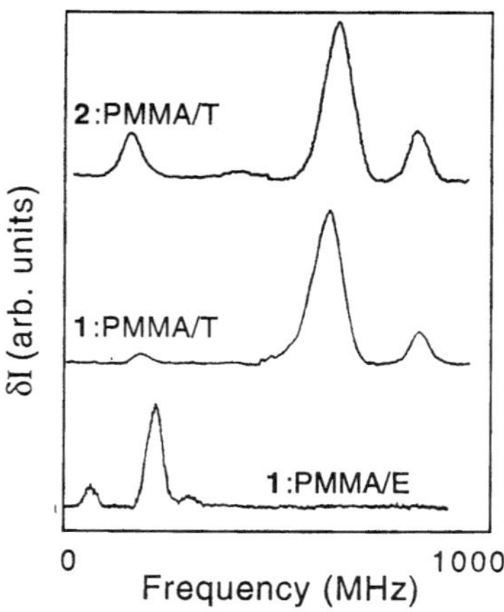

Figure 32 The zero field PLDMR spectrum of functionalized C_{60} derivatives (1 and 2) embedded in PMMA matrics. (*From Ref. 51; used with permission.*)

Table 3   Zero-Field Splitting Parameters of $^3C_{60}$ Single Crystals, Films, and Derivatives

Reference	Sample	Technique	$\lvert D \rvert$ $(10^{-4}\,\mathrm{cm}^{-1})$	$\lvert E \rvert$ $(10^{-4}\,\mathrm{cm}^{-1})$
Groenen et al [42,44]	Single crystal	Pulsed LESR and	70	21
		Zero-field PLDMR	86	16
Lane et al [18,27]	Film	PLDMR	89	*
			310	100
Agostini et al [51]	Derivative	Zero-field PLDMR	96	14
			248	33

* The polycrystalline nature of the film makes an unambiguous determination of E difficult.

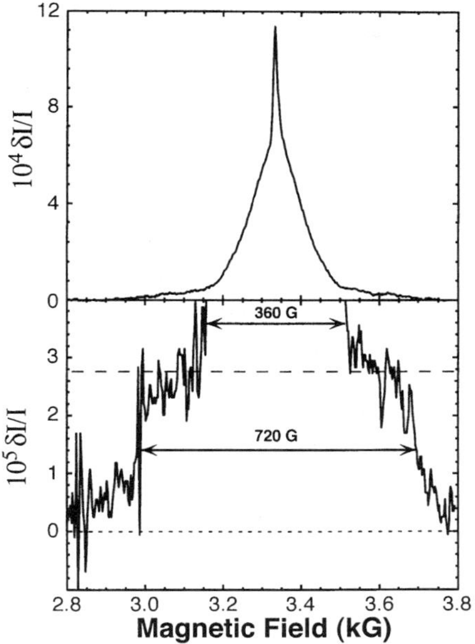

Figure 33   The full-field PLDMR spectrum of C_{70} film, measured at various temperatures. (*From Ref. 27.*)

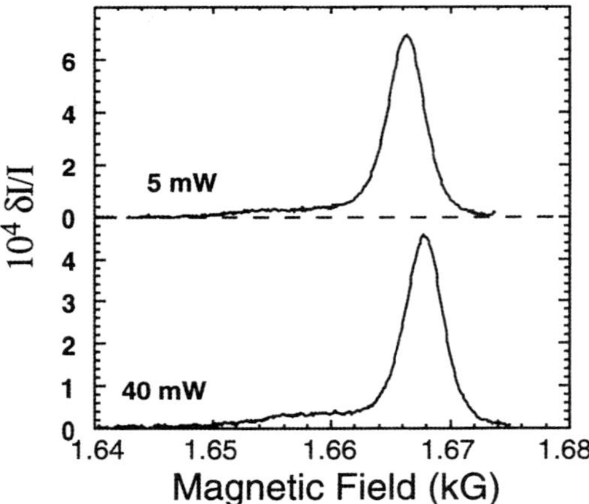

Figure 34 The half-field PLDMR spectrum of C_{70} film, measured at various temperatures. (*From Ref. 27.*)

C_{70}

The full-field PLDMR spectrum of a C_{70} film evaporated from solution, shown in Figure 33, contains the same three features seen in the C_{60} film. The delocalized triplet powder pattern of C_{70} is narrower than that of C_{60}, due to the larger molecule and consequently greater delocalization of the triplet. This powder pattern is virtually featureless, which may be due to a greater distribution of ZPS parameters. Zero-field measurements would be able to confirm whether or not this is the case. Whereas the half-field powder pattern of C_{60} was dominated by the localized triplet exciton, the half-field powder pattern of C_{70} (Figure 34) is dominated by the delocalized triplet.

3.3 Vacuum Sublimed Fullerene Films

C_{60}

Sublimed C_{60} thin films were deposited on quartz or sapphire substrates at 450 °C at a rate of 1 Å/s from purified C_{60} powder. The film thickness was determined by absorption measurements to be ~0.1 mm [52]. Figure 35 shows the PA spectrum between 0.25 and 2.6 eV of a sublimed C_{60} film at 80 K and modulation frequencies of 20 Hz and (b) 20 kHz. The PA spectrum exhibits a derivative shaped feature with a zero crossing at 2.4 eV and four PA bands: C_1 at 0.8 eV, T_1 at 1.2 eV, T_2 at 1.8 eV, and C_2 at 2.0 eV. All PA bands increase as the square

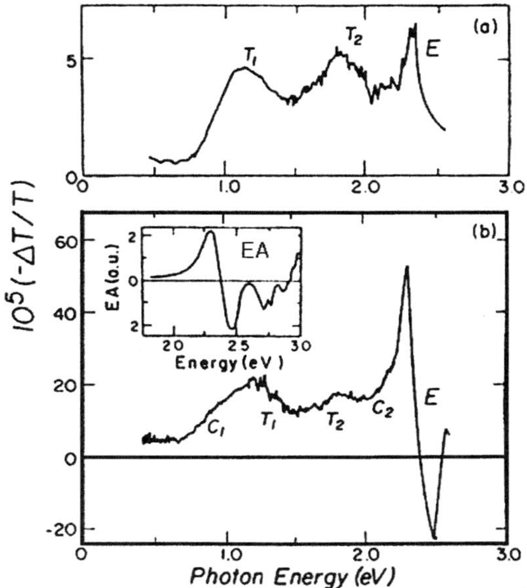

Figure 35 The PA spectrum of a vacuum sublimed C_{60} film measured at 80 K with modulation frequencies of (a) 20 Hz and (b) 20 kHz. Inset: the electroabsorption spectrum. (*From Ref. 52.*)

root of the photoexcitation intensity, indicating bimolecular recombination kinetics. The feature at 2.4 eV can be identified by electroabsorption spectroscopy, which measures changes in the absorption spectrum due to an external electric field. The EA spectrum of a C_{60} film, inset in Figure 35, shows strong derivative-shaped features caused by a Stark shift of the lowest excited state [53]. The EA spectrum is identical to the high-energy feature, enabling us to identify it as due to a photoinduced internal electric field. The source of the electric field in PA is photoinduced charge separation in the C_{60} film [54].

The various PA bands can be correlated with one another, due to different dependencies on temperature and the modulation frequency ν_c. C_1, C_2, and the EA depend strongly on V_c, dropping by an order of magnitude between 20 Hz and 20 kHz, whereas T_1 and T_2 depend weakly on ν_c. T_1 and T_2 also have a much stronger dependence on temperature, as demonstrated by their complete disappearance from the PA spectrum at 300 K (Figure 36). As the EA appears to be correlated with bands C_1 and C_2, we tentatively attribute these PA bands to photoinduced charged excitations. T_1 and T_2 are due to long-lived neutral excitations, most likely triplet excitons. PADMR is an ideal technique to deconvolute

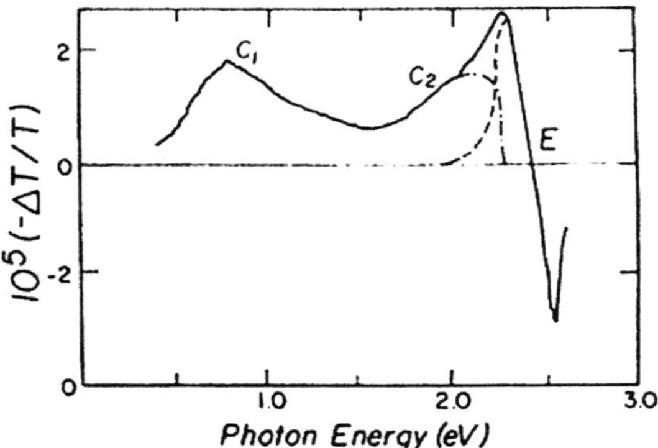

Figure 36 The PA spectrum of a sublimed C_{60} film measured at 300 K. (*From Ref. 52.*)

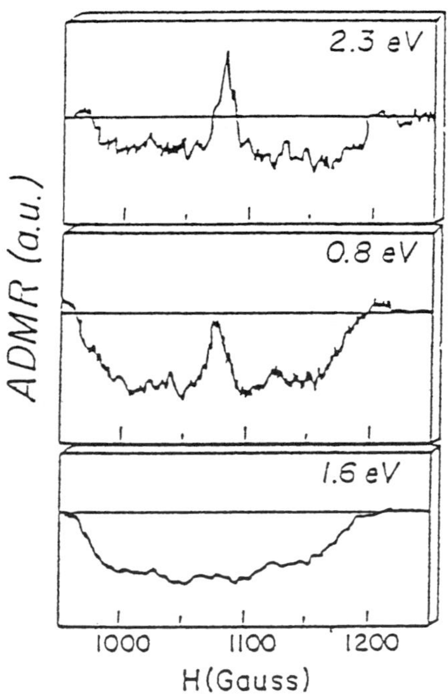

Figure 37 H-PADMR spectra of sublimed C_{60} film, measured at various probe energies. (*From Ref. 52.*)

such a complicated spectrum, as it combines the advantages of PA (spectral information) with those of magnetic resonance (spin selectivity).

Figure 37 shows three typical PADMR spectra of the sublimed C_{60} film, measured at 0.8, 1.5 and 2.3 eV. Each spectrum contains two components with opposite signs: a $\delta n < 0$ component with FWHM 180 G and a $\delta n > 0$ component with FWHM 15 G. Correlated with the broad component, there is a $\delta n < 0$ half-field signal at 531 G (not shown here). The negative component is clearly due to enhanced recombination of triplet excitons with $|D| \approx 90$ G, in agreement with the PLDMR measurements described above. The narrow component cannot be due to radicals as these are removed by vacuum sublimation [55]. This component is also much broader (15 G) than that of the radicals (5 G). We therefore identify the narrow component as due to spin $^1/_2$ polarons ($C_{60}^{\pm}$) with $g \cong 2.000$.

The different dependence of the two magnetic resonance signals on the probe energy enabled us to unambiguously determine the origin of the various features in the PA spectrum shown in Figures 35 and 36. The λ-PADMR spectrum of the C_{60} film is shown in Figure 38. The triplet spectrum contains two $\delta n < 0$ bands at the same energies as the T_1 and T_2 PA bands. The spin-$^1/_2$ λ-PADMR spectrum contains three features: two $\delta n > 0$ bands with energies at

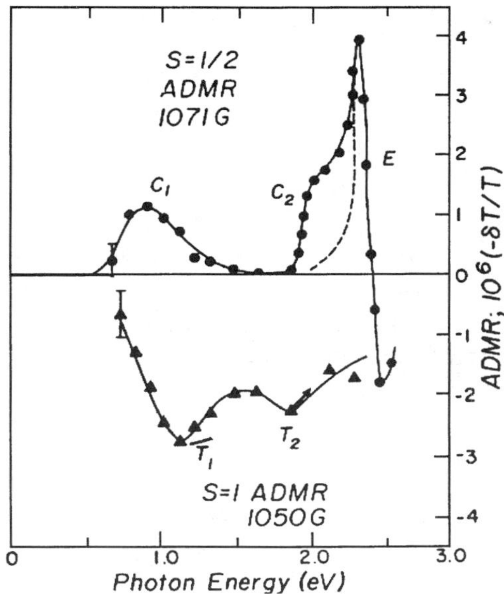

Figure 38   Spin-$^1/_2$ (top) and spin-1 (bottom) λ-PADMR spectra of sublimed C_{60} film. (*From Ref. 52.*)

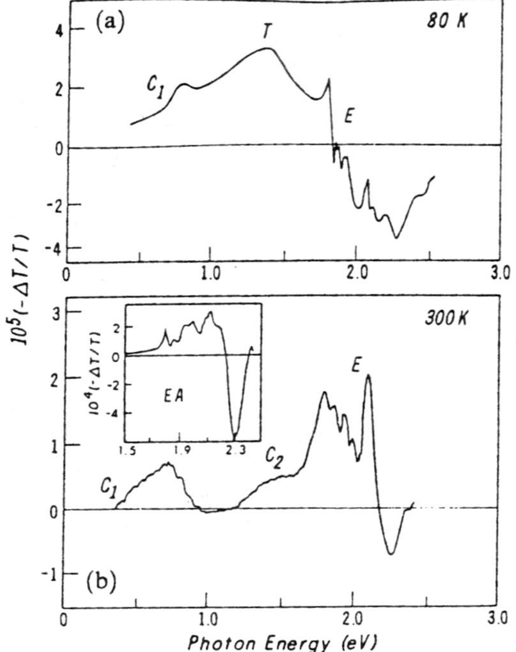

Figure 39 PA spectra of sublimed C_{70} film, measured at (a) 80 and (b) 300 K. Lower inset: the electroabsorption spectrum. (*From Ref. 58.*)

0.8 and 2.0 eV and a derivativelike band with zero crossing at 2.4 eV. In agreement with the assignment of PA bands C_1 and C_2 to photogenerated charged excitations, these PA bands were attributed to spin-$^1/_2$, singly charged polarons ($C_{60}^{\pm}$). Some photogenerated polarons are stable even at room temperature, in agreement with persistent photoconductivity measurements in C_{60} films at 300K [56].

C_{70}

C_{70} films were deposited at 450°C on sapphire substrates by evaporation at 5×10^{-6} torr [57]. The PA spectrum of the C_{70} film, measured at 80 and 300 K with $v_c = 500$ Hz, is shown in Figure 37 [58]. The PA spectrum of the C_{70} film contains at least four features: PA bands C_1 at 0.6 eV and C_2 at 1.8 eV, a broad PA band T_1 with a maximum at 1.4 eV, and a derivativelike feature with a zero crossing at 2.2 eV (E). From the similarity of the C_{70} PA spectrum to that of C_{60}, we identify the C_1 and C_2 bands as due to optical transitions of polarons whereas

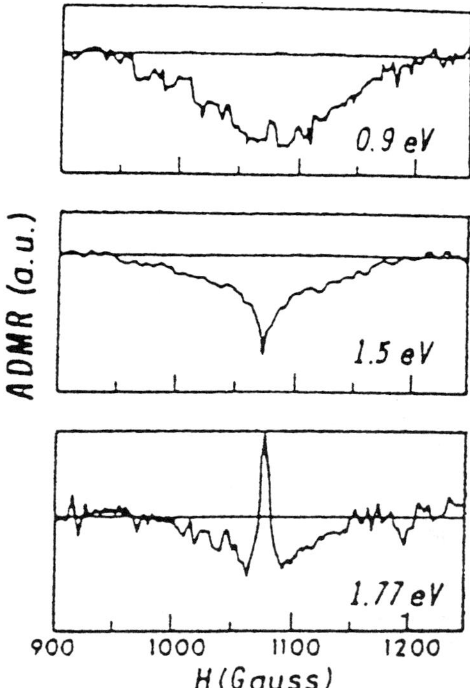

Figure 40   H-PADMR spectra of sublimed C_{70} film, measured at various probe energies. (*From Ref. 58.*)

the E band is due to electroabsorption associated with photoinduced electric fields in the film caused by photogenerated charged excitations.

The PADMR spectrum (Figure 37 inset) of the C_{70} film consists of a $\delta n < 0$ powder pattern with FWHM $\Delta H = 120$ G and a narrow $\delta n > 0$ band. The λ-PADMR spectra of the spin-$\frac{1}{2}$ and spin-1 PADMR components are shown in Figure 38. The triplet spectrum contains a broad band peaked at 1.5 eV, identical to the T band in the PA spectrum. Hence we conclude that the long-lived neutral photoexcitations in C_{70} films are triplet excitons. By contrast, the spin-$\frac{1}{2}$ λ-PADMR spectrum contains three features: two $\delta n > 0$ PA bands identical to C_1 and C_2 and a derivativelike feature with zero crossing at 2.2 eV.

4. SUMMARY

Magnetic resonance studies of photoexcited fullerenes have revealed rich photophysics. The zero-field splitting parameters of $^3C_{60}$ show that the excited

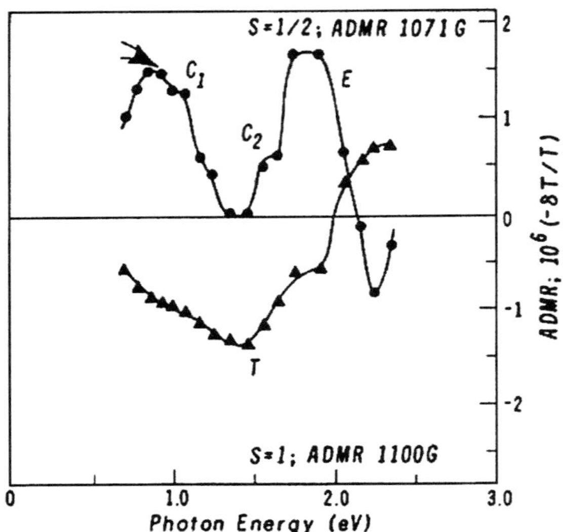

Figure 41 Spin-½ (top) and spin-1 (bottom) λ-PADMR spectra of sublimed C_{70} film. (*From Ref. 58.*)

molecule does not have icosohedral symmetry. This has been attributed to static and dynamic Jahn-Teller effects which lower the symmetry of the excited molecule from I_h to D_{2h}. Two overlapping triplet species have been detected in high field LESR and ODMR measurements of C_{60} molecules. By combining zero-field ODMR with hole burning (double resonance) measurements, the ZFS parameters of the two triplets were determined. Measurements of the temperature dependence of the ODMR spectrum showed that a thermally activated process is involved, which was modeled by a hopping of the triplet orientation amongst the six degenerate principal axes. The excited state dynamics of C_{70} differ from C_{60}; it has much longer triplet lifetimes and is consequently phosphorescent. The PLDMR spectrum of C_{70} has a strong dependence on temperature and microwave modulation frequency. This is due to the interplay between sublevel lifetimes, spin-lattice relaxation, and dynamic Jahn-Teller effects. Two distinct triplet species were observed in C_{70} molecules, which can be observed by monitoring changes in the flourescence or the phosphorescence by ODMR.

Transient LESR and zero-field measurements on C_{60} single crystals detected three distinct triplet species. Two of these triplets have the same ZFS parameters, but different orientations of their principal axes. Hence, these triplets could be distinguished by rotating the single crystal with respect to the applied magnetic field. As the orientation of the principal Z axis corresponded to the

[001] or [110] axes of the cubic lattice, it was concluded that the triplets are delocalized along dimers or chains of C_{60} molecules. High-field ODMR measurements of a C_{60} film evaporated from solution detected half- and full-field powder patterns due to triplets detected above. Furthermore, a broad full-field triplet powder pattern was detected and attributed to fullerene molecules residing at defects in the polycrystalline film. The ZFS parameters of this triplet indicate that it is localized ($\langle \bar{r} \rangle \cong 3.5$ Å), most likely to a face of the fullerene molecule. A narrow spin-$\frac{1}{2}$ resonance was also detected, which λ-PADMR and transient LESR measurements show is due to radical/triplet interactions. Similar features were detected in evaporated C_{70} films. The PA spectrum of fullerene films revealed multiple spectral features with different dynamics and temperature dependence. Magnetic resonance spectroscopy showed that the bands are correlated with excitations having spin $S = \frac{1}{2}$ or $S = 1$. The PA spectrum was thus attributed to charged photocarriers (polarons), triplet excitons, and electroabsorption due to electric fields associated with polarons.

ACKNOWLEDGMENTS

We would like to thank Xing Wei, Leland Swanson, and Jonathan Partee for valuable contributions to the research reported here. Ames Laboratory is operated by Iowa State University for the U.S. Department of Energy under Contract W-7405-Eng-82. The work at Iowa State University was supported by the Director for Energy Research, Office of Basic Energy Sciences, USDOE. The work at the University of Utah was supported by the Department of Energy under grants DE-FG 03-93 ER 45490 and DE-FG 02-89 ER 45409, the National Science Foundation under grant No. DMR-92-22947, and by the Office of Naval Research under grant No. N00014-91-C-0104.

REFERENCES

1. M. M. Khaled, R. T. Carlin, P. C. Trulove, G. R. Eaton, and S. S. Eaton, *J. Am. Chem. Soc.* **116**, 3465 (1994).
2. J. Stinchcombe, A. Penicaud, P. D. W. Boyd, and C. A. Reed, *J. Am. Chem. Soc.* **115**, 5212 (1993).
3. T. Kato, T. Kodama, and T. Shida, *Chem. Phys Lett.* 205, 405 (1993).
4. P. D. W. Boyd, P. Byrappa, P. Paul, J. Stinchcombe, R. D. Bolskar, Y. Sun, and C. A. Reed, *J. Am. Chem. Soc.* **117**, 2907 (1995).
5. Y. Maniwa, M. Nagasaka, A. Ohi, K. Kume, K. Kikuchi, K. Saito, I. Ikemoto, S. Suzuki and Y. Achiba, *Japan J. Appl. Phys.* **33**, L173 (1994).

6. M. R. Wasielewski, M. P. O'Neill, K. R. Lykke, M. J. Pellin, and D. M. Gruen, *J. Am. Chem. Soc.* **113**, 2774 (1991).

7. C. P. Poole, in *Electron Spin Resonance*, J. Wiley & Sons, New York, 1983.

8. A. Matsushita, A. M. Frens, E. J. J. Groenen, O. G. Poluektov, J. Schmidt, G. Meijer, and M. A. Verheijen, *Chem. Phys. Lett.* **214**, 349 (1993).

9. Y. Zeng, L. Biczok, and H. Linschitz, *J. Phys. Chem.* **96**, 5237 (1992).

10. N. Minami, S. Kazaoni, and R. Ross, *Mol. Cryst. Liq. Cryst.* **256**, 233 (1994).

11. C. Reber, L. Yee, J. McKiernan, J. I. Zink, X. Williams, W. M. Tong, D. A. A. Ohlberg, R. L. Whetten, and F. Diederich, *J. Phys. Chem.* **95**, 2127 (1991).

12. A. Andreoni, M. Bondani, and G. Consolati, *Phys. Rev. Lett.* **72**, 844 (1994).

13. Y. Zeng, L. Biczok, and H. Linschitz, *J. Phys. Chem.* **96**, 5237 (1992).

14. D. J. van den Heuvel, I. Y. Chan, E. J. J. Groenen, J. Schmidt, and G. Meijer, *Chem. Phys. Lett.* **231**, 111 (1994).

15. R. R. Hung and J. J. Grabowski, *J. Phys. Chem.* **95**, 6073 (1991).

16. X. Wei and Z. V. Vardeny, *Phys. Rev.* **B52**, R2317 (1995).

17. X. Wei, Ph.D. Thesis, 1992, University of Utah (unpublished).

18. P. A. Lane, L. Swanson, Q.-X. Ni, J. Shinar, J. P. Engel, T. J. Barton, and L. Jones, *Phys. Rev. Lett.* **68**, 887 (1992).

19. M. Bennati, A. Grupp, M. Mehring, K. P. Dinse, and J. Fink, *Chem. Phys. Lett.* **200**, 440 (1992).

20. A. Angerhofer, J. U. von Schütz, D. Widmann, W. H. Müller, H. U. ter Meer, and H. Sixl, *Chem. Phys. Lett.* **217**, 403 (1994).

21. X. Wei and Z. V. Vardeny, *Phys. Rev.* **B52**, R2317 (1995).

22. N. Koga and K. Morokuma, *Chem. Phys. Lett.* **196**, 191 (1992).

23. W. Z. Wang, C. L. Wang, Z. B. Su, and L. Yhu, *Phys. Rev. Lett.* **72**, 3550 (1994).

24. W. Z. Wang, A. R. Bishop, and L. Yu, *Phys. Rev.* **B50**, 5016 (1994).

25. A. Auerbach, *Phys. Rev. Lett.* **72**, 2931 (1994); A. Auerbach et al., *Phys. Rev.* **B49**, 12998 (1994).

26. W. Z. Wang, C. L. Wang, A. R. Bishop, L. Yu, and Z. B Su, *Phys. Rev.* **B51**, 10209 (1995).

27. P. A. Lane and J. Shinar, *Phys. Rev.* **B51**, 10028 (1995).

28. J. W. Greis, A. Angerhofer, J. R. Norris, H. Scheer, A. Struck, and J. U. von Schütz, *J. Chem. Phys.* **100**, 4820 (1994)

29. P. W. Anderson, *J. Phys. Soc. Jpn.* **9**, 316 (1954).

30. H. Levanon, V. Meiklyar, A. Michaeli, S. Michaeli, and A. Regev, *J. Chem. Phys.* **96**, 6128 (1992).

31. A. Regev, D. Gamliel, V. Meiklyar, S. Michaeli, and H. Levanon, *J. Chem. Phys.* **97**, 3671 (1993).

32. S. Alexander, *J. Chem. Phys* **37**, 967 (1962).

33. X. Wei, Z. V. Vardeny, D. Moses, V. I. Srdanov, and F. Wudl, *Synth. Met.* **54**, 273 (1993).

34. P. A. Lane, X. Wei, Z. V. Vardeny, J. Partee and J. Shinar, *Phys. Rev.* **B53**, R7580 (1996).

35. W. Guss et al, *Phys. Rev. Lett.* **72**, 2644 (1994).

36. M. R. Fraelich and R. B. Weisman, *J. Phys. Chem.* **97**, 11145 (1993).

37. H. T. Etheridge and R. B. Weisman, *J. Phys. Chem.* **99**, 2782 (1995).

38. C. Saal, N. Weiden, and K.-P. Dinse, *Appl. Mag. Reson.* **11**, 335 (1996).

39. P. W. Stephens, L. Mihaly, P. L. Lee, R. L. Whetten, S. M. Huang, R. Kaner, F. Diederich, and K. Holczer, *Nature* **351**, 632 (1991).

40. C. S. Yannoni, R. D. Johnson, G. Meijer, D. S. Bethune, and J. R. Salem, *J. Phys. Chem.* **95**, 9 (1991).

41. R. Tycko, R. C. Haddon, G. Dabbagh, S. H. Glarum, D. C. Douglass, and A. M. Mujsce, *J. Phys. Chem.* **95**, 518 (1991).

42. E. J. J. Groenen, O. G. Poluektov, M. Matsushita, J. Schmidt, J. H. van der Waals, and G. Meijer, *Chem. Phys. Lett.* **197**, 314 (1992).

43. M. S. Dresselhaus, G. Dresselhaus, and P. C. Eklund, *Science of Fullerenes and Carbon Nanotubes*, Academic Press, San Diego, 1996.

44. M. Matsushita, A. M. Frens, E. J. J. Groenen, O. G. Poluektov, J. Schmidt, G. Meijer, and M. A. Verheijen, Chem. *Phys. Lett.* **214**, 349 (1993).

45. M. D. Pace, T. C. Christidis, J. J. Yin and J. Milliken, *J. Phys. Chem.* **96**, 6855 (1992).

46. C. A. Steren, H. van Willigen, and M. Fanciulli, *Chem. Phys. Lett.* **245**, 244 (1995).

47. C. Blattler, F. Jent, and H. Paul, Chem. *Phys. Lett.* **166**, 375 (1990).

48. G. H. Goudsmit, H. Paul, and A. I. Shushin, *J. Phys. Chem.* **97**, 13243 (1993).

49. X. Wei and Z. V. Vardeny, unpublished results.

50. N. Atherton, *Electron Spin Resonance*, Halstead, New York, 1973.

51. G. Agostini, C. Corvaja, M. Maggini, L. Pasimeni, and M. Prato, *J. Phys. Chem.* **100**, 13416 (1996).

52. D. Dick, X. Wei, S. Jeglinski, R. E. Benner, Z. V. Vardeny, D. Moses, V. I. Srdanov, and F. Wudl, *Phys. Rev. Lett.* **73**, 2760 (1994).

53. K. Pichler et al. *J. Phys. C Condensed Matter* **3**, 9259 (1991).

54. D. D. C. Bradley and O. M. Gelsen, *Phys. Rev. Lett.* **67**, 2589 (1991).

55. Y. Maniwa, M. Nagasaka, A. Ohi, K. Kume, K. Kikuchi, K. Saito, I. Ikemoto, S. Suzuki and Y. Achiba, *Japan J. Appl. Phys.* **33**, L173 (1994).

56. A. Hamed, L. L. Rasmussen, and P. H. Hor, *Phys. Rev. B* **48**, 14760 (1993).

57. A. F. Hebard, R. C. Haddon, R. M. Flemming, and A. R. Kortan, *Appl. Phys. Lett.* **59**, 2109 (1991).

58. X. Wei, S. Jeglinski, O. Paredes, Z. V. Vardeny, D. Moses, V. I. Srdanov, G. D. Stucky, K. C. Khemani, and F. Wudl, *Solid State Comm.* **85**, 455 (1993).

8
Electrons and Phonons in Fullerenes and Carbon Nanotubes

M. S. Dresselhaus and G. Dresselhaus
Massachusetts Institute of Technology
Cambridge, Massachusetts

P. C. Eklund
University of Kentucky
Lexington, Kentucky

R. Saito
University of Electro-Communications
Tokyo, Japan

1. ELECTRONS AND PHONONS IN FULLERENES

In this section a brief review is given of electrons and phonons in fullerenes. Detailed discussions of these topics are now available [1] and should be consulted.

1.1 Vibrational Modes

Because of the high symmetry of the C_{60} molecule (point group I_h), there are only 46 distinct mode frequencies corresponding to the $180 - 6 = 174$ degrees of freedom for the isolated C_{60} molecule, and of these, only four are infrared-active (all with T_{1u} symmetry) and 10 are Raman-active (two with A_g symmetry and eight with H_g symmetry). Raman and infrared spectroscopy provide sensitive methods for distinguishing C_{60} from higher fullerenes with lower symmetry (e.g.,

217

C_{70} has D_{5h} symmetry). Since most of the higher molecular weight fullerenes have lower symmetry as well as more degrees of freedom, they have many more infrared- and Raman-active modes. Fully interpreted IR and Raman spectra thus far are limited to C_{60} and C_{70}, and closely related materials.

The first-order infrared (IR) spectrum of solid C_{60} (see Figure 1) shows four prominent spectral lines. The simplicity of this spectrum provides a convenient method for characterizing the compositional purity of C_{60} samples, especially with regard to their contamination with other fullerenes which have more complex spectra. The IR spectrum of solid C_{60} remains almost unchanged relative to the isolated C_{60} molecule, with only the addition of a few weak features. This is indicative of the molecular nature of solid C_{60}. Higher-order features in the IR spectra provide information about the vibrational frequencies of many of the silent modes [1–8].

Likewise, Raman spectroscopy provides valuable information about the intra- and intermolecular bonding in solid C_{60} and C_{60}-related compounds [3]. The Raman spectrum in Figures 1 and 2 for solid C_{60} shows 10 strong Raman lines, the number of Raman-allowed vibrational modes expected for the free molecule. These spectral lines are therefore assigned to intramolecular modes. The normal modes in molecular C_{60} above 1000 cm^{-1} involve carbon atom displacements that are predominantly tangential to the C_{60} molecule surface, while the modes below $\sim$800 cm^{-1} involve predominantly radial motion. The A_g "pentagonal pinch" mode (1469 cm^{-1}) corresponds to breathing-type tangential displacements of the 5 carbon atoms around each of the 12 pentagons. The other A_g mode is found at 492 cm^{-1} and corresponds to a radial breathing mode. Under high laser flux from an Ar ion laser, the spectral feature associated with the pentagonal pinch mode downshifts irreversibly to 1458 cm^{-1}, which has been interpreted as a signature of a photoinduced transformation [9].

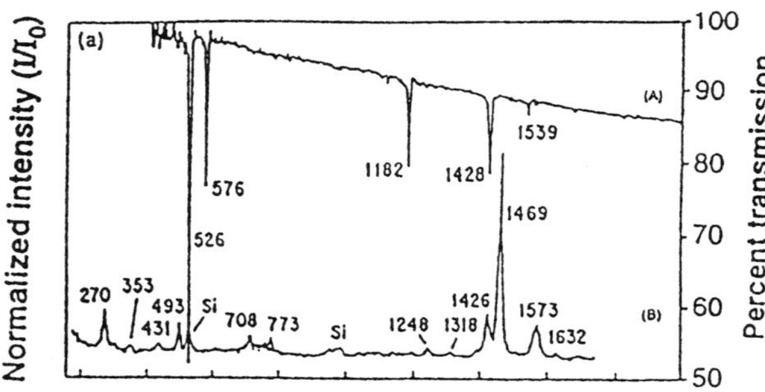

Figure 1 First-order infrared (A) and Raman (B) spectra for C_{60} taken with low incident optical power levels (<50 mW/mm^2). (*From Ref. 1.*)

In the solid phase, cubic crystal field interactions are expected to lower the symmetry of the lattice modes and to cause a splitting of the five-fold degenerate H_g-derived molecular modes, into three-fold (T_g) and two-fold (E_g) lattice modes [10]. The study of spectral line splittings and their symmetry in solid C_{60} requires the use of single crystals, polarization studies, and studies of the temperature dependence of the Raman spectra. Both Raman and IR spectroscopy have been used to monitor the effect of the 250 K phase transition involving the orientational ordering of the fullerenes. This phase transition affects the spectral line shifts, linewidths, line splittings, and their symmetries [1].

The addition of alkali metal dopants, even to the saturation stoichiometry M_6C_{60} (where M = K, Rb, Cs) perturbs the Raman spectra only slightly [7], with very little change from one alkali metal dopant to another. For metallic K_3C_{60} [see Figure 2a] and Rb_3C_{60}, the coupling between the phonons and a low energy electronic continuum in these metallic systems strongly broadens the H_g-derived modes [3] and gives rise to modifications in the Raman lineshape of some of the low frequency modes.

As a result of alkali metal doping, electrons are transferred to the π-electron orbitals on the surface of the C_{60} molecules, elongating the C—C bonds and downshifting the intramolecular tangential modes. Referring to Figure 2b, the dependence of the Raman frequencies on K concentration x for three dominant modes in K_xC_{60} is presented [11]. The softening of the 1469 cm^{-1} pentagonal pinch A_g mode by alkali metal doping is used as a convenient method to characterize the stoichiometry x of K_xC_{60} samples. The radial breathing mode, on the other hand, stiffens slightly due to competing effects associated with a mode softening arising from charge transfer effects (similar to the situation for the tangential modes) and a larger mode stiffening effect due to electrostatic interactions between the charged C_{60} molecule and the surrounding charged alkali atoms as their atomic separations change during a normal mode vibrational displacement [7]. Models for the intramolecular modes have been used to study the electron-phonon interaction, with particular emphasis on its connection to the observed superconductivity in the M_3C_{60} compounds. The effects of phototransformation on the intramolecular modes have been studied in detail [12]. Time t resolved pump-probe measurements of the reflectivity $\Delta R(t)/R$ in K_3C_{60} and Rb_3C_{60} [13] have revealed a highly damped mode near 150 cm^{-1} which is identified with an A_g symmetry vibration between the C_{60} anion and an alkali metal cation, where $\Delta R(t)$ is the change in the reflectivity at the probe frequency (1.55 eV) stimulated by an intense, short femtosecond pump pulse (20 fs duration) at the same frequency. The high degree of damping and relatively low frequency of this mode (see Figure 3) makes observation of this mode very difficult by Raman scattering techniques. The $A_g(1)$ radial breathing mode near 490 cm^{-1} has also been observed in K_3C_{60} and Rb_3C_{60} by pump-probe techniques [13], and the very long dephasing time of 5 ps for the small reflectivity oscillations in Figure 4 allow observation of

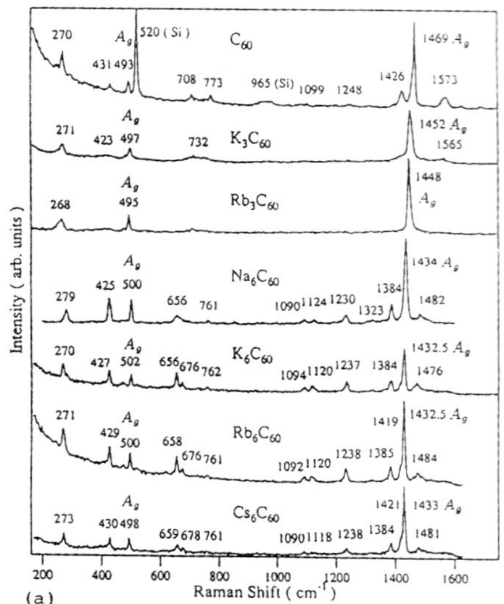

(a)

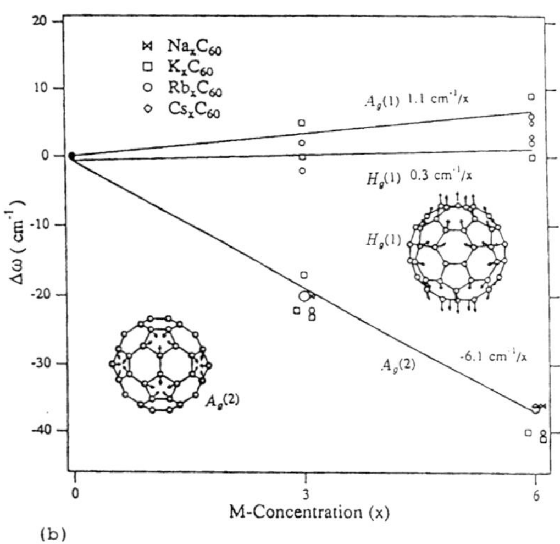

(b)

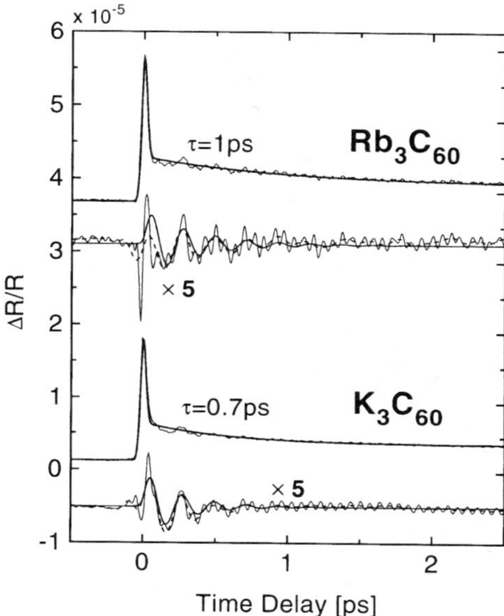

Figure 3 The pump-probe traces of changes in the reflectivity $\Delta R(t)/R$ versus delay time between the pump and probe pulses for K_3C_{60} and Rb_3C_{60} using pump and probe frequencies of 1.55 eV. Fits of the filtered data are made to fast decay sinusoids which have a frequency of ~150 cm^{-1}, and the results are shown by the thick solid curves. (*From Ref. 13.*)

Figure 2 (a) Unpolarized Raman spectra ($T = 300$ K) for solid C_{60}, K_3C_{60}, Rb_3C_{60}, Na_6C_{60}, K_6C_{60}, Rb_6C_{60} and Cs_6C_{60} [3, 4]. The tangential and radial modes of A_g symmetry are identified, as are the features associated with the Si substrate. From these spectra it is concluded that the molecule-molecule interactions between the C_{60} molecules are weak, as are the interactions between the molecules and the alkali metal ions. (b) Dependence of the frequencies of the $A_g(1)$, $A_g(2)$, and $H_g(1)$ modes on alkali metal concentration x in M_xC_{60}, where $M = $ Na, K, Rb, Cs. The frequency shifts of these Raman-active modes are plotted relative to the frequencies $\omega[A_g(1)] = 493$ cm^{-1}, $\omega[A_g(2)] = 1469$ cm^{-1}, and $\omega[H_g(1)] = 270$ cm^{-1} for C_{60} at $T = 300$ K. A schematic of the displacements for the eigenvectors for the $A_g(2)$ pentagonal pinch and $H_g(1)$ modes are shown. All the C atom displacements for the $A_g(1)$ modes are radial and of equal magnitude. Data from Refs. 5 and Ref. 6 are artificially displaced to the left and right, respectively. The centered data are from Ref. 7.

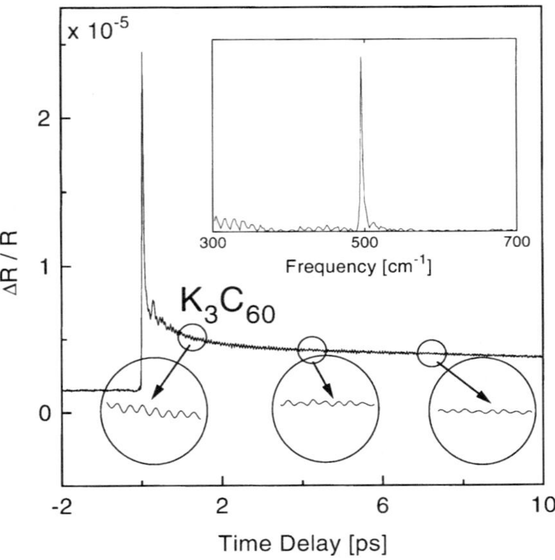

Figure 4   Coherent phonon oscillations in K$_3$C$_{60}$ at 300 K. The pump-probe data were taken in reflectivity using a single-wavelength pump-probe setup with a time resolution of about 20 fs. The three circular insets show the small oscillations superimposed on the decay. The inset shows that the Fourier transform power spectrum of the pump-probe data has a sharp peak at 492.5 ± 0.25 cm^{-1} obtained for K$_3$C$_{60}$, corresponding to the $A_g(1)$ radial breathing mode of the C$_{60}$ anion. (*From Ref. 13.*)

more than 200 oscillations of the impulsively induced intramolecular $A_g(1)$ breathing mode, thereby confirming the strongly molecular character of the K$_3$C$_{60}$ and Rb$_3$C$_{60}$ solid phase [13].

The low frequency acoustic and rotational-librational modes associated with the displacements and rotations of the C$_{60}$ molecules in a cubic lattice have been calculated [14], and experimental low frequency Raman [15] and inelastic neutron scattering measurements [16] on single crystals of C$_{60}$ have been performed.

1.2   Electronic Structure of Fullerenes

Based on the spectroscopic studies discussed in Section 1.1, we conclude that the fullerenes and even alkali metal derived fullerenes form molecular solids. Thus we expect their electronic energy band structures to be closely related to the electronic levels of the isolated molecules. Each carbon atom in C$_{60}$ has two single bonds along adjacent sides of a pentagon and one double bond between two adjoining hexagons. If these bonds were coplanar, they would be very similar to sp^2 trigonal bonding in graphite. However, the curvature of C_{60} causes the planar-derived trigonal orbitals to hybridize with the remaining p orbitals, thereby giving rise to different electronic states than in graphite. The shortening of the double bonds and lengthening of the single bonds in the Kekulé arrangement of the C_{60} molecule also strongly influence the electronic structure.

The most extensive calculations of the electronic structure of fullerenes thus far have been done for C_{60}. To obtain the molecular orbitals for C_{60}, the multielectron problem must be solved [18]. Representative results for C_{60} are shown in Figure 5a for the free molecule based on a one-electron approximation [17]. Because of the molecular nature of solid C_{60}, the electronic structure for the solid phase would be expected to be closely related to that of the free molecule, and some authors have followed this approach. A large amount of effort has also been given to the local density functional (LDF) band structure approach, an example of which is shown in Figure 5b where we see results for FCC solid C_{60}, based on total energy calculations [17]. The various band calculations for C_{60} yield a narrow band ($\sim$0.4–0.6 eV bandwidth) solid, with a HOMO-LUMO-derived direct band gap of $E_{HL} \approx 1.5$ eV separating the lowest unoccupied bands from the highest occupied bands. Since the HOMO (h_u) and LUMO (t_{1u}) levels have the same odd parity, electric dipole transitions between these levels are symmetry forbidden in the free C_{60} molecule. In the solid, transitions between the direct bandgap states at the Γ and X points in the FCC Brillouin zone are

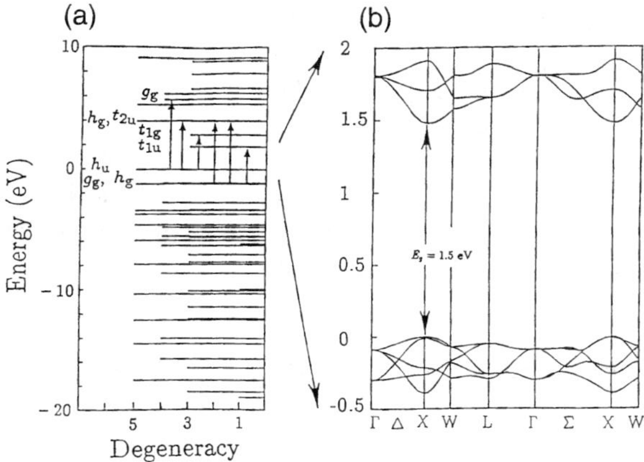

Figure 5 Calculated electronic structure for (a) an isolated C_{60} molecule and (b) the FCC solid C_{60} where the direct band gap at the X-point is 1.5 eV. (*From Ref. 17.*)

also forbidden. The allowed electric dipole transitions for the free C_{60} molecule are indicated by arrows in Figure 5a. Since the lowest energy transition is symmetry forbidden for electric dipole transitions in both the free molecule and the cubic solid, experiments, such as the temperature dependence of the electrical resistivity of undoped fullerenes [19], which are sensitive to the magnitude of the bandgap E_g and not to the symmetries of the initial and final states, would be expected to yield lower values for E_g than would be obtained from optical studies which are dominated by electric dipole-allowed transitions. Other experiments, such as luminescent emission also show different photon threshold energies, since luminescent emission is particularly sensitive to excitonic processes which also involve a Coulomb attraction between the electrons and holes participating in the recombination process.

Photoemission and inverse photoemission experiments reported by Weaver and coworkers on $M_x C_{60}$ for $M = K$, Rb, and Na and their alloys [19,20] show direct evidence for filling of the LUMO levels upon alkali metal doping, consistent with optical and transport measurements, but the implied energy gap in the density of states for these experiments is sensitive to screening and correlation effects. These experiments are discussed in more detail in Section 1.3.

Doping C_{60} with an alkali metal transfers electrons to the LUMO levels which, because of their t_{1u} symmetry, can accommodate three spin up and three spin down electrons. Total energy calculations on $K_x C_{60}$ show that the interstitial sites with tetrahedral symmetry are occupied with K^+ ions before the octahedral interstitial sites [17]. Assuming one electron to be transferred to the C_{60} molecule per alkali-metal atom dopant, the LUMO levels are expected to be half occupied at the alkali metal stoichiometry M_3C_{60} and totally full at M_6C_{60}, leading to a filled shell configuration with A_g symmetry. Thus M_6C_{60} would be expected to be semiconducting with a band gap between the t_{1u}, and t_{2g} levels (see Figure 5), while M_3C_{60} should be metallic provided that no bandgap is introduced at the Fermi level by a Peierls distortion. This simple picture is borne out by experiment.

Fermi surface calculations for K_3C_{60} indicate a small hole Fermi surface around the Γ-point and a larger and more complicated Fermi surface consisting of both electron and hole regions [21]. No direct measurements of the Fermi surface topology for any of the M_3C_{60} materials are yet available. Because of the weak interaction of the fullerenes with each other and with the alkali metal dopants, solid M_3C_{60} should also be viewed as a molecular solid having energy levels with little dispersion, giving rise to a very high density of states near the Fermi level, which is important in understanding superconductivity in M_3C_{60} materials.

1.3 Optical Properties

During early research on the separation of C_{60} from impurities and higher fullerenes using liquid chromatography [22], it became apparent that fullerenes in solution with organic solvents exhibited characteristic colors which would be useful for identifying and purifying these fullerenes. For example, C_{60} and C_{70} appear magenta and reddish-orange, respectively, in toluene and benzene solutions.

More detailed studies show that the optical properties of fullerenes are complex and provide important insights into the electronic structure of the fullerenes. Since the HOMO-LUMO optical transitions in C_{60} and C_{70} are symmetry forbidden, a great deal of attention has focused upon understanding the optical behavior near the fundamental absorption edge, which is very sensitively studied by optical absorption spectra. Such spectra are shown in Figure 6a and b vs. wavelength in the ultraviolet (UV)-visible range (200–700 nm) for C_{60} and C_{70}, respectively,

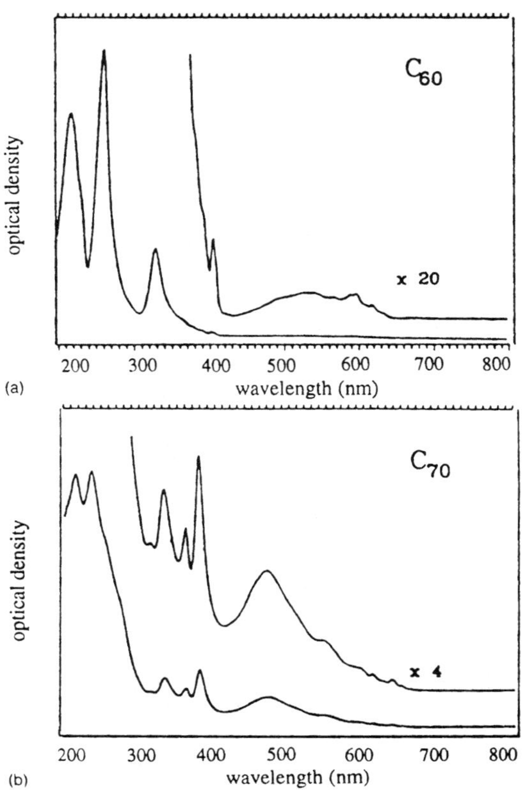

Figure 6 Optical density $\log_{10}$ (transmission coefficient) vs. wavelength for (a) C_{60} and (b) C_{70} in hexane solutions. For both of these molecular solution spectra the data near the absorption edge is shown more clearly by using solutions containing higher concentrations of fullerenes as indicated. (*From Ref. 22 and 23.*)

in hexane solution [22]. The strong absorption bands in the solution spectra are identified with electric dipole-allowed transitions between occupied (bonding) and empty (antibonding) sp^2-hybridized molecular orbitals. As mentioned above, electric dipole transitions between the one-electron HOMO level with h_u symmetry and the one-electron LUMO level with t_{1u} symmetry in C_{60} are forbidden. Using many-electron molecular states [18], we equivalently identify the fully occupied ground-state HOMO level with a singlet S_0 state having A_g symmetry. Allowed electric dipole transitions can likewise be described in a many-electron (exciton) picture. In terms of a many-electron picture the weak absorption between 490 and 640 nm for C_{60} in solution is identified with transitions from a singlet S_0 ground state to a singlet exciton state denoted by S_1 (Figure 6a). In the one-electron description, the h_u to t_{1u} transitions are activated by vibronic coupling. This vibronic coupling is also necessary to activate the $S_0 \rightarrow S_1$ transition. From the long wavelength limit ($\sim$640 nm) of this weak absorption, the energy of the S_1 state can be estimated to be $\sim$1.9 eV [19]. Referring to Figure 6a, the broad weak absorption band is identified with the $h_u \rightarrow t_{1u}$ forbidden one-electron transition between multiplets associated with the S_0 and S_1 singlet excitonic states. The strong transition starting at $\sim$2.9 eV relates to the dipole-allowed transition between the h_u and the t_{1g} one-electron states (see Figure 5). An analysis of the C_{70} optical density spectrum (see Figure 6b) is more difficult than for C_{60} since molecular orbital calculations for C_{70} [17] reveal a large number of closely spaced orbitals both above and below the HOMO-LUMO gap $E_{HL} = 1.65$ eV [17]. A number of theoretical works have clarified detailed aspects related to the optical absorption edge of C_{60} and C_{70} [24–26].

Further insight into the electronic structure is provided by pulsed laser studies of C_{60} and C_{70}, which probe the photodynamics of the optical excitation spectra (see Figure 7). The importance of these dynamic studies is to show that photoexcitation in the long wavelength portion of the UV-visible spectrum leads to the promotion of the system from the singlet S_0 state into a singlet S_1 state, which decays quickly with a nearly 100% efficiency via an "intersystem" crossing to the lowest excited triplet state T_1 (1.7 eV) which has the same orbital quantum numbers as the S_1 state, and therefore the S_1 and T_1 states are strongly coupled. Weak luminescence between S_1 and S_0 is nevertheless observed and is decorated by vibrational sidebands at 0.18 eV and 0.09 eV and associated with the two vibrational A_g modes of C_{60} (see Figure 7). Once in the lower T_1 triplet excitonic state, a nonradiative transition to the ground states takes place. Furthermore, once C_{60} reaches the T_1 triplet state, efficient electronic dipole excitation to other excited triplet states can occur with high efficiency, thereby greatly increasing the absorption coefficient, and providing a mechanism for achieving efficient optical limiting [27]. In the presence of oxygen, the lifetime of T_1 becomes very short.

To date, most of the optical studies on pristine fullerene solids have been carried out in transmission on thin solid films deposited on various substrates

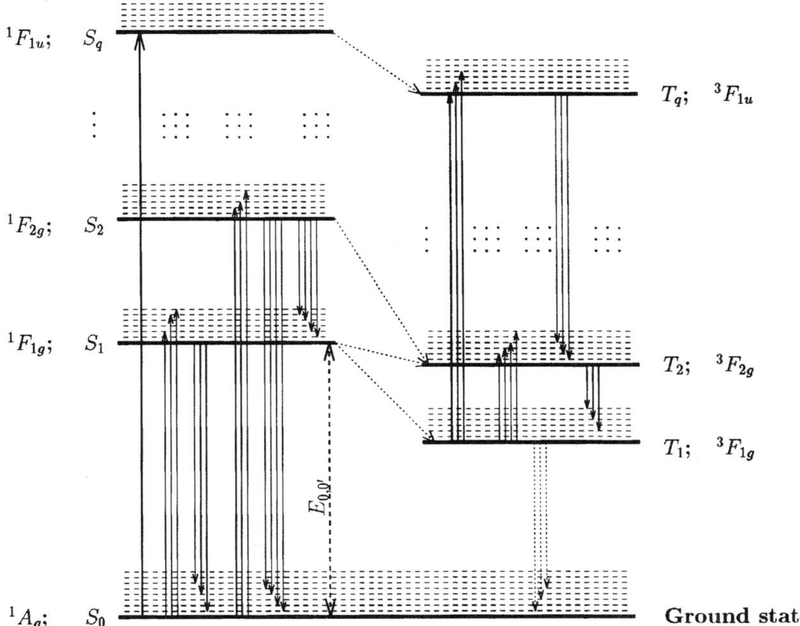

Figure 7 Schematic diagram for two of the lowest excitonic levels (boldface lines) associated with the $h_u^9 t_{1u}^1$ excitonic configuration. Associated with each excitonic level are a number of vibronic levels some of which couple to the ground state by an absorption or emission (luminescence) process. Also shown schematically are intersystem crossings between the singlet (left) and triplet (right) vibronic levels, as well as transitions from a triplet exciton state to the electronic ground state (or vibronic state) in a phosphorescence transition. The S_0 ground state together with its associated vibronic levels are shown across both left and right columns for convenience. The ($^1F_{1u}$; S_q) and ($^3F_{2u}$; T_q) states in the figure are associated with the $h_g^9 h_u^{10} t_{1u}^1$ excitonic configuration of the HOMO-1 and LUMO exciton and represent states for allowed optical transitions (heavier full vertical lines). The transition to the ($^1F_{1u}$; S_q) state is expected to occur in C_{60} at about 3 eV. (*From Ref. 1.*)

(such as quartz or KBr). The UV-visible transmission spectra for C_{60} and C_{70} solid films are observed to be remarkably similar to the respective solution spectra, thus providing further evidence for the molecular nature of fullerene solids. The optical properties of solid C_{60} and C_{70} have been studied in the UV-visible frequency range using the variable angle spectroscopic ellipsometry (VASE) technique. From these VASE and more straightforward transmission/reflection studies, the complex refractive index $\tilde{n}(\omega) = n(\omega) + ik(\omega)$ is determined, which is related

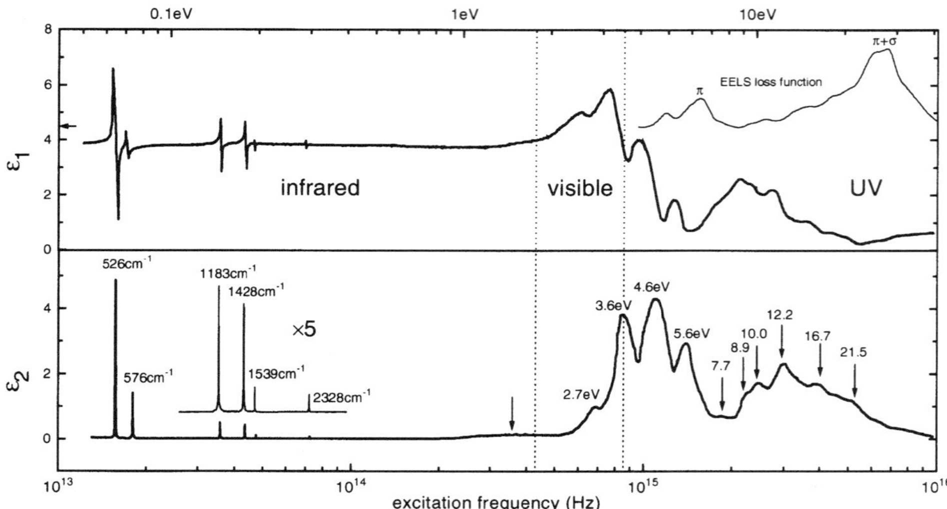

Figure 8 Real and imaginary parts of the dielectric function $\varepsilon(\omega)$ determined from variable angle spectroscopic ellipsometry (VASE) and Fourier transform infrared FTIR measurements for C_{60}. (*From Ref. 1.*)

to the optical dielectric function $\varepsilon(\omega) = \varepsilon_1(\omega) + i\varepsilon_2(\omega) = \tilde{n}^2(\omega)$. In Figure 8 we show the results for $\varepsilon_1(\omega)$ and $\varepsilon_2(\omega)$ obtained from VASE and transmission-FTIR studies on thin solid films of C_{60} on KBr at $T = 300$ K [1]. The strong, sharp structure at low energy is identified with infrared-active optic phonons and at higher energies the structure is due to electronic transitions.

Photoemission and inverse photoemission experiments have been especially useful [19,28] in providing information on the density of states within a few eV of the Fermi level for both undoped and doped fullerenes. Typical photoemission spectra ($E < E_F$, where E_F is the Fermi energy and E is the electronic state energy) and inverse photoemission spectra ($E > E_F$) are shown in Figure 9 for C_{60} and K_xC_{60} ($0 \leq x \leq 6$) [19], where the intensity maxima correspond to peaks in the density of states. This identification is established on the basis of a one-electron picture and a constant optical matrix element approximation. Photoemission and inverse photoemission spectra (PES and IPES) provide convincing evidence for charge transfer and for band filling as x in K_xC_{60} increases, though strictly speaking, x is not a continuous variable, unless taken to denote the average stoichiometry. Five crystallographic phases, with narrow stability ranges, have been identified ($x = 0, 1, 3, 4, 6$). At x values between these stable phases the material is in a mixed phase. Referring to Figure 9, we

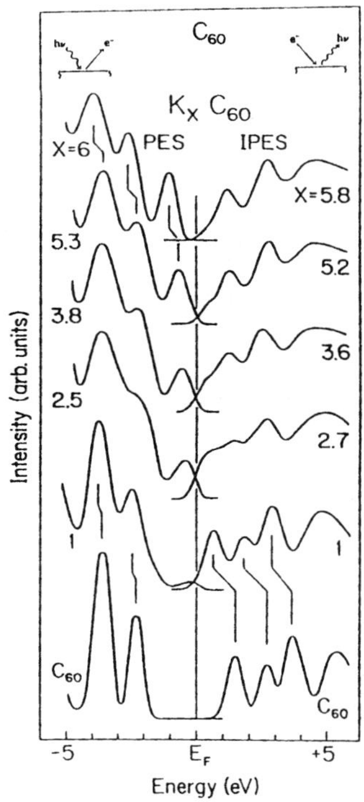

Figure 9 Photoemission (PES) for $E < E_F$ and inverse photoemission (IPES) spectra for $E > E_F$ taken on K$_x$C$_{60}$ (see text). (*From Ref. 20.*)

can see that the peak in the density of states, associated with the t_{1u}-derived band just above the Fermi level E_F in the C$_{60}$ trace, moves closer to E_F as K is added and some K$_3$C$_{60}$ phase is introduced. Upon further addition of potassium beyond $x = 3$, the t_{1u}-related peak crosses the Fermi level as the K$_4$C$_{60}$ and K$_6$C$_{60}$ phases form, and eventually for $5.8 \leq x \leq 6.0$, the t_{1u}-derived peak falls below E_F, indicating complete filling of the t_{1u}-derived level; the data for this composition further show a small bandgap to the next higher lying t_{1g}-derived level. No absorption threshold was detected down to 0.5 eV in the M$_6$C$_{60}$ materials, establishing an experimental upper bound for the bandgap between the t_{1u}-derived and t_{1g}-derived one-electron bands [1]. Photoemission and inverse photoemission [29] studies have confirmed strong similarities between the electronic structure of K$_x$C$_{60}$ and both Rb$_x$C$_{60}$ and Cs$_x$C$_{60}$. However, significant differences have also been clearly demonstrated between the density of states for K$_x$C$_{60}$ and for both

Na_xC_{60} and Ca_xC_{60}, for which multiple metal ions can be accommodated in the same octahedral interstitial sites in the lattice [1].

2. CARBON NANOTUBES

The electron and phonon dispersion relations for carbon nanotubes differ from those in fullerenes in a fundamental way, arising from the one-dimensional characteristic symmetry of carbon nanotubes. The first observation of carbon nanotubes in 1991 involved multi-wall nanotubes [30]. A multiwall carbon nanotube is a collection of coaxial single-wall nanotubes, each nanotube being one atomic monolayer in thickness (see Figure 10). This early work on multiwall nanotubes stimulated a large number of theoretical studies on the structure and properties of the simpler single-wall carbon nanotube constituents (see Figure 10). The experimental discovery of single-wall carbon nanotubes in 1993 [31,32] further

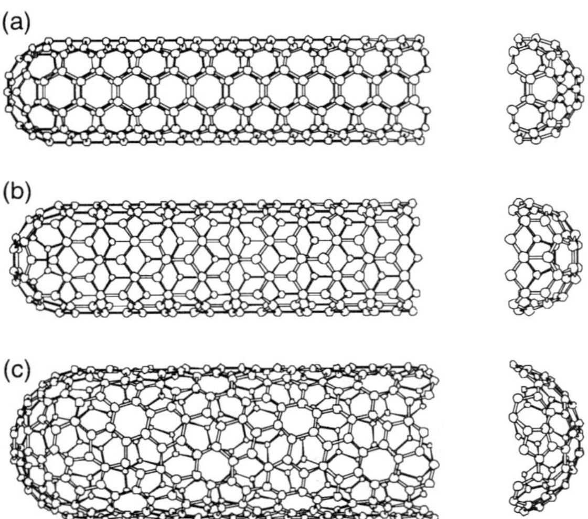

Figure 10 Schematic models for a single-wall carbon nanotube with the nanotube axis normal to: (a) the $\theta = 30°$ direction [an ''armchair'' (n, n) nanotube], (b) the $\theta = 0°$ direction [a ''zigzag'' $(n, 0)$ nanotube], and (c) a general direction, such as OB (see Figure 11), with $0 < \theta < 30°$ [a ''chiral'' (n, m) nanotube]. The actual nanotubes shown here correspond to (n, m) values of (a) (5, 5), (b) (9, 0), and (c) (10, 5). The nanotubes in this figure are capped by one half of a fullerene molecule. Single-wall nanotubes currently being synthesized have a typical aspect ratio (length/diameter) of $10^3 - 10^4$. (*From Ref. 34.*)

stimulated work in the field, but because of their short supply, most of the experimental studies continued to be done on multiwall nanotubes. This situation continued until the discovery in 1996 of a much more efficient synthesis route. Laser vaporization of graphite, in a catalyzed reaction [33], was used to prepare *bundles* of weakly coupled single-wall nanotubes, as discussed further below. The availability of single-wall nanotubes with a relatively narrow distribution of diameters [33] greatly stimulated systematic experimental studies of carbon nanotubes and has facilitated comparisons to theoretical calculations.

2.1 Structure of Carbon Nanotubes

The basic carbon nanotube is a cylinder of carbon atoms arranged on a honeycomb lattice, as in a single layer of graphite, and with almost the same nearest-neighbor C—C spacing (1.421 Å in graphite and 1.44 Å in carbon nanotubes) [35,36]. Carbon nanotubes can have any one of the three basic geometries shown in Figure 10. Carbon nanotubes are usually capped at either end by half of a fullerene (see Figure 10), so that the smallest diameter nanotube that is observed corresponds to the smallest diameter fullerene obeying the isolated pentagon rule, C_{60}, which has a diameter of 7.1 Å.

The structure of the nanotube can be understood by referring to Figure 11, which demonstrates the rolling of a segment of a single graphite layer (called a graphene sheet) into a seamless cylinder. In this figure we see that points O and A are crystallographically equivalent on the graphene sheet. The points O and A can be connected by a chiral vector $\vec{C}_h = n\hat{a}_1 + m\hat{a}_2$, where $\hat{a}_1$ and $\hat{a}_2$ are unit vectors for the honeycomb lattice of the graphene sheet. Lines OB and AB' are perpendicular to $\vec{C}_h$ at points O and A. If we now roll the graphene sheet and superimpose OB onto AB', then we obtain a cylinder of carbon atoms which constitutes a carbon nanotube. A single-wall carbon nanotube is thus uniquely determined by the integers (n, m) which specify the chiral vector $\vec{C}_h$. The nanotube can also be uniquely specified by its diameter d_t and chiral angle θ, where

$$d_t = \frac{C_h}{\pi} = \frac{\sqrt{3}a_{C-C}(m^2 + mn + n^2)^{1/2}}{\pi} \tag{1}$$

in which a_{C-C} is the nearest-neighbor C—C distance, and C_h is the length of the chiral vector $\vec{C}_h$. The chiral angle θ is defined as the angle that $\vec{C}_h$ makes with the "zigzag" direction ($\hat{a}_1$), and is given by

$$\theta = \tan^{-1}\left(\frac{\sqrt{3}m}{m + 2n}\right) \tag{2}$$

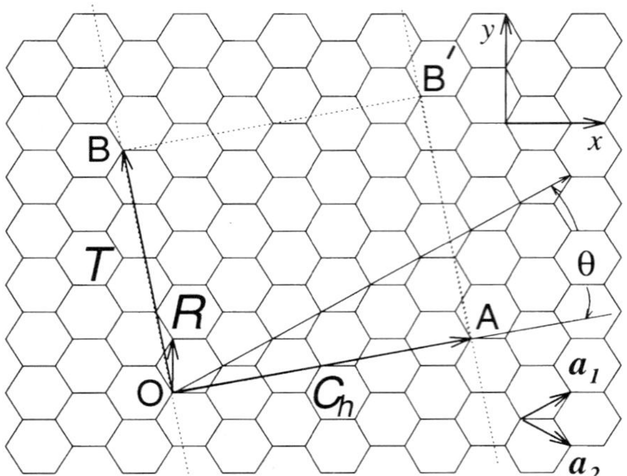

Figure 11 The chiral vector $\vec{C}_h = n\hat{a}_1 + m\hat{a}_2$ is defined on the flat honeycomb lattice of carbon atoms by the graphene unit vectors $\hat{a}_1$ and $\hat{a}_2$ and (n, m) are integers. The chiral angle θ is defined as the angle between $\vec{C}_h$ and $\hat{a}_1$. Along the "zigzag axis" ($\hat{a}_1$), we have $\theta = 0°$. Also shown is the lattice vector $\vec{OB} \equiv \vec{T}$ of the 1D nanotube unit cell. The vector $\vec{R}$ denotes the basic symmetry operation for the carbon nanotube. The diagram is constructed for $(n, m) = (4, 2)$. The area defined by the rectangle $(OAB'B)$ contains the carbon atoms that lie within the 1D unit cell of the nanotube. (*From Ref. 1.*)

as shown in Figure 11. The unit cell of the 1D carbon nanotube is, therefore, a rectangle formed by the vectors $\vec{C}_h$ and $\vec{T}$ shown in Figure 11, where $\vec{T}$ is the smallest lattice vector from O in the direction normal to $\vec{C}_h$, and parallel to the nanotube axis. Since the basis vectors $\vec{C}_h$ and $\vec{T}$ of the 1D unit cell are large compared with the basis vectors $\hat{a}_1$ and $\hat{a}_2$ of the unit cell for the graphene sheet (see Figure 11), the reciprocal space 1D unit cell (Brillouin zone) is small compared to that for the graphene sheet.

The as-prepared ropes (or bundles) of single-wall nanotubes produced by either the laser vaporization method [33] or the carbon arc method [37] appear in a scanning electron microscope (SEM) image as a mat of carbon "ropes" 10–20 nm in diameter and up to 100 µm or more in length, as shown in Figure 12 [37]. Under transmission electron microscope (TEM) examination, each rope is found to consist of a bundle of single-wall carbon nanotubes aligned along a common axis and arranged on a triangular lattice. X-ray diffraction (which views many ropes at once) and transmission electron microscopy (which views a single rope) show [33] that the diameters of the single-wall nanotubes have a strongly peaked narrow distribution of diameters. For both the laser vaporization and arc discharge synthesis conditions that were used in early work [33,37], the diameter

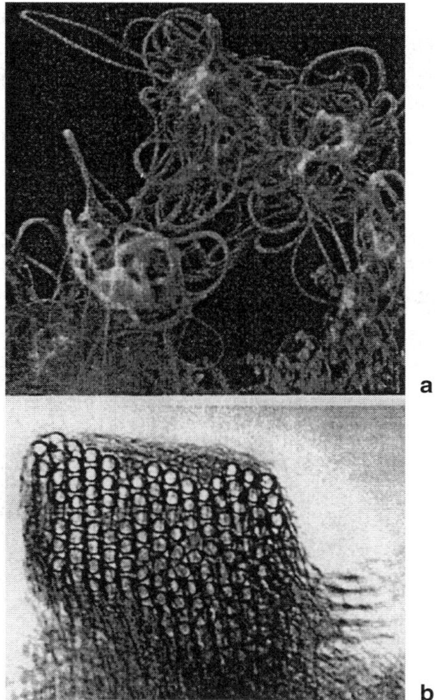

Figure 12   (a) Ropes of single-wall carbon nanotubes observed by scanning electron microscopy (SEM). The ropes are 10–20 nm thick and ∼100 μm long. (b) At higher magnification, the TEM image shows that each rope contains a bundle of single-wall nanotubes with diameters of ∼1.4 nm, arranged in a triangular lattice (with lattice constant 1.7 nm). The image in (b) is seen when the rope bends through the image plane of the transmission electron microscope (TEM). (*From Ref. 33.*)

distribution was found to be strongly peaked at 1.38±0.02 nm [33], very close to the diameter of an ideal armchair (10, 10) nanotube. X-ray diffraction measurements [33] showed that, within a bundle or rope, these single-wall nanotubes form a triangular lattice with a lattice constant of 1.7 nm, and an intertube separation of 0.315 nm at closest approach within a rope, in good agreement with prior theoretical modeling [38]. This lattice has been observed directly in TEM images of the rope cross section.

   Whereas multiwall carbon nanotubes do not require any catalyst for their growth, a catalyst is necessary for the growth of single-wall nanotubes [33]. Mixed metal catalysts (e.g., Ni-Co or Ni-Y) seem to be needed to synthesize ropes of single-wall carbon nanotubes efficiently by either the laser vaporization or arc methods [33,37,39]. The role of the catalyst is not yet well understood although it must simply prevent premature capping of the nanotube during growth. Variations in the most probable diameter and in the width of the diameter distribution is sensitively controlled by the composition of the catalyst, the growth temperature and other growth conditions [40]. Characterization of nanotubes with regard to their diameters and chiral angles can be carried out either by transmission electron microscopy (TEM) or by scanning tunneling microscopy (STM) (see Section 2.2). There is presently a large effort ongoing worldwide to produce narrower diameter distributions and to gain better control and understanding of the growth process.

2.2 Electronic Properties

Calculation of electronic energy band structure shows that for small diameter graphene nanotubes, about one-third of the nanotubes are metallic and two-thirds are semiconducting, depending on their nanotube diameter d_t and chiral angle θ. This unusual property follows from zone folding of the 2D dispersion relations $E_{g_{2D}}(k_x, k_y)$ for electrons on a graphene sheet [41]:

$$E_{g_{2D}}(k_x, k_y) =$$

$$\pm \gamma_0 \left\{ 1 + 4\cos\left(\frac{\sqrt{3}k_x a_0}{2}\right)\cos\left(\frac{k_y a_0}{2}\right) + 4\cos^2\left(\frac{k_y a_0}{2}\right) \right\}^{1/2} \quad (3)$$

where $a_0 = 1.421 \times \sqrt{3}$ Å is the lattice constant for a 2D graphene sheet and γ_0 is the nearest-neighbor C—C transfer integral [42] which has a value of about 2.7–2.8 eV for an isolated graphene sheet. A set of 1D energy bands is obtained from Eq. (3) by considering the quantum condition in the circumferential direction that $(\pi d_t / \lambda) = p$, where p is an integer and λ is the wavelength of the electron. The simplest cases to consider are the armchair and zigzag nanotubes, which have the highest symmetry.

For the case of the armchair (n, n) and zigzag $(n, 0)$ nanotubes, it is convenient to choose a real space rectangular unit cell in accordance with Figure 13. As shown for simplicity, the area of each real space unit cell for both the armchair and zigzag nanotubes contains only two hexagons or four carbon atoms, and is defined by the vectors $\vec{T}$ and $\vec{C}_h$. The area of the cylindrical belt around a zigzag or armchair nanotube is actually n times larger than the real space unit cells shown in Figure 13a and b. The appropriate periodic boundary conditions used to obtain the energy eigenvalues for the (N_x, N_x) armchair nanotube define the small number of allowed wave vectors $k_{x,q}$ in the circumferential direction

$$N_x \sqrt{3} a_0 k_{x,q} = q2\pi \qquad q = 1, \ldots, N_x \qquad (4)$$

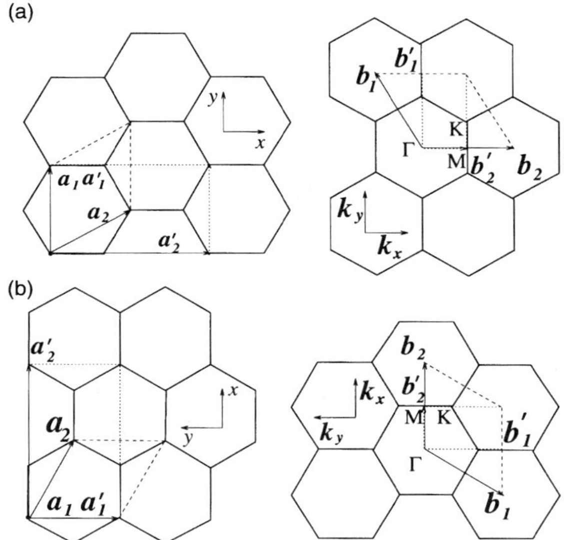

Figure 13 Real space unit cell and Brillouin zone for (a) armchair and (b) zigzag nano-tubes (dotted lines). Those for a 2D graphene sheet (dashed lines) are shown for comparison. See further explanation in the text. (*From Ref. 43.*)

Substitution of the discrete allowed values for $k_{x,q}$ given by Eq. (4) into Eq. (3) yields the energy dispersion relations $E_q^a(k)$ for the armchair nanotube [44]:

$$E_q^a(k) = \pm\gamma_0\left\{1 \pm 4\cos\left(\frac{q\pi}{N_x}\right)\cos\left(\frac{ka_0}{2}\right) + 4\cos^2\left(\frac{ka_0}{2}\right)\right\}^{1/2} \tag{5}$$

$$-\pi < ka_0 < \pi \qquad q = 1, \ldots, N_x$$

where the superscript a refers to the armchair nanotube, and k is a one-dimensional vector perpendicular to $k_{x,q}$ and directed along the nanotube axis [1].

Application of periodic boundary conditions associated with the rolling of the graphene sheet into a cylinder also yields the 1D dispersion relations for zigzag and chiral carbon nanotubes (see Figure 14). In general, metallic conduction in a carbon nanotube (n, m) is achieved when $|n - m| = 3q$, where n and m are integers specifying the nanotube diameter and chiral angle, and q is an integer. The calculations predict that all armchair (n, n) nanotubes are metallic (Figure 14a), but only one-third of the possible zigzag nanotubes are metallic (Figure 14b) [43,48,49], and likewise for the chiral nanotubes.

Scanning tunneling microscopy/spectroscopy (STM/STS) studies [50,51]

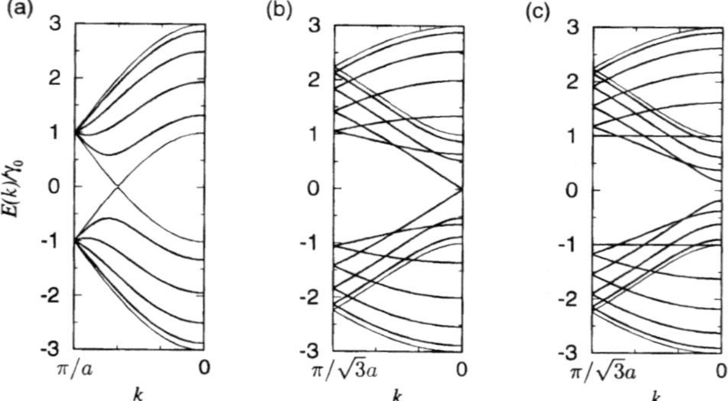

Figure 14 One-dimensional energy dispersion relations for (a) an armchair (5, 5) nanotube, (b) a zigzag (9, 0) nanotube, and (c) a zigzag (10, 0) nanotube, with the Fermi level placed at the zero of energy $E_F = 0$ for all three cases [45]. Here k is the 1D electron wave vector along the nanotube axis.

and resonant Raman studies [52] are sensitive to the electronic density of states of carbon nanotubes. Shown in Figure 15 are the 1D electronic density of states predicted for a metallic (9) zigzag nanotube (for which the 1D density of states is non-zero at the Fermi level $E_F = 0$), and for a semiconducting (10, 0) zigzag nanotube (for which the 1D density of states vanishes at the Fermi level which is located at $E_F = 0$). The spikes in these curves are associated with the $\rho(E) \approx |E - E_0|^{-1/2}$ singularities expected in the 1D density of states where E_0 denotes the energy of the edge of each subband. Also shown in Figure 15 for comparison is the density of states for a 2D graphene sheet (dashed lines). Combined STM/STS studies confirm that some nanotubes are metallic and some are semiconducting [50,51], and the current-voltage (I-V) curves show that the bandgap for the semiconducting nanotubes is proportional to the reciprocal of the nanotube diameter ($1/d_t$), independent of nanotube chirality, in agreement with the theoretical predictions, as shown in Figure 16 [1]. Measurements in the scanning tunneling mode (STM) [51,53,54] allow characterization of individual single-wall nanotubes according to their diameter and chiral angle [or equivalently their (n, m) indices could be found from d_t and θ by solving Eqs. (1) and (2)]. Measurements of the 1D density of states (which is proportional to dI/dV) are made in the scanning tunneling spectroscopy mode (as shown in Figure 17). The combined STM/STS studies [51,53,54] indicate that: (1) about two-thirds of the nanotubes are semiconducting, one-third metallic; (2) the density of states exhibits singularities characteristic of expectations for 1D systems; (3) E_g for the semiconducting nanotubes is proportional to $1/d_t$ with $E_g = 2\gamma_0 a_{C-C}/d_t$ and γ_0 is found to be 2.7 eV from the STS data [51], in good agreement with theory; (4) the density

(a) *(n,m)=(9,0)*

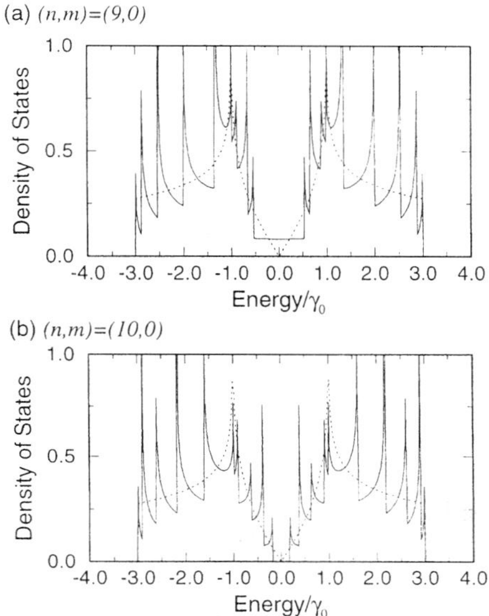

(b) *(n,m)=(10,0)*

Figure 15 Electronic 1D density of states (DOS) per unit cell vs. dimensionless energy normalized to γ_0, the nearest neighbor overlap energy of a 2D graphene sheet for two zigzag nanotubes: (a) the (9, 0) nanotube which has metallic behavior, (b) the (10, 0) nanotube which has semiconducting behavior. Also shown in the figures as dotted lines is the density of electronic states for a 2D graphene sheet. The dependence of the electronic density of states on nanotube index (n, m) is illustrated in Figure 30, where DOS curves for armchair (n, n) nanotubes are shown for different values of n in the range $8 \leq n \leq 11$. (*From Ref. 46.*)

of states near $E_F = 0$ is zero for semiconducting nanotubes, and nonzero for metallic nanotubes [54]; and (5) the energy separation between the two spikes on either side of the Fermi level is much larger for the case of metallic nanotubes than for semiconducting nanotubes, as shown in Figures 15 and 17 [51]. These 1D density of states curves are also important for explaining the quantum effects observed in the resonant Raman experiments on carbon nanotubes discussed in Section 2.5. More detailed future STM/STS studies should provide more quantitative information on the band model and the band parameters governing the 1D electronic structure for carbon nanotubes.

2.3 Transport Properties

The transport properties of carbon nanotubes, of course, depend strongly on whether they are metallic or semiconducting. Differences in transport properties

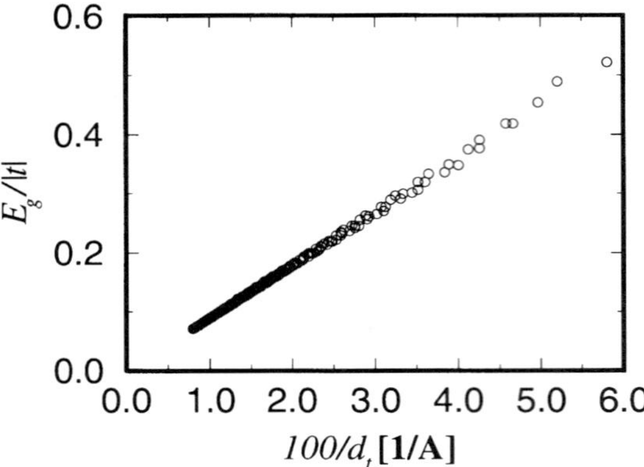

Figure 16   The energy gap E_g scaled by the transfer integral $|t| = \gamma_0$ for a semiconducting, chiral carbon nanotube as a function of $100/d_t$, where d_t is the nanotube diameter in Å and a value of $\gamma_0 = |t| = 2.5$ eV was taken for the transfer integral. It is noted that the relationship between E_g and $1/d_t$ becomes linear for large values of d_t and is given by $E_g = 2\gamma_0 a_{C-C}/d_t$ [47]. Note that the dependence of E_g on $1/d_t$ is independent of chiral angle and that about one-third of the nanotubes are metallic and therefore do not pertain to this plot.

are expected based on the nanotube symmetry (n, m). To date, transport properties studies have focused on the electrical resistance of carbon nanotubes. The samples have taken the form of mats of nanotube bundles, individual bundles or ropes, and individual multiwall nanotubes. Even individual single wall nanotubes have been studied. What is measured is really the current per unit electric field. For a single nanotube this quantity should be viewed as the conductance of the nanotube. The conversion of this experimental quantity into a volume conductivity or resistivity provides limited insight into the physical conduction mechanism. This view is even more relevant for studies on collections of nanotubes. The interpretation of resistance or conductance measurements on mats of nanotubes, multiwall nanotubes and even individual nanotube bundles require some assumptions regarding the nature of internanotube and interbundle electrical contacts, as well as the nature of the external contacts to the nanotubes.

The first transport measurement on an individual isolated single-wall carbon nanotube was made on a nanotube 1.3 nm in diameter resting across Pt electrodes on a Si/SiO_2 substrate. Two Pt electrodes (15 nm thick and 140 nm wide) were used to measure the current vs. bias voltage V_{bias}, where V_{bias} denotes the difference in the chemical potential between the two electrodes, μ_1 and μ_2, as shown in Figure 18 [55]. A field effect transistor was constructed by the appli-

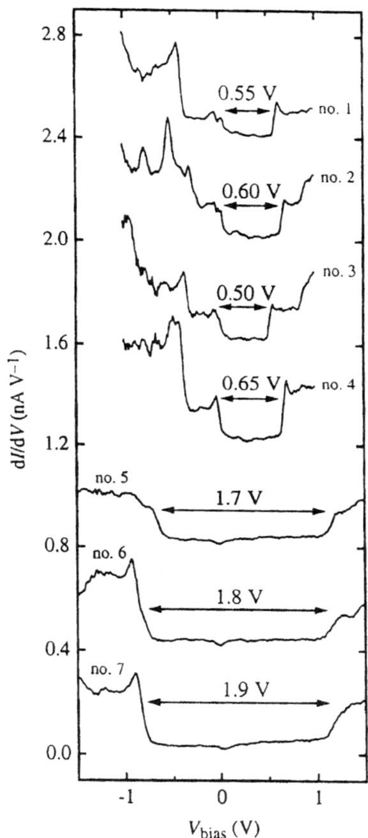

Figure 17 Derivative of the current-voltage curves (dI/dV) obtained by scanning tunneling spectroscopy on various nanotubes. Using the STM image mode, it was found that nanotubes labeled #1–6 are chiral and #7 is zigzag. The bias voltage V_{bias} applied to the sample corresponds to the tunneling electron energy relative to the Fermi level. Peaks in the tunneling conductance (dI/dV) in the figure correspond to peaks (singularities) in the 1D electronic density of states (see Figures 15 and 30). Energy differences between the peaks in the 1D density of states of the highest valence band and the lowest conduction band are indicated by the arrows. Nanotubes #1–4 typically have values of ~0.5–0.6 V between these 1D density of states peaks (see Figure 15), and are identified as semiconducting, whereas nanotubes #5–7 have larger values of ~1.7–1.9 V, indicating that these are metallic. The small dips at zero bias present in some of the curves are due to an experimental artifact. (*From Ref. 51.*)

cation of a gate voltage V_{gate}, to change the local electrostatic potential seen by the nanotube (see top of Figure 18). As the gate voltage V_{gate} was changed, the positions of the energy levels of the nanotube were varied relative to the chemical potentials μ_1 and μ_2 at the platinum current contacts. Because of the finite length ($L = 3\ \mu m$) of the nanotube, the one-dimensional energy band is split into propagating energy states whose spacing is $\Delta E = h v_F/2L$. Figure 19 shows that tunneling occurs when one of these quantum dot levels falls in the range of the tunneling window shown on the bottom right of Figure 19a. The results shown in Figure 19 yield a value of $\Delta E \sim 0.6$ meV, where the Fermi velocity is taken as $v_F \sim 8.1 \times 10^5$ m/s [55]. Single-wall carbon nanotubes thus provide a unique system for studying single molecule transistor effects [56,57].

The I versus V_{bias} curves obtained at a gate voltage of 88.2 mV (curve A), 104.1 mV (curve B), and 120.0 mV (curve C) at 5 mK are shown in Figure 19a [55]. Curves A, B, C all show plateaus of nonzero current, which present clear evidence for ballistic transport in which a conducting channel is in the tunneling window range of $V_{bias} = (\mu_1 - \mu_2)/e$ (see lower right of the figure). The position of the steps in the I versus V_{bias} curves is changed by increasing the gate voltage. Coulomb charging effects of the nanotube were also observed at low temperature, whereby the nanotube, considered as a capacitor with a capacitance C, has a thermal energy k_BT smaller than the charging energy E_c for a single electron, where $k_BT < E_c = e^2/2C$. A Coulomb blockade is observed when current flow

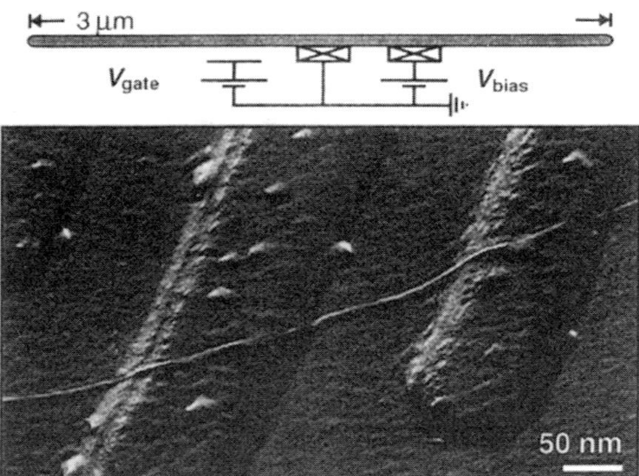

Figure 18   AFM tapping-mode image of a thin (~ 1 nm) carbon nanotube lying across two 15 nm thick and 140 nm wide Pt electrodes deposited using modern lithographic techniques on a Si/SiO$_2$ substrate. A gate voltage was used to control the current passed down the nanotube. A schematic circuit diagram for the transport experiments shown in Figure 19 is given at the top. (*From Ref. 55.*)

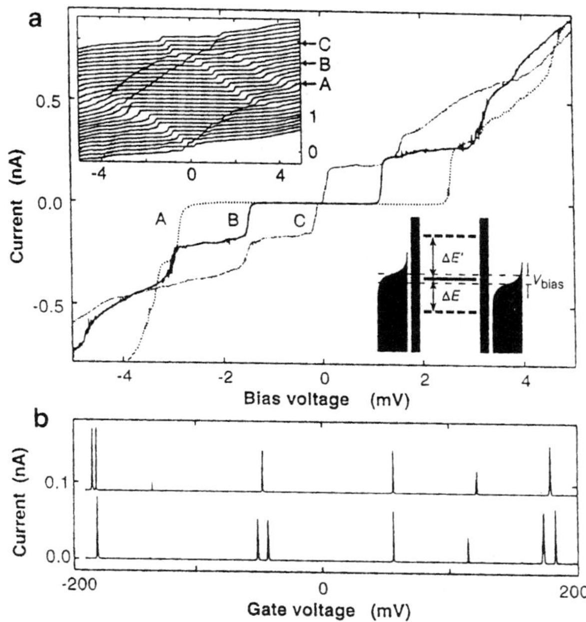

Figure 19   (a) Low temperature ($\sim$5 mK) current-voltage curves of a carbon nanotube at a gate voltage of 88.2 mV (curve *A*), 104.1 mV (curve *B*), and 120.0 mV (curve *C*). In the upper left inset, a number of I versus V_{bias} curves are shown with V_{gate} ranging from 50 mV (bottom curve) to 136 mV (top curve). In the lower right, the wide window for tunneling is shown to occur when a quantum-dot level lies within this window of width ($\mu_2 - \mu_1$) which is equal to V_{bias}. (b) The two traces of current versus gate voltage at V_{bias} = 30μV were performed under the same conditions and represent a bistability of the system. (*From Ref. 55.*)

is blocked by the energy shift E_c which shifts levels out of the bias window between μ_1 and μ_2. Thus, current flow in Figure 19b appears only when $V_{\text{bias}} > E_c$. Systematic studies of the Coulomb blockade phenomena as a function of nanotube diameter in armchair nanotubes and in other metallic nanotubes are of great current interest [57]. Since the position of the nanotube energy levels can be modified by V_{gate}, the step positions shown in the inset of Figure 19a can be changed smoothly as a function of V_{gate}. The phenomena observed in Figure 19 can be understood by steps in the quantum conductance, and charging effects associated with the microcapacitances of the nanotube and the gate.

Transport measurements (current-voltage) on a single 12 nm diameter rope, containing about 60 single-wall nanotubes of 1.4 nm diameter [58], are shown

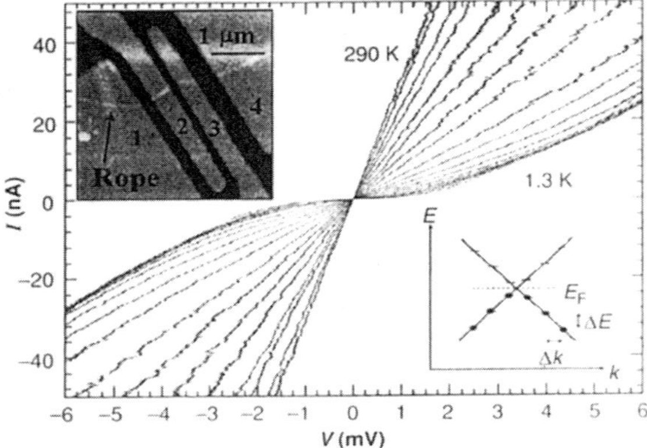

Figure 20 I versus V characteristics of a single 12-nm-diameter rope made up about 60 single-wall nanotubes with diameters of about 1.4 nm. The I versus V curves are taken between contacts #2 and #3 (left inset) and at various temperatures between 1.3 K and 290 K. The right inset schematically shows energy levels near the Fermi energy. These energy levels are quantized (quantum dot levels) because of the finite length of the nanotube. (*From Ref. 58.*)

in Figure 20 for the rope segment between contacts 2 and 3 (left inset). At room temperature, the I versus V characteristic exhibits a linear (Ohmic) behavior, while the reduction in conductance near $V = 0$ for $T < 10K$ is interpreted in terms of a band gap associated with the charging energy of single electron tunneling (i.e., Coulomb blockade effect). These data further show an increase in resistance with decreasing temperature over a wide temperature range from 1.3 to 290 K. In another experiment on a single rope, about a 10% increase of the resistivity was reported from 50 K to 280 K [59], while a decrease in the resistivity with increasing T was found below 50 K. The reasons for these discrepancies in the temperature-dependent resistivity of carbon nanotubes within a single rope are not presently understood. Since electrical contact may not have been made to all the individual nanotubes of a nanotube bundle, the transport of electrons between single-wall nanotubes in a rope may also make an important contribution to the observed transport properties.

Consistent with early theoretical predictions, the resistance measurements for various kinds of nanotube samples by a number of different groups generally show that there are metallic and semiconducting nanotubes [60–62] (see Section 2.2). Absolute values of resistivity are usually difficult to deduce quantitatively, for reasons stated above, so that the actual measurements are, at best, proportional to a volume resistivity. With this caution in mind, resistivity values in the range 10^{-4}–10^{-3} Ωcm have been reported for metallic nanotubes, while the room temperature resistivity of semiconducting nanotubes is much higher, $\sim 10^{1}$ Ωcm. Semiconducting nanotubes exhibit a slope in a plot of log R (where R is the sample resistance) versus $1/T$, which indicates an energy gap in the range 0.1–0.3 eV, which is roughly consistent with theoretical values of the energy gap for the corresponding nanotube diameters. On the other hand, for the metallic nanotubes, large fluctuations from sample to sample are observed in the absolute values of the resistance and in the temperature dependence of the resistance over a wide temperature range from 4 to 300 K. By plotting the temperature dependent data as $\Delta R(T)/R_{300} = [R(T) - R_{300}]/R_{300}$, where R_{300} is the room temperature resistance the comparison between the measurements of different groups could perhaps be facilitated. Differences in behavior from sample to sample, and from group to group, may be associated with the different (n, m) distribution of nanotubes in samples under investigation, to structural defects, differences in the diameters and chiralities of the constituent nanotubes, difference in the methods used to make contacts, and in the electrical fields associated with the contacts, and differences in the temperature range for the experiments.

In the low temperature regime, weak localization and universal conductance phenomena have been reported for an individual multiwall nanotube with an outer diameter of 20 nm [62]. Research opportunities remain for systematic studies of weak localization and universal conductance phenomena in the small diameter range where 1D phenomena are expected. There is particular value in studying the dependence of weak localization and universal conductance fluctuation phenomena on nanotube diameter and chirality. Theoretical models are needed to explain weak localization and universal conductance phenomena for carbon nanotubes as a function of nanotube diameter and chirality.

The junction between a semiconducting and a metallic nanotube has been of interest from a scientific point of view and in terms of device applications, should it be possible to synthesize such structures in a controlled way [56, 64–66]. The theoretical progress that has been made in understanding the structure and conductance properties of the junction region is encouraging.

2.4 Phonon Modes

The phonon dispersion relations for graphite are shown in Figure 21 and form the basis for understanding the spectra for carbon nanotubes. Because of the very weak coupling between carbon atoms on different graphite layers, it is customary to use the phonon dispersion relations in the basal plane for bulk graphite to describe the phonons on a 2D graphene sheet, which is defined as an isolated layer of crystalline graphite.

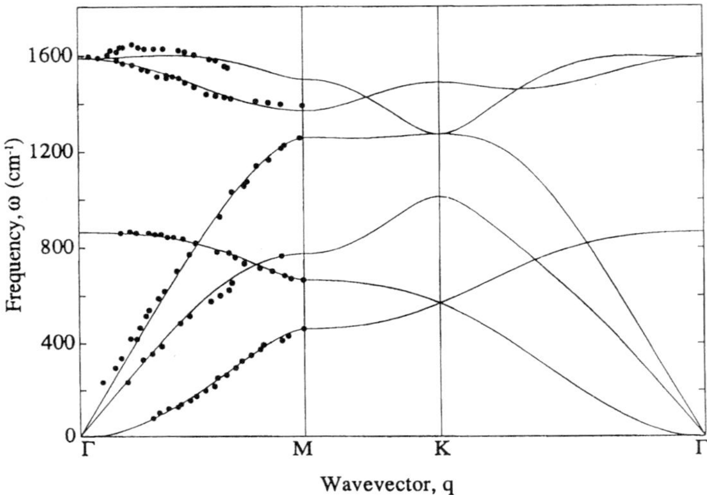

Figure 21 The phonon dispersion relations for graphite plotted along high-symmetry directions. Experimental points from neutron scattering and electron energy loss spectra were used to obtain values for the in-plane and out-of-plane carbon-carbon force constants and to determine the phonon dispersion relations throughout the Brillonin zone. (*From Ref. 63.*)

The phonon dispersion relations for a single-wall carbon nanotube can be obtained from those of the 2D graphene sheet by using the same zone folding approach as was used to obtain the 1D electronic dispersion relations in Section 2.2 [63, 67]. These 1D phonon dispersion relations are given by

$$\omega_{1D}(k) = \omega_{2D}(k\hat{K}_2 + \mu\hat{K}_1) \qquad \mu = 0, 1, 2, \ldots N - 1 \tag{6}$$

where the subscripts 1D and 2D refer, respectively, to the one-dimensional nanotube and the two-dimensional graphene sheet, k is the phonon wave vector along $\hat{K}_2$ the nanotube axis direction, μ is a nonnegative integer used to label the wave vectors or the states along the $\hat{K}_1$ reciprocal space direction normal to the nanotube axis, and N denotes the number of hexagons in the 1D unit cell of the nanotube. The boundary condition for periodicity in the circumferential direction is applied in writing the second term on the right-hand side of Eq. (6).

The proper application of the 1D dispersion relations [Eq. (6)] requires that the nanotubes be much longer than their diameters. In this limit, the k values are essentially continuous, and the contributions from the carbon atoms in the caps can be neglected. We give explicit results here for the symmetry analysis for the vibrational modes for the various types of carbon nanotubes. For armchair (n, n) nanotubes with even $n/2$, the vibrational modes at $k = 0$ are decomposed according to the following irreducible representations of the point group D_{nh}:

$$
\begin{aligned}
\Gamma^{\text{vib}}_{(n,n)} = {} & 4A_{1g} + 2A_{1u} + 4A_{2g} + 2A_{2u} + 2B_{1g} \\
& + 4B_{1u} + 2B_{2g} + 4B_{2u} + 4E_{1g} + 8E_{1u} + 8E_{2g} \\
& + 4E_{2u} + \cdots + 4E_{(n/2-1)g} + 8E_{(n/2-1)u} \qquad n/2 = \text{even integer}
\end{aligned} \tag{7}
$$

where the $\cdots$ denote the contributions from the E_{qg} and E_{qu} terms for $2 < q < n/2 - 1$. The zero frequency mode which describes a rigid rotation about the cylindrical axis has A_{2g} symmetry and the zero frequency rigid translation of the center of mass along the cylinder axis has A_{2u} symmetry, while the two rigid nanotube translations along the directions perpendicular to the nanotube axis have E_{1u} symmetry. The remaining modes at $k = 0$ are optical modes (i.e., $\omega \neq 0$). As an example of an (n, n) armchair nanotube where $n/2$ is an even integer, we consider the (8,8) armchair nanotube which has the number of hexagons in the 1D unit cell $N = 16$ and the resulting 32 carbon atoms have 96 degrees of freedom, of which 92 are optical modes and 4 have $\omega \to 0$ as $k \to 0$, as discussed above. Equation (7) shows that there are 60 distinct mode frequencies for the (8, 8) nanotube at $k = 0$ of which 24 are nondegenerate, 36 are doubly degenerate and 57 correspond to non-zero mode frequencies. From the character table for the symmetry group D_{nh}, the modes that transform according to the A_{1g}, E_{1g}, or E_{2g} irreducible representations are Raman active, while those that transform as A_{2u} or E_{1u} are infrared active. Hence there are only 16 distinct Raman-active mode frequencies $(4A_{1g} + 4E_{1g} + 8E_{2g})$, only 8 distinct infrared-active non-zero frequencies $(A_{1u} + 7E_{1u})$, and 33 optically silent nonzero mode frequencies for (8, 8) nanotubes with D_{nh} symmetry.

If $n/2$ is an odd integer [such as for $(n, n) = (10, 10)$], the 4 and 8 are interchanged in the last two terms in Eq. (7). Although the number of vibrational modes increases as the diameter of the carbon nanotube increases, the number of Raman-active and infrared-active modes remains constant, yielding the same number of Raman-active and infrared-active modes for the (6, 6) and (8, 8) armchair nanotubes. The $4E_{3g} + 8E_{3u}$ modes, present in the (8, 8) nanotube but absent in the (6, 6) nanotube, are all optically silent modes. The constant number of Raman-active and infrared-active phonon nanotube modes for a given symmetry group, or for a given chirality, is common to each category of carbon nanotube. As an example of calculated phonon dispersion relations $\omega(k)$, we show the results calculated by a fully 3D model for a (10, 10) armchair carbon nanotube in Figure 22 [35,69]. For the $N = 20$ hexagons per circumferential strip, we have 120 vibrational degrees of freedom, but because of mode degeneracies there are only 72 distinct normal mode frequencies at $k = 0$, of which 24 modes are nondegenerate and 48 are doubly degenerate.

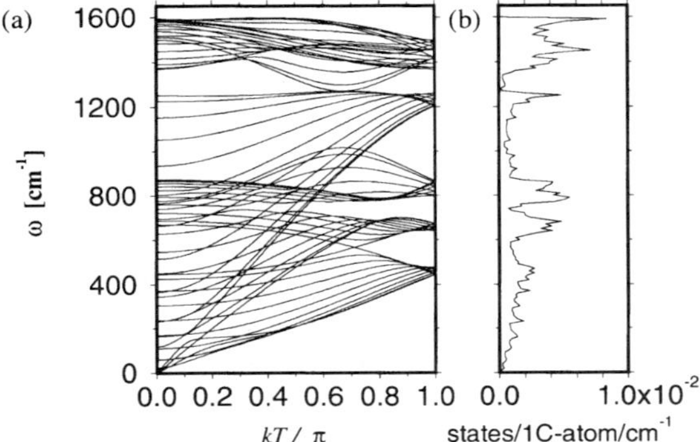

Figure 22   (a) The calculated phonon dispersion relations of an armchair carbon nanotube with $\vec{C}_h = (10, 10)$. (b) Phonon density of states for a $(10, 10)$ nanotube [68]. For the $(10, 10)$ nanotube, $N = 20$, and there are 120 phonon branches with 72 distinct frequencies or irreducible representations. Of these 16 are Raman-active $(4A_{1g} + 4E_{1g} + 8E_{2g})$ at the Γ-point $(k = 0)$, 8 are infrared-active $(A_{2u} + 7E_{1u})$, and 45 are optically silent $(\omega \neq 0)$. The acoustic modes and the rotational mode with $\omega = 0$ at $k = 0$ have symmetries $(A_{2g} + A_{2u} + E_{1u})$. (*From Ref. 35.*)

Armchair nanotubes (n, n) for which n is an odd integer exhibit D_{nd} symmetry and have vibrational modes with the following symmetries:

$$\Gamma^{\text{vib}}_{(n,n)} = 3A_{1g} + 3A_{1u} + 3A_{2g} + 3A_{2u}$$
$$+ 6E_{1g} + 6E_{1u} + 6E_{2g} + 6E_{2u} \tag{8}$$
$$+ \cdots + 6E_{[(n-1)/2]g} + 6E_{[(n-1)/2]u} \qquad n = \text{odd integer}$$

in which $+ \cdots +$ denotes modes associated with irreducible representations $6E_{qg}$ and $6E_{qu}$ where $2 < q < (n - 1)/2$. From the character table for group D_{nd}, the zone center $k = 0$ vibrational frequencies are zero for one A_{2u} mode (displacement of carbon atoms along the nanotube axis) and for the doubly degenerate E_{1u} mode (displacement of carbon atoms perpendicular to the nanotube axis), while the rotational mode with $\omega \to 0$ as $k \to 0$ has A_{2g} symmetry. Furthermore, there are 15 distinct optical Raman-active mode frequencies $(3A_{1g} + 6E_{1g} + 6E_{2g})$ and 7 distinct optical infrared-active mode frequencies $(2A_{1u} + 5E_{1u})$ for armchair nanotubes with D_{nd} symmetry.

The idea of zone folding is applicable for almost all the phonon modes of a carbon nanotube. However, it has been pointed out [63] that zone-folding alone does not always give the correct dispersion relations for a carbon nanotube, especially in the low frequency region, and some additional physical concepts must be introduced. For example, the out-of-plane tangential acoustic (TA) modes of

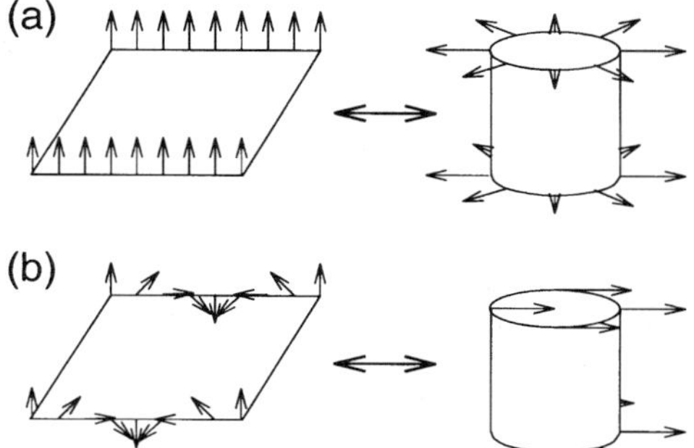

Figure 23   (a) The zero frequency out-of-plane transverse acoustic modes at k = 0 (left) in a single layer of graphite give rise to a radial breathing mode in the carbon nanotube with non-zero frequency (right). (b) An acoustic mode of a carbon nanotube whose vibration is perpendicular to the nanotube axis (right) corresponds to a linear combination of both in-plane and out-of-plane C-atom displacements on a graphene sheet (left). (*From Ref. 35.*)

a graphene sheet shown in Figure 23a on the left do not give zero energy at the Γ point when rolled into a nanotube, as shown on the right. Here, at $k = 0$, all the carbon atoms of the nanotube move radially with the same amount of displacement. This nanotube mode corresponds to an out-of-plane radial acoustic vibration in a graphene sheet. However, the breathing mode exhibits a nonzero frequency [63] (see Figure 23a on the right), and is featured prominently in the vibrational spectra of carbon nanotubes. In the breathing mode, only the out-of-plane force constant ϕ_r and the force constant ϕ_{ti} in the circumferential direction of the nanotube are related as a first approximation to the nanotube breathing mode, and this results in a finite frequency at the Γ point ($k = 0$). Since there is no vibration in the direction of the nanotube axis in the breathing mode, the bond angles of the hexagon network are modified during the breathing mode, so that it is necessary to also include more distant neighbor force constant in describing the nanotube radial breathing mode.

On the other hand, when we consider the vibrations of a carbon nanotube in the context of three-dimensional space, we generally expect three acoustic modes which correspond to rigid translations of the center of mass of the nanotube in the x, y, z directions in the limit $k \to 0$ as $\omega \to 0$. However, the two directions which are perpendicular to the nanotube axis do not correspond to any two-dimensional graphite phonon modes. In a graphene sheet, the in-plane and out-of-plane modes are decoupled from each other. However, when the graphene strip is rolled up into a nanotube, the graphite-derived in-plane and out-of-plane modes do couple to each other, as shown on the left-hand side of Figure 23b, to form the acoustic mode of the nanotube shown on the right. Furthermore, when the calculation for the phonon dispersion relations is carried out directly in 3D space, the proper symmetry of the nanotube is included in the construction of the dynamical matrix, as was, for example, done in Figure 22. In general, the phonon dispersion relations for a carbon nanotube depend on the (n, m) indices for the nanotube. As we shall see below, some modes are more sensitive to (n, m) or to the nanotube diameter than others (see Figure 24).

All zigzag nanotubes $(n, 0)$ have $3A_{1g}$, $3A_{2u}$, $6E_{1g}$, $6E_{1u}$, and $6E_{2g}$ optically active modes, irrespective of whether they have D_{nh} (n even) or D_{nd} (n odd) symmetries. For odd n, the vibrational modes of zigzag nanotubes are described by the D_{nd} point group and have symmetries

$$\Gamma^{\text{vib}}_{(n,0)} = 3A_{1g} + 3A_{1u} + 3A_{2g} + 3A_{2u}$$
$$+6E_{1g} + 6E_{1u} + 6E_{2g} + 6E_{2u} \tag{9}$$
$$+ \cdots + 6E_{[(n-1)/2]g} + 6E_{[(n-1)/2]u} \qquad n = \text{odd integer}$$

while zigzag nanotubes with even n are described by the D_{nh} point group and have symmetries

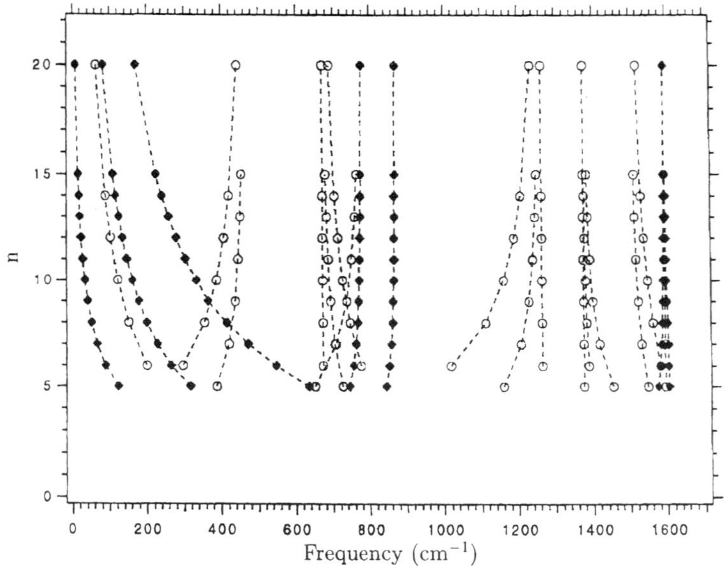

Figure 24 The armchair index n versus mode frequency for the Raman-active modes of single-wall armchair (n, n) carbon nanotubes [52]. For the armchair nanotubes the diameter is given by $d_t = 1.38 \times n$ Å.

$$\Gamma^{\text{vib}}_{(n,0)} = 3A_{1g} + 3A_{1u} + 3A_{2g} + 3A_{2u}$$
$$+ 3B_{1g} + 3B_{1u} + 3B_{2g} + 3B_{2u}$$
$$+ 6E_{1g} + 6E_{1u} + 6E_{2g} + 6E_{2u}$$
$$+ \cdots + 6E_{[(n-2)/2]g} + 6E_{[(n-2)/2]u} \qquad n = \text{even integer.} \tag{10}$$

For any zigzag nanotube $(n, 0)$ there are 15 optical Raman-active symmetry types $(3A_{1g} + 6E_{1g} + 6E_{2g})$ and 7 optical infrared-active symmetry types $(2A_{2u} + 5E_{1u})$ with distinct nonzero frequencies. The symmetries of the $\omega = 0$ modes are A_{2u} and E_{1u} for the acoustic branches and A_{2g} for the rotational $\omega = 0$ mode. As an example, consider the $(9, 0)$ zigzag nanotube which has $N = 18$ and 108 degrees of freedom and 60 distinct mode frequencies from Eq. (9). Of these distinct mode frequencies, 15 are Raman active, 7 are infrared active, and 35 are optically silent $(\omega \neq 0)$, while three $(A_{2u} + E_{1u} + A_{2g})$ correspond to the acoustic and rotational modes for which $\omega = 0$ at $k = 0$.

For chiral nanotubes (n, m), the symmetries of the vibrational modes are

$$\Gamma^{\text{vib}}_{N} = 6A + 6B + 6E_1 + 6E_2 + \cdots + 6E_{N/2-1} \tag{11}$$

Of these modes, the Raman-active modes are those that transform as A, E_1, and E_2 and the infrared-active modes are those that transform as A or E_1. This gives 15 distinct optical Raman-active mode frequencies $(3A + 6E_1 + 6E_2)$ and 9 distinct optical infrared-active mode frequencies $(3A + 6E_1)$, after subtracting the modes associated with acoustic translations $(A + E_1)$ and with rotation of the cylinder (A). Because of the large size of the 1D unit cell for chiral nanotubes, they have many degrees of freedom. For example, the $(n, m) = (7, 4)$ nanotube has a relatively small diameter $d_t = 0.75$ nm but a large number of hexagons in the unit cell $N = 62$, so that the 1D unit cell has 372 degrees of freedom. The phonon modes will include branches with $6A + 6B + 6E_1 + 6E_2 + \cdots + 6E_{30}$ symmetries, 192 branches in all. Of these, there are 15 distinct optical Raman-active mode frequencies at $k = 0$, while 9 distinct optical mode frequencies are infrared active, and 162 are optically silent at $k = 0$. Again the acoustic modes and rotational mode corresponding to $\omega = 0$ at $k = 0$ have symmetries $(2A + E_1)$.

The zone-folding method yields a large number of phonon modes, but only a few of the $k = 0$ vibrational modes are Raman active or infrared active, the others being optically silent for a first-order process. Since the number of Raman-active and infrared-active modes for a given symmetry category is independent of nanotube diameter, the dependence of a particular vibrational mode on nanotube diameter can be investigated. Many of the mode frequencies and their Raman cross sections are found to be highly sensitive to the nanotube diameter, while others are not. Figure 24 shows the dependence of the frequency of the Raman-active modes on the nanotube diameter for armchair nanotubes, expressed in terms of their (n, n) indices [70]. A similar diagram can be constructed for the infrared-active modes for the armchair nanotubes, and also for Raman and infrared active modes for zigzag and chiral nanotubes [1,70].

2.5 Raman Spectra

Raman spectroscopy has provided a particularly valuable tool for examining the mode frequencies of carbon nanotubes with specific diameters and hence evaluating the merits of theoretical models for the 1D phonon dispersion relations, for characterizing nanotube samples in terms of the diameter distribution of the nanotubes in the sample, and for studying the 1D electron density of states in resonant Raman experiments through the electron-phonon coupling mechanism [52].

The mode frequencies for some of the Raman-active modes for armchair nanotubes have been determined by Raman scattering experiments on a sample containing several ropes of single-wall carbon nanotubes prepared by laser vaporization techniques [33], and a room-temperature spectrum is shown in Figure 25

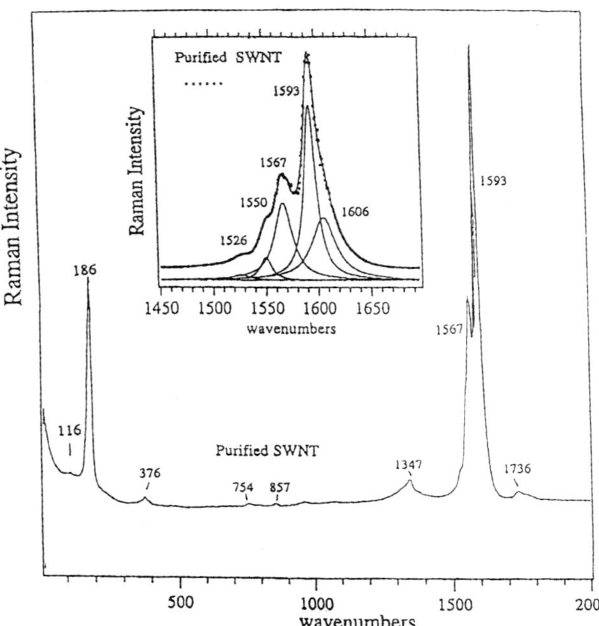

Figure 25 Experimental room temperature Raman spectrum from a sample consisting primarily of single-wall nanotubes with diameters near that of the (10, 10) nanotube. The spectrum was excited with 514.5 nm radiation from an Ar ion laser. (From Ref. 52.)

[52]. Prominent in this spectrum are a number of modes near 1580 cm^{-1}, which exhibit a weak dependence on nanotube diameter (see Figure 24), and a strong band at $\sim$186 cm^{-1}, which is identified with A_{1g} radial breathing modes for the constituent nanotubes in the sample. We note that the A_{1g} breathing mode is strongly dependent on the nanotube diameter, as shown in Figure 24. The atomic displacements associated with the most prominent of these modes are shown in Figure 26.

Most of the early experiments on the vibrational spectra of carbon nanotubes were carried out on multiwall carbon nanotubes which were of too large diameter to observe detailed quantum effects associated with the 1D dispersion relations discussed above [71,72]. Spectral features observed in the first- and second-order Raman spectra could be identified with features previously reported for disordered graphite, but somewhat shifted in frequency because of the presence of some small diameter nanotube constituents. An example of first- and second-order Raman spectra from a multiwall carbon nanotube sample is shown in Figure 27 [72] in comparison with spectra on a graphite powder sample, and the similarities between these spectra are clearly seen.

The first work to show clear evidence for 1D quantum effects in carbon

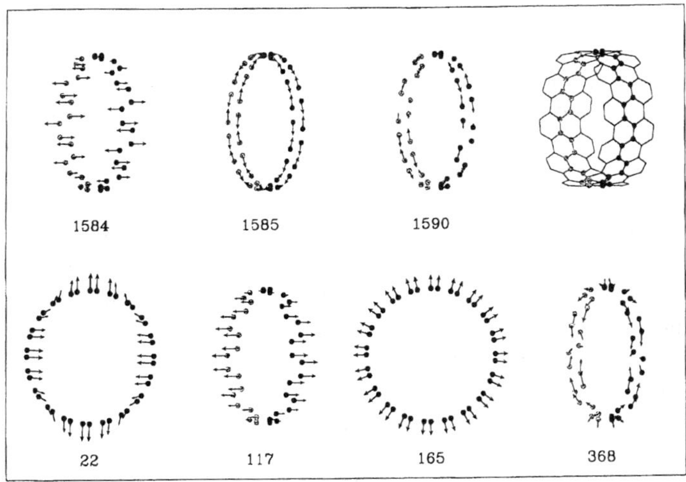

Figure 26 Normal mode carbon atom displacements for several of the modes prominent in the Raman spectra for single-wall nanotubes. Upper right: unit cell of a (10, 10) nanotube showing C-atoms as solid circles. The calculated C-atom mode frequencies listed below the mode displacements are pertinent to (10, 10) armchair nanotubes. (*From Ref. 52.*)

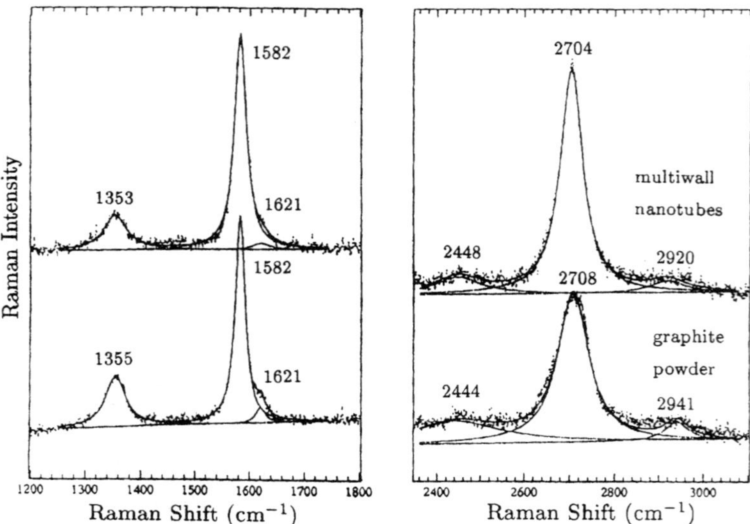

Figure 27   Experimental first-order (left) and second-order (right) Raman spectra from a sample consisting primarily of multiwall nanotubes shown in comparison with graphite powder, which can be considered to be weakly disordered graphite. (*From Ref. 72.*)

nanotubes [73] was carried out on samples containing only a small concentration of single-wall nanotubes, and having a wide distribution of diameters and chiralities, so that only the 1D-related features pertinent to phonon modes (near 1580 cm^{-1}) which are very weakly dependent on nanotube diameter, could be observed. No Raman features that could be identified with the A_{1g} radial breathing mode were found. Second-order harmonics and combination modes showed very little phonon dispersion shift relative to their related first-order constituents [73].

A major advance in Raman spectroscopy studies on carbon nanotubes occurred through the availability of ropes of single-wall carbon nanotubes [33] with a narrow distribution of diameters and chiral angles [74]. Raman spectra taken on such samples (see Figure 25) show a number of well-resolved Raman features, including intense features near 1580 cm^{-1}, similar to what had been previously reported [73,75], and a strong feature at 186 cm^{-1}, which is associated with the A_{1g} radial breathing mode of the nanotube. The dependence of the mode frequency of the radial breathing mode on nanotube radius is given in Figure 28 and this mode is not sensitive to the chiral angle θ.

Quantum effects are observed in the Raman spectra of single-wall carbon nanotubes through the resonant Raman enhancement effect, which is seen experimentally by measuring the Raman spectra at a number of laser excitation ener-

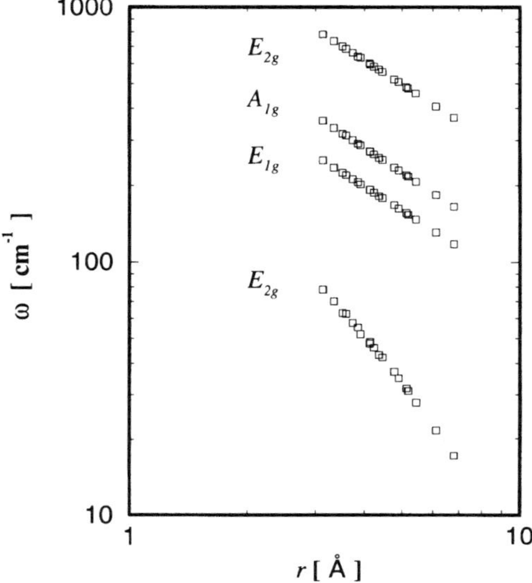

Figure 28   Log-log plot of the lower Raman mode frequencies as a function of carbon nanotube radius. Calculations show that the vibrational frequencies for these modes are not sensitive to the chiral angle θ. (*From Ref. 68.*)

gies, as shown in Figure 29. Resonant enhancement in the Raman scattering intensity from carbon nanotubes occurs when the laser excitation frequency corresponds to a transition between the sharp features in the one-dimensional electronic density of states of the carbon nanotube, as shown in Figure 30 for armchair nanotubes (8,8), (9,9), (10,10), and (11,11) [52]. Since the energies of these sharp features in the density of states are strongly dependent on the nanotube diameter, a change in the laser frequency brings into resonance a carbon nanotube with a different diameter. For example, the model calculation which gives rise to the electronic density of states in Figure 30 shows that the (10, 10) armchair nanotube would be expected to be resonant at a laser frequency of 1.28 eV, while the (9, 9) nanotube would be resonant at 1.42 eV. However, nanotubes with different diameters have different vibrational frequencies for the A_{1g} breathing mode (see Figures 24 and 28), independent of their chiral angle. By comparing the various Raman spectra in Figure 29, we see large differences in the vibrational frequencies of the strong A_{1g} mode, consistent with a resonant Raman effect involving nanotubes of different diameters. The dependence of the Raman intensity of the A_{1g} mode on laser excitation frequency (see Figure 30) is also consistent with a resonant Raman scattering mechanism. These quantum effects thus lend strong

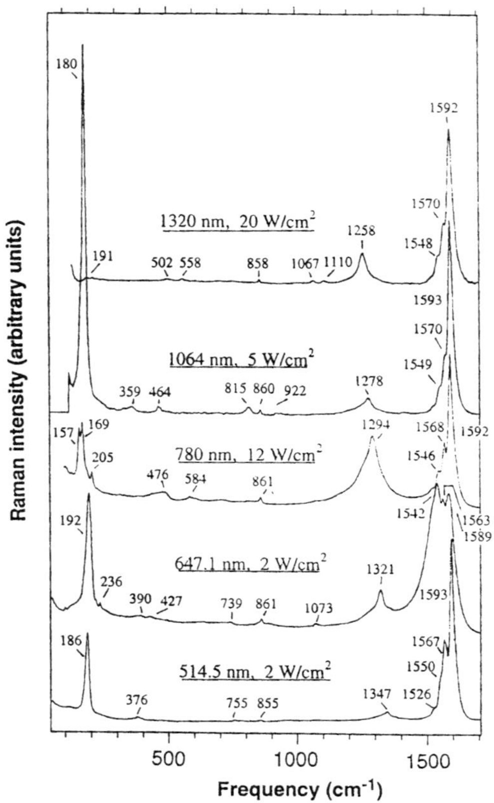

Figure 29 Experimental room temperature Raman spectra for purified single-wall carbon nanotubes excited at five different laser excitation wavelengths. The laser wavelength and power density for each spectrum are indicated, as are the vibrational frequencies (in cm⁻¹). The equivalent photon energies for the laser excitation are: 1320 nm (0.94 eV), 1064 nm (1.17 eV), 780 nm (1.58 eV), 780 nm (1.58 eV), 647.1 nm (1.92 eV), and 514.5 nm (2.41 eV). (*From Ref. 52.*)

credibility to the 1D aspects of the electronic and phonon structure of single-wall carbon nanotubes. In fact, these resonant Raman phenomena (Figure 29) provided the first clear confirmation for the theoretical predictions about the singularities in the 1D electronic density of states of carbon nanotubes. More detailed measurements of the 1D electronic density of states by STM/STS spectroscopy [51] (see Section 2.2) will be very helpful for the interpretation of the

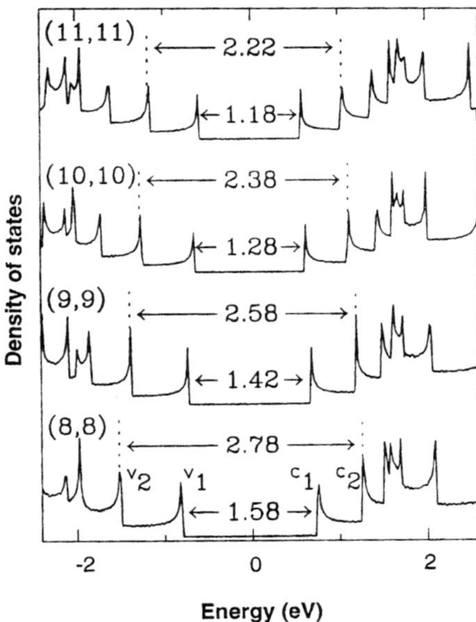

Figure 30 Electronic 1D density of states (DOS) calculated in a tight binding model for (8, 8), (9, 9), (10, 10), and (11, 11) armchair nanotubes. Wavevector conserving optical transitions can occur between mirror image peaks in the 1D density of states, i.e., $v_1 \rightarrow c_1$ and $v_2 \rightarrow c_2$, etc., and these optical transitions are responsible for the resonant Raman effect shown in Figure 29. (*From Ref. 52.*)

resonant Raman spectra and for the role of nanotube curvature in determining the detailed 1D electronic structure of carbon nanotubes. The distribution of nanotube diameters within the nanotube sample may also be responsible for some of the details of the spectral features near 1580 cm^{-1} in the inset to Figure 25.

Raman spectra have also been reported on ropes of single-wall carbon nanotubes doped with the alkali metals K and Rb and with the halogen Br$_2$ [76]. These spectra show effects similar to the effect of alkali metal and halogen intercalation into graphite [42], exhibiting upshifts in the frequency of the 1580 cm^{-1} band associated with the donation of electrons from graphite to halogens in the case of acceptors, and downshifts in mode frequencies associated with donor charge transfer to graphite in the case of alkali metal intercalation (see Figure 31). Just as for the case of the graphite intercalation compounds [42], these frequency shifts of the nanotube modes can, in principle, be calibrated and used to characterize the amount of intercalate uptake that has occurred on the nanotube wall.

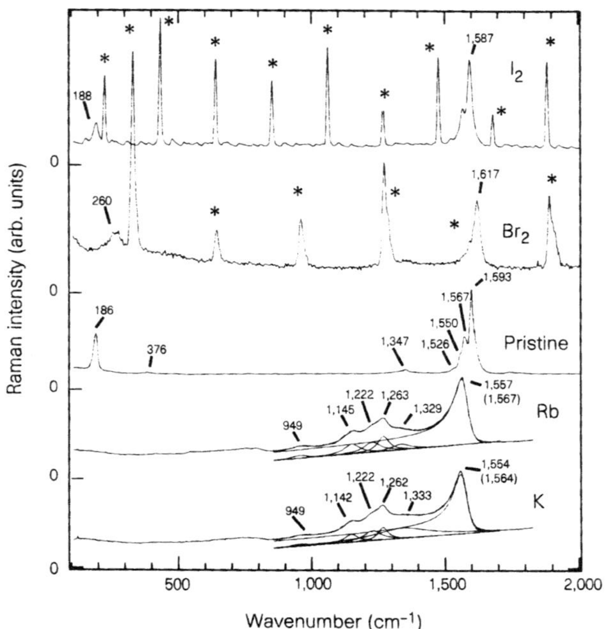

Figure 31　Raman spectra for pristine single-wall nanotube bundles reacted with the vapors of various donor and acceptor reagents. From top to bottom: I_2, Br_2, pristine single-wall nanotube, Rb, and K. The backscattering spectra were taken at $T = 300$ K using 514.5 nm radiation. In the spectra for both of the halogen-doped single-wall nanotube bundles, a harmonic series of peaks (indicated by an asterisk) are observed, which are identified with the fundamental stretching frequency ω_s: of $\sim$220 cm^{-1} (I_2) and of $\sim$324 cm^{-1} (Br_2). In the vicinity of the strongest high-frequency mode around 1550 cm^{-1}, the Raman spectra for single-wall nanotube bundles doped with K or Rb are fitted with a superposition of Lorentzian functions and an asymmetric Breit-Wigner-Fano lineshape on a linear continuum. Peak frequencies are indicated; the values in parentheses are renormalized phonon frequencies. (*From Ref. 76.*)

3.　CONCLUDING REMARKS

The last few years have seen the emergence of a variety of nanostructured π-bonded carbon materials, most notably the closed cage fullerene molecules, and the single-wall carbon nanotubes. The high present level of research activity on these carbon nano-materials is likely to lead to important scientific advances in the nanostructures field generally, and hopefully useful applications will soon follow. Because of the complexity of the carbon phase diagram and the unique locations of carbon in the periodic table, it is likely that other interesting nano-structured forms of carbon will be discovered, and that these new carbon materials will also lead to interesting science.

ACKNOWLEDGMENTS

We gratefully acknowledge the helpful discussions with Professor M. Endo and Dr. A.M. Rao. We are also in debt to many other colleagues for assistance. The research at MIT is funded by NSF grant DMR-98-04734. The work at UK was supported by NSF grant OSR-9452895. PCE was supported by the University of Kentucky Center for Applied Energy Research and the NSF Grant No. EHR-91-08764. The authors (RS, GD, MSD) thank the International Joint Research Program of the New Energy and Industrial Technology Organization (NEDO), Japan for their support. Part of the work by RS is supported by a Grant-in Aid for Scientific Research (No. 08454079 and No. 9243211) from the Ministry of Education and Science of Japan.

REFERENCES

1. M. S. Dresselhaus, G. Dresselhaus, and P. C. Eklund, *Science of Fullerenes and Carbon Nanotubes*, Academic Press, New York, 1996.
2. W. Krätschmer, L. D. Lamb, K. Fostiropoulos, and D. R. Huffman, *Nature (London)* **347**, 354–358 (1990).
3. K. A. Wang, Y. Wang, Ping Zhou, J. M. Holden, S. L. Ren, G. T. Hager, H. F. Ni, P. C. Eklund, G. Dresselhaus, and M. S. Dresselhaus, *Phys. Rev.* **B45**, 1955 (1992).
4. Ping Zhou, Kai-An Wang, Ying Wang, P. C. Eklund, M. S. Dresselhaus, G. Dresselhaus, and R. A. Jishi, *Phys. Rev.* **B46**, 2595 (1992).
5. T. Pichler, M. Matus, J. Kürti, and H. Kuzmany, *Phys. Rev.* **B45**, 13841 (1992).
6. S. H. Glarum, S. J. Duclos, and R. C. Haddon, *J. Am. Chem. Soc.* **114**, 1996–2001 (1992).
7. P. C. Eklund, Ping Zhou, Kai-An Wang, G. Dresselhaus, and M. S. Dresselhaus, *J. Phys. Chem. Solids* **53**, 1391 (1992).
8. K. A. Wang, A. M. Rao, P. C. Eklund, M. S. Dresselhaus, and G. Dresselhaus, *Phys. Rev.* **B48**, 11375–11380 (1993).
9. P. Zhou, K.-A. Wang, A. M. Rao, P. C. Eklund, G. Dresselhaus, and M. S. Dresselhaus, *Phys. Rev.* **B45**, 10838 (1992).
10. G. Dresselhaus, M. S. Dresselhaus, and P. C. Eklund, *Phys. Rev.* **B45**, 6923 (1992).
11. Ping Zhou, K. A. Wang, A. M. Rao, P. C. Eklund, G. Dresselhaus, and M. S. Dresselhaus, *Phys. Rev.* **B45**, 10838 (1992).
12. A. M. Rao, P. Zhou, K.-A. Wang, G. T. Hager, J. M. Holden, Ying Wang, W. T.

Lee, Xiang-Xin Bi, P. C. Eklund, D. S. Cornett, M. A. Duncan, and I. J. Amster, *Science* **259**, 955–957 (1993).

13. S. B. Fleischer, B. Pevzner, D. J. Dougherty, H. J. Zeiger, G. Dresselhaus, M. S. Dresselhaus, E. P. Ippen, and A. F. Hebard, *Appl. Phys. Lett.* **71**, 2734–2736 (1997).

14. Y. Wang, D. Tománek, and G. F. Bertsch, *Phys. Rev.* **B44**, 6562 (1991).

15. M. Matus and H. Kuzmany, *Appl. Phys.* **A56**, 241 (1993).

16. L. Pintschovius, B. Renker, F. Gompf, R. Heid, S. L. Chaplot, M. Haluška, and H. Kuzmany, *Phys. Rev. Lett.* **69**, 2662 (1992).

17. S. Saito and A. Oshiyama, *Phys. Rev. Lett.* **66**, 2637 (1991).

18. Gerhard Herzberg, Molecular Spectra and Molecular Structure, Van Nostrand, Princeton, New Jersy, 1945.

19. *Accounts Chem. Res.* **25**, 3 (1992). Edited by F. W. McLafferty and R. E. Smalley. A compilation of reviews with numerous references.

20. J. H. Weaver, *Accounts Chem. Res.* **25**, 143 (1992).

21. S. C. Erwin and M. R. Pederson, *Phys. Rev. Lett.* **67**, 1610 (1991).

22. H. Ajie, M. M. Alvarez, S. J. Anz, R. D. Beck, F. Diederich, K. Fostiropoulos, D. R. Huffman, W. Krätschmer, Y. Rubin, K. E. Schriver, D. Sensharma, and R. L. Whetten, *J. Phys. Chem.* **94**, 8630–8633 (1990).

23. T. W. Ebbesen, K. Tanigaki, and S. Kuroshima, *Chem. Phys. Lett.* **181**, 501 (1991).

24. K. Harigaya and S. Abe, *Phys. Rev.* **B49**, 16746–16752 (1994).

25. K. Harigaya and S. Abe, *Synth. Metals* **70**, 1415 (1995).

26. See chapter 1 of this volume.

27. See chapter 5 of this volume.

28. J. H. Weaver, J. L. Martins, T. Komeda, Y. Chen, T. R. Ohno, G. H. Kroll, N. Troullier, R. E. Haufler, and R. E. Smalley, *Phys. Rev. Lett.* **66**, 1741–1744 (1991).

29. M. B. Jost, N. Troullier, D. M. Poirier, J. L. Martins, J. H. Weaver, L. P. F. Chibante, and R. E. Smalley, *Phys. Rev.* **B44**, 1966 (1991).

30. S. Iijima, *Nature (London)* **354**, 56 (1991).

31. S. Iijima and T. Ichihashi, *Nature (London)* **363**, 603 (1993).

32. D. S. Bethune, C. H. Kiang, M. S. de Vries, G. Gorman, R. Savoy, J. Vazquez, and R. Beyers, *Nature (London)* **363**, 605 (1993).

33. A. Thess, R. Lee, P. Nikolaev, H. Dai, P. Petit, J. Robert, C. Xu, Y. H. Lee, S. G. Kim, A. G. Rinzler, D. T. Colbert, G. E. Scuseria, D. Tománek, J. E. Fischer, and R. E. Smalley, *Science* **273**, 483–487 (1996).

34. M. S. Dresselhaus, G. Dresselhaus, and R. Saito, *Carbon* **33**, 883–891 (1995).

35. R. Saito, J. Dresselhaus, and M. S. Dresselhaus, *Physical Properties of Carbon Nanotubes*, Imperial College Press, London, 1998.

36. T. Ebbesen, in *Fullerenes and Nanotubes* (Pierre Delhaès and P. M. Ajayan, eds.), Gordon and Breach, Paris, 1998. Series: *World of Carbon*, vol. 2.

37. C. Journet, W. K. Maser, P. Bernier, A. Loiseau, M. Lamy de la Chapelle, S. Lefrant, P. Deniard, R. Lee, and J. E. Fischer, *Nature (London)* **388**, 756 (1997).

38. J. C. Charlier, X. Gonze, and J. P. Michenaud, *Europhys. Lett.* **29**, 43–48 (1995).

39. T. W. Ebbesen, in *Carbon Nanotubes: Preparation and Properties* (T. Ebbesen, ed.) CRC Press, Inc., Boca Raton, Florida, 1997, chap. IV, p. 139.

40. S. Bandow, S. Asaka, Y. Saito, A. M. Rao, L. Grigorian, E. Richter and P. C. Eklund, *Phys. Rev. Lett.*, *80*, 3779–3782 (1998).

41. P. R. Wallace, *Phys. Rev.* **71**, 622 (1947).

42. M. S. Dresselhaus and G. Dresselhaus, *Advances Phys.* **30**, 139–326 (1981).

43. R. Saito, M. Fujita, G. Dresselhaus, and M. S. Dresselhaus, *Appl. Phys. Lett.* **60**, 2204–2206 (1992).

44. Riichiro Saito, Mitsutaka Fujita, G. Dresselhaus, and M. S. Dresselhaus, *Phys. Rev.* **B46**, 1804–1811 (1992).

45. M. S. Dresselhaus, G. Dresselhaus, R. Saito, and P. C. Eklund, in C_{60}-*Related Balls and Fibers*, (J. L. Birman, C. Sébenne, and R. F. Wallis, eds.) (Elsevier Science Publishers, New York, 1992, chap. 18, pp. 387–417.

46. R. Saito, G. Dresselhaus, and M. S. Dresselhaus, *J. Appl. Phys.* **73**, 494 (1993).

47. M. S. Dresselhaus, R. A. Jishi, G. Dresselhaus, D. Inomata, K. Nakao, and Riichiro Saito, *Molecular Materials* **4**, 27–40 (1994).

48. J. W. Mintmire, B. I. Dunlap, and C. T. White, *Phys. Rev. Lett.* **68**, 631–634 (1992).

49. T. Hamada, M. Furuyama, T. Tomioka, and M. Endo, *J. Mater. Res.* **7**, 1178–1188 (1992), 2612–2620 (1992).

50. C. H. Olk and J. P. Heremans, *J. Mater. Res.* **9**, 259–262 (1994).

51. J. W. G. Wildöer, L. C. Venema, A. G. Rinzler, R. E. Smalley, and C. Dekker, *Nature (London)* **391**, 59–62 (1998).

52. A. M. Rao, E. Richter, S. Bandow, B. Chase, P. C. Eklund, K. W. Williams, M. Menon, K. R. Subbaswamy, A. Thess, R. E. Smalley, G. Dresselhaus, and M. S. Dresselhaus, *Science* **275**, 187–191 (1997).

53. K. Sattler, *Carbon* **33**, 915 (1995).

54. T. W. Odom, J. L. Huang, P. Kim, and C. M. Lieber, *Nature (London)* **391**, 62–64 (1998).

55. S. J. Tans, M. H. Devoret, H. Dai, A. Thess, R. E. Smalley, L. J. Geerligs, and C. Dekker, *Nature (London)* **386**, 474 (1997).

56. M. S. Dresselhaus, *Physics World* **9**(5), 18–19 (1996).

57. L. Kouwenhoven, *Science* **275**, 1896 (1997).

58. M. Bockrath, D. H. Cobden, P. L. McEuen, N. G. Chopra, A. Zettl, A. Thess, and R. E. Smalley, *Science* **275**, 1922–1924 (1997).

59. C. L. Kane and E. J. Mele, *Phys. Rev. Lett.* **78**, 1932 (1997).

60. H. Dai, E. W. Wong, and C. M. Lieber, *Science* **272**, 523–526 (1994).

61. T. W. Ebbesen, H. J. Lezec, H. Hiura, J. W. Bennett, H. F. Ghaemi, and T. Thio, *Nature (London)* **382**, 54–56 (1996).

62. L. Langer, V. Bayot, E. Grivei, J. P. Issi, J. P. Heremans, C. H. Olk, L. Stockman, C. Van Haesendonck, and Y. Bruynseraeda, *Phys. Rev. Lett.* **76**, 479–482 (1996).

63. R. A. Jishi, L. Venkataraman, M. S. Dresselhaus, and G. Dresselhaus, *Chem. Phys. Lett.* **209**, 77–82 (1993).

64. R. Saito, G. Dresselhaus, and M. S. Dresselhaus, *Phys. Rev.* **B53**, 2044–2050 (1996).

65. Ph. Lambin, A. Fonseca, J. P. Vigneron, J. B. Nagy, and A. A. Lucas, *Chem. Phys. Lett.* **245**, 85–89 (1995).

66. L. Chico, V. H. Crespi, L. X. Benedict, S. G. Louie, and M. L. Cohen, *Phys. Rev. Lett.* **76**, 971–974 (1996).

67. R. A. Jishi, D. Inomata, K. Nakao, M. S. Dresselhaus, and G. Dresselhaus, *J. Phys. Soc. Jpn.* **63**, 2252–2260 (1994).
68. R. Saito, T. Takeya, T. Kimura, G. Dresselhaus, and M. S. Dresselhaus, *Phys. Rev. B57*, 4145–4153 (1998).
69. E. Richter and K. R. Subbaswamy, *Phys. Rev. Lett.* **79**, 2738 (1997).
70. P. C. Eklund, J. M. Holden, and R. A. Jishi, *Carbon* **33**, 959 (1995).
71. H. Hiura, T. W. Ebbesen, K. Tanigaki, and H. Takahashi, *Chem. Phys. Lett.* **202**, 509 (1993).
72. N. Chandrabhas, A. K. Sood, D. Sundararaman, S. Raju, V. S. Raghunathan, G. V. N. Rao, V. S. Satry, T. S. Radhakrishnan, Y. Hariharan, A. Bharathi, and C. S. Sundar, *PRAMANA—Journal of Phys.* **42**, 375–385 (1994).
73. J. M. Holden, Ping Zhou, Xiang-Xin Bi, P. C. Eklund, Shunji Bandow, R. A. Jishi, K. Das Chowdhury, G. Dresselhaus, and M. S. Dresselhaus, *Chem. Phys. Lett.* **220**, 186 (1994).
74. J. M. Cowley, P. Nikolaev, A. Thess, and R. E. Smalley, *Chem. Phys. Lett.* **265**, 379–384 (1997).
75. K. Tohji, T. Goto, H. Takahashi, Y. Shinoda, N. Shimizu, I. Matsuoka, Y. Saito, A. Kasuya, T. Ohsuna, K. Hiraga, and Y. Nishina, *Nature (London)* **383**, 679 (1996).
76. A. M. Rao, P. C. Eklund, S. Bandow, A. Thess, and R. E. Smalley, *Nature (London)* **388**, 257 (1997).

9

Photoconductivity in Fullerene Thin Films and Solids

Nobutsugu Minami and Said Kazaoui
National Institute of Materials and Chemical Research (AIST)
Tsukuba, Japan

1. INTRODUCTION

Ever since fullerenes (C_{60}, C_{70}, etc.) became available in macroscopic quantity, there have been an exploding number of works exploring various aspects of this fascinating and intriguing third allotrope of solid carbon. One of the unique features of fullerenes is the rich π-conjugated electrons covering their entire spherical surface. It was natural that optical and electrical properties inherent in solid fullerene became the targets of intense studies, because π-conjugated electrons are the very origin of a variety of electronic processes occurring in organic molecular solids. Studies of the photoconductivity of fullerene thin films or solids can be rationalized just in this trend. The uniqueness of fullerene as an electronic material lies in its molecular shape and structure that bring about exceptional behavior such as the formation of fcc (or hcp) closed packed crystal lattice, the fast molecular rotation in the solid state, and the relevant phase transition near room temperature (Heiney, 1992). Such features were expected to contribute to the occurrence of some specific photoconductive properties and their possible applications.

Photoconductivity is a photoelectric process that involves two important steps: photogeneration of charge carriers and their transport through solid. Since optical excitation is the first event preceding any photogeneration step, electronic structure of ground and excited states should be of great relevance for photoconductive properties. As will be discussed later, the importance of excited states delocalized over adjacent C_{60} molecules has recently been found in close connection with the photoconductivity and other optical processes in solid C_{60}. Besides, such states are thought to intrinsically originate from the spherical π-conjugation constituting the closed packed lattice. For the carrier transport step, connection with the unique structure of solid fullerene has yet to be established because of its susceptibility to various uncontrollable disorders in solid such as impurities, dislocations, adsorbed gases. Nonetheless, experimental challenges on the basis of the measurements of fast transient photoconductivity have demonstrated some new aspects and features in carrier transport processes through fullerene solids.

The earliest works on the photoconductivity of fullerene were reported by several groups in mid to late 1991, about one year after macroscopic quantity of C_{60} became available. Minami (1991) found sizable photoconductivity in thin film of C_{60}/C_{70} mixture in the surface cell configuration. The spectral dependence (400–900 nm) essentially agreed with its absorption spectrum, although at longer wavelength (around 620 nm) an extra peak was observed where no absorption structure could be detected; later it was found that it originated from intramolecular dipole forbidden transition. Mort et al. (1991) reported photoconductivity in C_{60}/C_{70} films sandwiched between aluminum and/or gold electrodes. Their results showed that the photoconductivity onset extended to the near infrared region (850 nm); this was ascribed to the onset of the lowest direct gap transition which was optically dipole forbidden. The peak photoefficiency (photocarrier per absorbed photon) was estimated to be 10^{-4}. Kaiser et al. (1992) studied steady-state photoconductivity of C_{60} thin films in the surface cell configuration. They measured the action spectrum and temperature dependence of the photocurrent and proposed different transport mechanisms for low- and high-temperature regime, suggesting a situation similar to amorphous silicon. These were the earliest pieces of evidence for the existence of the photoconductivity in fullerene solids. While they did not necessarily give comprehensive views of the photoconductive processes therein, these were pioneering works that triggered the later development in the field of fullerenes as novel photoelectric materials.

The present review is intended to be a general survey of the developments achieved in the study of the photoconductivity of fullerene for the last eight years. Various experimental results will be summarized with their established (and sometimes tentative) interpretations. Because the field is ever developing and the electronic structure in solid fullerene remains to be fully understood, different aspects and views will be introduced broadly at the slight expense of strong coherence between the subjects. While this might not give a picture as focused as those for more established fields, it should reveal the overall situation about what has been achieved up to now and what still awaits further challenges. Several attempts at the possible technological applications of fullerene's photoconductivity will be described in a later section, including xerographic photoreceptors and photovoltaic devices.

2. PHOTOCARRIER GENERATION AND RECOMBINATION

2.1 Photocarrier Generation and Photocurrent Action Spectra

Spectral dependence of photocurrent in connection with absorption spectra is an important factor to consider when studying photoconductivity in a new material. By comparing both spectra, we obtain information as to whether or not photogeneration of carriers originates from electronic excitation of the material being considered and what electronic excited states play important roles. From the earliest stage, it was known that photocurrent action spectra of C_{60} are dominated by or in close connection with the absorption spectra of molecular as well as thin film C_{60}. Since electronic absorption in molecular solids is usually dominated by excitonic transition, main routes for photocarrier generation in fullerene solids were thought to be the ones involving excitons, as in traditional π-conjugated molecular solids like anthracene. On the other hand, some groups tended to consider optical excitation in fullerene solids in terms of interband transition, where free electrons and holes are directly generated by absorption of photons. However, considering the fact that the overall features of absorption spectrum of C_{60} thin film closely resemble those of the isolated molecule (see Section 2.3), direct excitation of free electrons and holes seems less supposable at least at the lower excitation energy. This review basically takes the former viewpoint, that is, ''via exciton'' mechanism, which seems more plausible for solid fullerene these days, but different views are occasionally referred to.

In general, photocarrier generation in π-conjugated molecular solids such as fullerene can take place by the following mechanisms (Wright, 1995, p. 183): (1) charge separation from excitons at the surface or in the bulk, (2) exciton-exciton collision leading to the formation of one ground-state molecule and one pair of separated charges, (3) photoionization of excitons, (4) optical detrapping of trapped charges. In case 1, charge separation sites may be impurities or defects in the bulk, crystal surfaces, interfaces with electrodes, or adsorbed gas. If it takes place at the interface with electrodes, either an electron or hole is generated in the sample, which is often called photoinjection of carriers. In this case, only one type of carriers contribute to photoconductivity (unipolar photoconduction). On the other hand, if exciton separation takes place in the bulk, both types of charges can exist in the sample and thus can be involved in the photoconduction process. As will be discussed later, bimolecular recombination can take place under this condition, leading to considerable alteration of photocurrent action spectra depending on experimental configurations adopted (film thickness, surface or sandwich cell, etc.). Alternatively, charge separation could occur in a more intrinsic way, where charge carriers are generated from bound electron-hole pairs having some spatial distance that are formed by autoionization from molecular, higher excited states; in such a case defects or impurities may not be necessary. In addition, such bound charge pairs can be formed by direct photo-excitation called intermolecular charge transfer excitation, which turned out to be an important process in fullerene solids, as will be discussed in later sections. In the cases 2 and 3, since two photons are involved in the generation process, it should critically depend on excitation intensity. For case 4, the spectral behavior of photoconductivity may largely depend on samples, because the nature of traps can vary as their preparation and measurement conditions are different. Little has been studied about such aspects of fullerene's photoconductivity. Since works so far published mostly studied photocarrier generation from single excitons formed by one-photon excitation, this review deals mainly with case 1.

For the photoconductivity of fullerene solids or thin films, various types of photocurrent action spectra have been reported by different groups. On the whole, they can be divided into two categories: spectra going parallel with and opposite to the absorption spectrum, often called symbatic and antibatic relationships, respectively. According to the results of the reviewer's group, where measurements were made on interdigitated surface gap cells of vapor deposited C_{60} films (Kazaoui et al., 1994), symbatic relationship was observed for thin films (Fig. 1a), while it tended to become antibatic as the thickness increased (Fig. 1c). Such antibatic behavior is prominent around the absorption peak at 350 nm. An action spectrum reported by Mort et al. (1991) for a sandwich cell of C_{60}/C_{70} film also showed antibatic trend. Giro et al. (1993) reported antibatic relationship for vapor-deposited C_{60} film sandwiched between aluminum electrodes. They analyzed their results in terms of a model involving bimolecular recombination of bulk-generated photocarriers (here the term "bimolecular" refers to electron and hole whose concentration can be assumed to be the same because carriers are generated in the bulk). In the next section, the outline of the model is described.

2.2 Bimolecular Recombination

As mentioned above, depending on the experimental conditions, the relationship between absorption and photocurrent action spectra is not always straightforward. This is because, under the condition of steady-state photoconductivity measurements as have been often employed, not only the photogeneration but recombination of carriers are important factors determining the steady-state carrier concentration. Therefore, depending on the carrier recombination mechanism, its spectral and light-intensity dependence alter. In other words, the analysis of these properties can give us information about what mechanisms are dominant in the fullerene solids.

Giro et al. (1993) analyzed the photocurrent action spectra of C_{60} thin films sandwiched between aluminum electrodes on the basis of a model involving the bimolecular recombination of bulk-generated photocarriers, as discussed below.

In general, the time dependence of carrier density $n(x)$ at a distant x from the surface is expressed as

$$\frac{dn(x)}{dt} = I_0 \alpha \phi \exp(-\alpha x) - \beta n(x) - \gamma n^2(x) \tag{1}$$

where I_0 is the number of incident photons, α the absorption coefficient, ϕ the probability that the absorption of a photon leads to production of a carrier pair, β the monomolecular decay constant, and γ the bimolecular recombination coefficient for the created charge carriers. It is presumed here that the density of electrons and holes are both $n(x)$ and therefore their recombination rate should be proportional to $n^2(x)$. β represents carrier decay processes such as discharge at the electrodes or trapping. Under the steady-state condition $[dn(x)/dt = 0]$ and the predominance of carrier decay due to the bimolecular term $[\gamma n^2(x) \gg \beta n(x)]$, Eq. (1) yields

$$n(x) = \left(\frac{\alpha \phi I_0}{\gamma}\right)^{1/2} \exp\left(\frac{-\alpha x}{2}\right) \tag{2}$$

Assuming that ϕ is independent of x and integrating over the sample thickness d, the average concentration of charge carriers is

$$n = \frac{1}{d} \int_0^d n(x)\, dx = \frac{2}{d} \cdot \left(\frac{\phi I_0}{\alpha \gamma}\right)^{0.5} \left[1 - \exp\left(\frac{-\alpha d}{2}\right)\right] \tag{3}$$

The photocurrent is thus expressed as

$$j = en\mu F = \frac{2e\mu F}{d} \cdot \left(\frac{\phi I_0}{\alpha \gamma}\right)^{0.5} \left[1 - \exp\left(\frac{-\alpha d}{2}\right)\right] \tag{4}$$

The validity of this model is shown by the experimental verification that the photocurrent was proportional to the square root of the incident photon flux and the action spectrum displayed antibatic relationship with the absorption spectrum, as expected from Eq. (4) for the film thickness larger than the penetration depth of light (Yonehara and Pac, 1992; Giro et al., 1993).

The above model was found to also apply to C_{60} thin films in the surface cell configuration, and in this case, it reasonably explained the alteration of the photocurrent action spectrum due to the variation of the sample thickness (Kazaoui et al., 1995). As shown in Figure 1, for a thin film ($d = 20$ nm) the action spectrum goes almost parallel with the absorption (symbatic) (a), while for a thicker film ($d = 200$ nm) a dip is shown in the action spectrum at 350 nm (c), where the absorption peak is located (antibatic). The solid lines (b, d) represent theoretical curves for $d = 20$ and 200 nm on the basis of Eq. (4), calculated

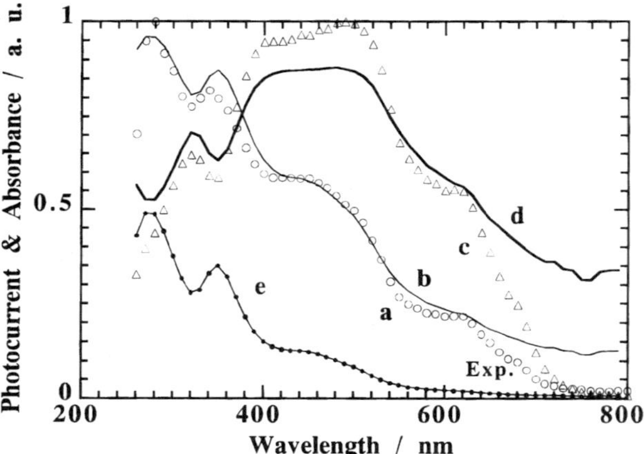

Figure 1 Experimental and calculated photocurrent action spectra of C_{60} films whose thickness $d = 20$ nm (a, b) and $d = 200$ nm (c, d), the absorption spectrum (e).

using experimentally determined $\alpha(\lambda)$. Both curves give reasonably good agreement with the corresponding experimental plots. The deviation at $\lambda > 610$ nm may indicate the decrease in ϕ for low-lying excited states, which, in the theoretical expressions, was assumed to be constant. Carrier loss by the bimolecular recombination mechanism was also shown to be prevailing by the measurement of the time-resolved transient photoconductivity of C_{60} thin films (Lee et al., 1995) (see Section 3.3).

2.3 Absorption and Luminescence in Connection with Photoconductivity

When considering optical excitation and photocarrier generation processes in fullerene solids, the examination of their absorption and luminescence behavior should give us important insight. In this section, such spectroscopic properties, especially those specific to solid-state C_{60}, are discussed. Figure 2 presents the UV-Vis absorption spectrum of a C_{60} thin film together with that of C_{60} solution. It is shown here that almost all the characteristic bands observed in the solution at <400 nm and at >560 nm are retained in the thin film, although red-shifted and broadened. This can be used as an indication that electronic states in the solid state are largely determined by those of isolated molecules. However, a significant difference arises around 400–540 nm, where the thin film has a broad band that has no straightforward counterpart in the solution; the existence of the

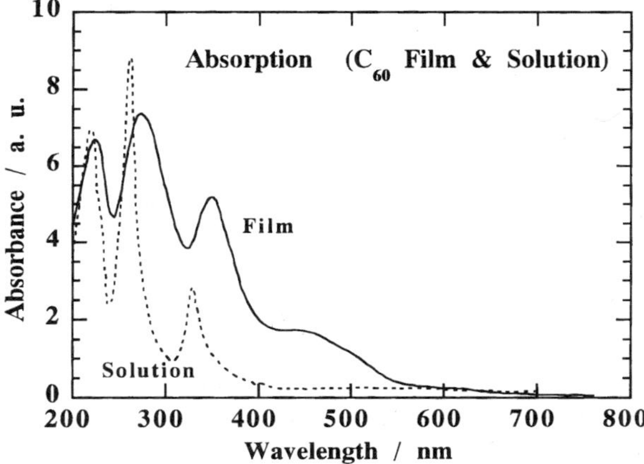

Figure 2   Absorption spectra of a C_{60} thin film and solution at room temperature.

band has been noticed since the early stage of fullerene research (Achiba et al., 1991).

By measurements at lower temperatures, two peaks have been resolved at 450 and 500 nm in this broad feature (Figure 3). As shown in Figure 4, the peak absorbance at 450 nm showed a break at 260 K below which the steeper rise was observed (Minami et al., 1995). On the other hand, the 350 nm peak, corre-

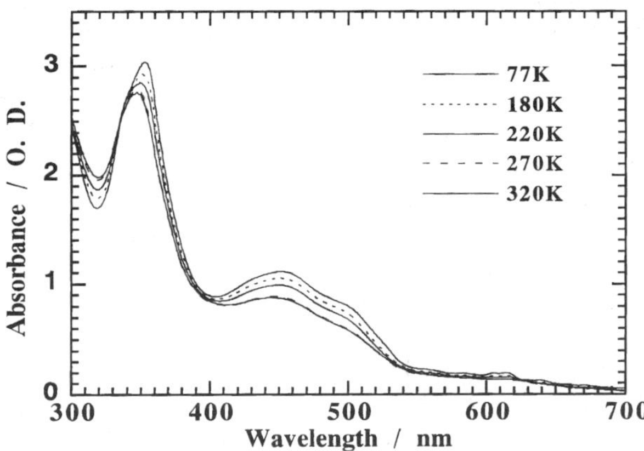

Figure 3   Absorption spectra of a C_{60} thin film at various temperatures.

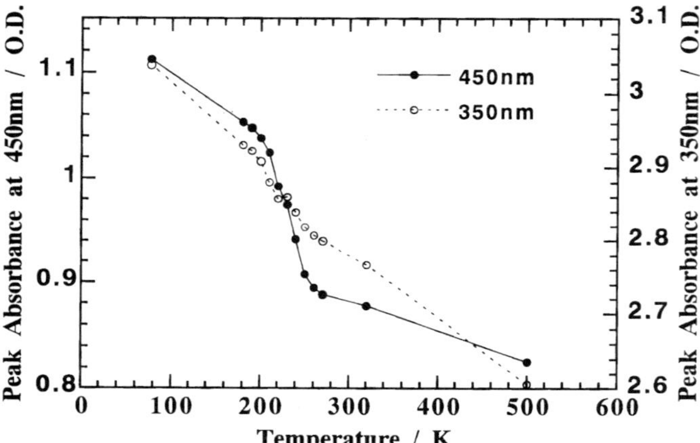

Figure 4 Temperature dependence of the peak absorbance at 450 and 350 nm for the C_{60} thin film. (Data from Figure 3.)

sponding to the lowest intramolecular allowed transition, underwent mostly monotonous increase throughout the temperature range used. The temperature at which the break appears coincides with that of the order-disorder phase transition from fcc to sc of the C_{60} crystal lattice, meaning that the feature is sensitive to the change in the lattice structure. These results suggest that this solid-state specific absorption should be related to intermolecular interaction between adjacent C_{60} molecules.

The consideration of luminescence excitation profile in connection with absorption spectra generally provides important information about relaxation processes of excited states. As shown in Figure 5a, the luminescence excitation spectrum of a C_{60} thin film has a peak at 3.5 eV coinciding with the lowest allowed electronic transition, a broad band in the range 2.3–3.0 eV corresponding to the solid-state specific absorption mentioned above, and fine structures below 2.3 eV (Kazaoui et al., 1995a). The last ones belong to the intramolecular electronic transitions coupled with vibrational modes (Schlaich et al., 1995). For the luminescence excitation spectrum of C_{60} solution, no spectral feature was observed around 2.3–3.0 eV (Williams and Verhoeven, 1994), again proving the existence of the solid-state specific excitation.

In Figure 5b is shown the calculated luminescence efficiency (emission monitored at 1.66 eV) assuming that it is proportional to the luminescence intensity divided by the number of absorbed photons $[1 - \exp(-\alpha d)]$ (α absorption coefficient, $d = 20$ nm, the film thickness). Below 2.3 eV, the spectral dependence

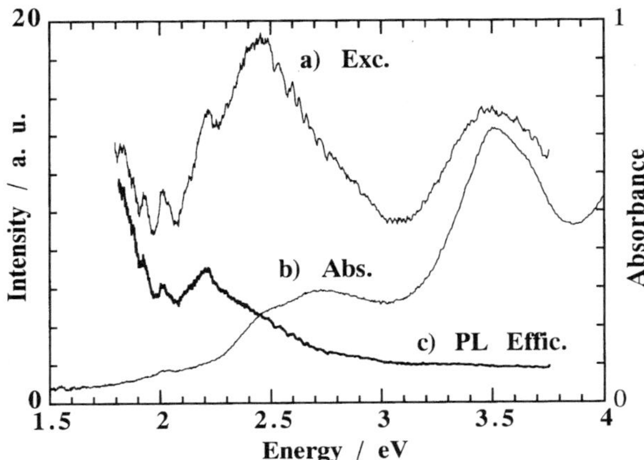

Figure 5 Luminescence excitation (monitored at $h\nu_{em} = 1.66$ eV) (a), absorption (b), and luminescence efficiency spectra (c) ($T = 77$ K, annealed C_{60} film, thickness ~20 nm).

of the efficiency reflects that of the luminescence excitation spectrum characterized by sharp peaks. An important observation here is that above 2.3 eV the efficiency is featureless and clearly decreases with increasing excitation energy. This means that, when the excitation energy exceeds 2.3 eV, a new decay route opens up that does not contribute to the luminescence being measured. In contrast, for isolated C_{60} molecules in solution the luminescence efficiency was shown to be constant irrespective of the excitation energy (Williams and Verhoeven, 1994). From these results, the decay route existing in C_{60} films should be regarded as a reflection of an intermolecular effect.

The drop of the luminescence efficiency above 2.3 eV suggests the existence of branching route in the relaxation process from the upper excited states, in competition with internal conversion to the lowest excited states from which the luminescence emission occurs. [The origins of the luminescence were discussed in terms of excitons and X-traps (Negri et al., 1992; van den Heuvel et al., 1994)]. Relevant to this is the presence of the solid-state specific absorption band that also emerges at 2.3 eV, as mentioned earlier. From these results, we propose intermolecular charge transfer (CT) excited states as a candidate for the branching route. The assumption of such states is consistent with electroabsorption studies (Jeglinski et al., 1992; Petelenz et al., 1994) and theoretical calculations (Tsubo and Nasu, 1994; Harigaya and Abe, 1994; Eilmes et al., 1997) that predicted the existence of distinct manifold of intermolecular CT states in this energy range.

2.4 Luminescence Quenching by Electric Field

The existence of the CT states just above 2.3 eV in solid C_{60} was further confirmed by the observation of electric field–induced luminescence quenching. It was found that the intensity of the luminescence emission peak at 1.7 eV with the excitation energy of 3.75 eV decreased by 12% under electric field ($E = 10^5$ V $\cdot$ cm^{-1}, $T = 77$ K) without appreciable shift or splitting. The integrated luminescence emission in the entire range 1.4–1.8 eV also showed a 10% decrease. A remarkable fact was that the quenching occurred only when the excitation energy was higher than 2.3 eV, below which electric field had no appreciable effects on luminescence, demonstrating that the luminescence quenching is excitation-energy-dependent. Figure 6a and b presents the luminescence excitation spectra before and after the application of electric field. The electric field–induced luminescence quenching efficiency (Q_{eff}) is defined as $Q_{eff} = 1 - I_{E\neq0}/I_{E=0}$ where $I_{E\neq0}$ and $I_{E=0}$ are the luminescence intensity with and without electric field, respectively. Figure 6c shows Q_{eff} as a function of excitation energy. The quenching occurred in the range 2.3–3.75 eV, whereas it was absent in the range 1.77–2.3 eV, within the experimental resolution.

It is should be stressed that the energy threshold of the electric field–induced luminescence quenching coincides with that of the drop in luminescence efficiency at 2.3 eV, as was mentioned in the previous section. This coincidence is again consistent with the proposal that the branching that occurs in the relaxation from the upper excited states (>2.3 eV) involves intermolecular CT states, because such states should strongly interact with the electric field due to their inherently large dipole moment. The absence of appreciable luminescence quenching in the range 1.77–2.3 eV appears consistent with the intramolecular transition (Frenkel-type excitons), whose interaction with electric field should be much weaker. These results are corroborated by the electroabsorption measurements that identified Frenkel exciton states at approximately 1.95 and 2.4 eV and CT states at 2.7 and 4 eV, except for some shift (Jeglinski et al., 1992; Petelenz et al., 1994).

Insight into the electric field–induced luminescence quenching was gained by studying its dependence on several parameters with an excitation energy fixed at 2.54 eV:

1. The luminescence quenching has been observed for both thin (thickness $\sim$20 nm) and thick films (thickness $\sim$200 nm) as compared to the light penetration depth (approximately 100 nm at 2.54 eV with absorption coefficient $\approx 10^5$ cm^{-1}). Moreover, the magnitude of the luminescence quenching is independent of the film thickness. The results show that the gradient in the distribution of excited states in the film should not be an important factor for the electric field–induced luminescence quenching.

2. In the range $E = 8 \times 10^4 - 2 \times 10^5 \, \text{V} \cdot \text{cm}^{-1}$, Q_{eff} increased as E^3 and $E^{4.5}$ at $T = 77$ and 300 K, respectively. Stark effect on the luminescence behavior can be discarded because the luminescence peaks were not split or shifted and because Q_{eff} did not obey E^2 dependence (Kalinowski et al., 1992).

3. In the excitation power range 0.25–5.0 mW $\cdot$ cm^{-2} (photon flux $\approx 6 \times 10^{14}$–10^{16} s$^{-1} \cdot$ cm^{-2}) the luminescence intensity increased linearly with excitation intensity either at $T = 77$ or 300 K. Q_{eff} was independent of the photon flux at constant electric field and temperature, discarding the effect of nonlinear processes as an origin of the luminescence quenching.

4. As a general trend Q_{eff} increased with the temperature. The quenching yield was weakly temperature dependent below 260 K with a bump at approximately 90 K, whereas above 260 K it was thermally activated with $E_a = 0.025$ eV. The features at 260 and 90 K might be related to the order-disorder phase transition and glass transition, respectively (Matsuo et al., 1992).

Photocarrier generation and radiative recombination (luminescence) should be complimentary phenomena, the balance between which can be modulated by external electric field. Expecting that there should be close relationship with the luminescence quenching, we measured a photoconductivity spectrum of C$_{60}$ thin

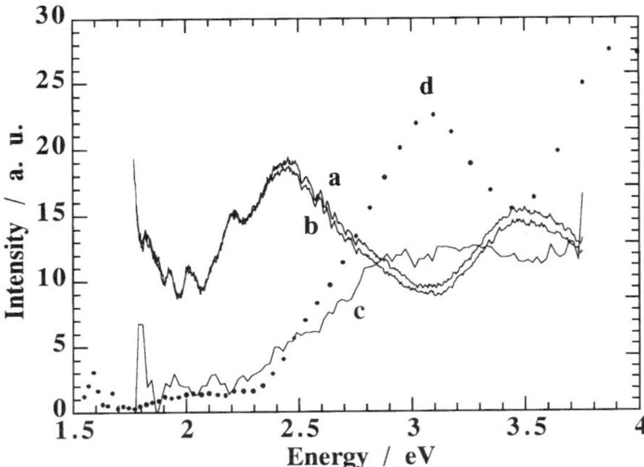

Figure 6 Luminescence excitation spectra before (a) and after (b) the application of electric field ($E = 10^5$ V/cm) (monitored at $h\nu_{\text{em}} = 1.66$ eV), electric field induced luminescence quenching efficiency (c) and photocurrent efficiency spectra (d) ($T = 77$ K, annealed C$_{60}$ film, thickness $\sim$20 nm).

film at 77 K, the same temperature at which the quenching experiments were performed. As shown in Figure 6d, photoconductivity was found to steeply increase above 2.3 eV, the threshold almost exactly coinciding with the onset of the luminescence quenching (Kazaoui et al., 1995a). Note that the lower-energy photoconductivity observed at room temperature could not be detected here (cf. Figure 1), because of the insufficient thermal energy for more tightly bound Frenkel excitons to dissociate. The coincidence in threshold energy between the luminescence quenching and photoconductivity is another experimental confirmation of the importance of the CT states in the photocarrier generation in fullerene solids.

2.5 Role of Charge Transfer Excitons

From these results, the photocarrier generation process in solid C_{60} are thought to be dominated by CT excitons existing just above 2.3 eV, especially at low temperatures. For a wide range of molecular solids and polymers, Onsager formalism of geminate recombination has been generally accepted as a model to describe photocarrier generation processes. (Borsenberger and Weiss, 1993; Silinsh and Cápek, 1994, p. 257; Wright, 1995, p. 185). The essence of this model is that photoexcitation of a molecule in a solid lattice leads, after a multistep process, to the formation of a bound electron hole pair (geminate charge pair) with some spatial distance, which is called Onsager radius. The pair of charges, which are bound to each other by Coulomb force, can be generated via different routes. One is the autoionization of high-lying Frenkel excitons, where electrons raised to higher excited states are ejected by the excessive kinetic energy and eventually thermalize to the charge pair. Another route is for lower-lying Frenkel excitons that do not have sufficient energy to autoionize, for which case the charge separation could occur via extrinsic processes like the effect of impurities, defects, or the interface with electrodes, as mentioned in Section 2.1. The third route is the direct formation of the pair by optical excitation. In either case, this pair is assumed to undergo Brownian motion under the effect of external electric field and thermal energy; some of them recombine to restore original state but others eventually escape recombination and generate free charge carriers contributing to photoconductivity.

If this picture is to be applied to solid C_{60}, CT excitons discussed in the preceding sections can be considered as the entity of the bound electron hole pair presumed in the Onsager formalism. As judged from the absorption, electroabsorption, and luminescence excitation spectra, photoexcitation in C_{60} thin films should involve both intramolecular and intermolecular process, depending on the excitation energy used. Above 2.3 eV, substantial part of the excitation leads to the formation of CT excitons; the process can be either direct (CT excitation) or indirect (autoionization of the upper Frenkel excitons to form the CT states). These states are regarded as precursors to free carriers. Although much work has yet to be done about the detailed mechanism by which these states contribute to free carrier generation, preliminary analysis by Mort et al. (1992) based on the xerographic discharge method (see Section 5.1) indicated that 2 nm would be a separation of primary electron hole pairs. Furthermore, an analysis based on time-resolved photoconductivity measurements yielded 1.3–1.7 nm as the Onsager radius for C_{60} films for photon energies between 2 and 4 eV (Könenkamp et al., 1996). These results should have some implications for understanding electronic states specific to solid C_{60}; hence more elaborate studies are desired, which would provide important information as to excited states delocalized over adjacent C_{60} molecules in the closed packed crystal lattice.

The role of the Frenkel excitons in the photocarrier generation should be briefly commented on here. At 77 K only very small, barely detectable response was found at excitation energy below 2.3 eV (Figure 6d), but at room temperature readily observable photoconductivity exists in the same energy range (Figure 1). This means that tightly bound Frenkel excitons can contribute to photocarrier generation with the assistance of thermal energy. Correspondingly, it was found that the activation energy for photoconductivity was substantially higher for <2.3 eV than for >2.3 eV (Kazaoui, 1995b).

It is remarkable that CT excitons can appear as such distinct absorption peaks, considering that for traditional π-conjugated molecular solids they normally give very weak spectral features and thus can only be detected by highly sensitive spectroscopic techniques such as electroabsorption. For one thing, this can be attributed to the fact that the lowest HOMO-LUMO transition in C_{60} is dipole forbidden, and thus the resultant absorption is not strong enough to conceal or obscure CT transitions existing in the same energy range. But more important would be the situation that a closed packed lattice constructed from the nearly perfect sphere of π-conjugated electrons inherently makes possible strong electronic interaction between adjacent molecules. In fact, for an anthracene crystal, it was shown that the relative contribution of the CT transition to the total absorption coefficient is less than 0.08 (Silinsh and Cápek, 1994, p. 155). Furthermore, if compared with the intensity of the absorption peak (mostly deriving from an intramolecular allowed transition), the CT transition is almost 2 orders of magnitude smaller even at its maximum position. This is contrasted with the result for the C_{60} thin film, where the CT transition is quite appreciable as compared with the intramolecular allowed ones (Figure 3). Thus the distinct appearance of the CT excitons should be a result of such unprecedented molecular and lattice structures. In that sense, these observations can be regarded as electronic properties unique to fullerene solids. It is natural to wonder what the situation will be for less symmetrical C_{70}'s solid. Several works have been oriented in that direction and the existence of CT excitons at somewhat different energies is now being revealed (Kazaoui and Minami, 1997; Kazaoui et al., 1998).

3. CARRIER TRANSPORT AND ITS DYNAMICS

3.1 Measurement of Carrier Mobilities

For the measurement of charge carrier mobility, use of single crystals is particularly important, because, under the usual conditions for available experimental techniques, mobility is a quantity largely defined for the macroscopic size (a few tens to hundreds µm) in any form of solid samples. When traversing such macroscopic distances, charge carriers are inevitably scattered or trapped at various disorders existing in the sample. For thin films, such disorders are abundant and generally hard to control; even epitaxially grown ones on crystalline substrates are made up of crystallites (domains) that are much smaller than the sample's dimensions. This brings about the involvement of numerous grain boundaries disturbing the smooth transport of carriers, making it difficult to extract intrinsic transport properties. Only by using single-crystal samples, one might be able to minimize the grain boundary problem. Nevertheless, single crystals still may not be free from defects that can influence carrier transport processes; thus one should recognize that the evaluation of mobility as materials' intrinsic property remains an experimental challenge.

To the best of our knowledge, there has been only one group that dealt with the measurement of charge carrier mobility in fullerene single crystals (Frankevich et al., 1993, 1996). It was performed by the time of flight (TOF) method. In this technique, samples are sandwiched between two electrodes and charge carriers are photogenerated by pulse laser excitation at one electrode. Depending on the polarity of the applied DC electric field, either electrons or holes start to migrate toward the other electrode. The photocurrent passing through the sample is monitored by an oscilloscope. In principle, a constant current signal appears while a sheet of carriers is traversing across the sample and then a sudden drop is observed, which represents the arrival of carriers at the other electrode. From this drop, one finds a transit time (T_{tr}), a time necessary for carriers to move across the sample. Dividing the sample thickness by T_{tr} yields a carrier drift velocity, division of which by the applied electric field then gives a mobility. Frankevich et al. (1993) made measurements on vapor phase–grown C_{60} crystals (thickness 55–120 µm) sandwiched between two electrodes. Applying either positive or negative voltage to the electrode on which light pulses from a nitrogen laser were irradiated, they were able to measure mobility of either holes or electrons, respectively.

Figure 7 shows the typical transient photocurrent by the TOF method for a C_{60} single crystal measured at room temperature. The arrows indicate the time at which carriers arrived at the other electrodes, from which T_{tr} values can be obtained. The traces A, B, and C represent the electron current, and D and E the hole current, for different bias voltages. Plot of the drift velocity thus obtained as a function of the applied electric field yielded a straight line, indicating the

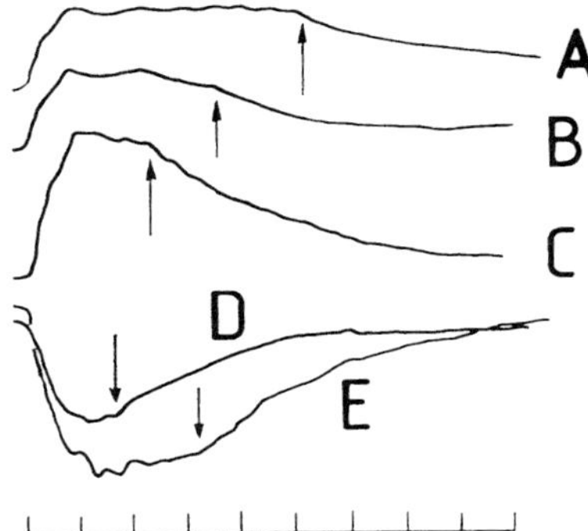

Figure 7 Typical pulses of transient current (ordinate) on the C_{60} single crystals grown from the vapor phase, at room temperature. The time scale (abscissa) is 50 ns/div; load resistance $R = 100\ \Omega$. (A) Electrons are moving, thickness of the crystal $d = 55\ \mu m$ ($\pm 10\%$), voltage $V = 200$ V; (B) the same as (A), but $V = 300$ V; (C) the same as (A), but $V = 500$V; (D) holes are moving, $d = 120\ \mu m$, $V = -1000$ V; (E) the same as (D), but $V = 500$ V. (*After Frankevich et al., 1993.*)

constant mobility independent of electric field. Values obtained are $\mu_h = 1.7 \pm 0.2$ cm$^2 \cdot$ V$^{-1} \cdot$ s^{-1} for holes and $\mu_e = 0.5 \pm 0.2$ cm$^2 \cdot$ V$^{-1} \cdot$ s^{-1} for electrons. These values are typical of π-conjugated molecular crystals such as anthracene, naphthalene, etc.

Figure 8 is the temperature dependence of the hole mobility. Two prominent features are observed here: the jump between 210 and 240 K and almost constant values below and above this temperature range. From the structural analysis by X-ray diffractometry, it is known that at 250 K the first-order phase transition occurs in C_{60} crystals. Above this point, molecules are thought to rotate freely and very rapidly, while below it they perform ratchet-type rotation on some fixed crystal axes. The coincidence of the jump of the mobility with the phase transition indicates some interplay between the charge carrier movement and the solid lattice structure of C_{60}. This should be derived from intermolecular electronic overlap that is critically influenced by the mode of the molecular rotation or by the change in intermolecular distance accompanying the phase transition. A number of previous works had found that the dark and photoconductivity of C_{60} crystals and thin films showed a jump at this phase transition temperature (Mort et al, 1992a; Alers et al., 1992; Wen et al., 1992). By the work of Frankevich et al. (1993; 1996) it was established that such jumps essentially originated from the steplike change in the carrier mobility.

The almost temperature-independent mobility is another characteristic feature in the C_{60} crystals. Similar behavior was reported for mobility along the *c*

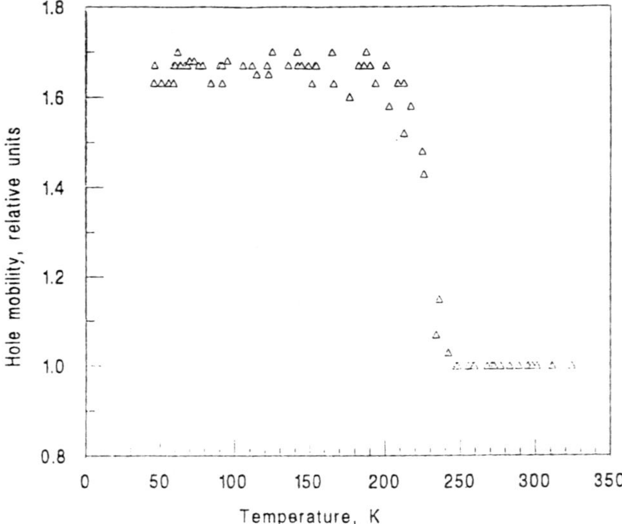

Figure 8   Dependence of the hole mobility in a C_{60} single crystal grown from the vapor phase on temperature. The mobility is normalized to the value at 300 K. (*After Frankevich et al., 1993.*)

axis of anthracene crystal and analyzed in terms of a model involving coherent motion in the *ab* plane but incoherent hopping along the *c* axis (Sumi, 1978). Although it remains to be seen if such formalism could be applied to highly symmetrical C_{60} crystals, this prominent result presents a strong impetus to theoretical works on electronic transport properties of this novel type of molecular solid. It can at least be said that such a constant mobility should be inconsistent with the simple band transport model, which predicts a mobility inversely varying with temperature (Wright, 1995, p. 157).

3.2   Carrier Mobility in MOS FET Configuration

Although it does not explicitly involve photoconductivity, works on field-effect transistors (FET) should be discussed here because it directly reflects carrier transport processes in fullerene solids. Several papers reported the field-effect mobility of C_{60} thin films on the basis of the measurement of FET parameters. The basic structure of the transistors used was similar to those commonly employed in inorganic semiconductor devices (Sze, 1969). Substrates used for thin film deposition were heavily *n*-type doped silicon wafers, whose surface was oxidized to form insulating SiO_2 layer. On top of this, a pair of coplanar metal electrodes were deposited with an interval of 10–100 µm between them, working as a source and drain of current. C_{60} thin films were vacuum deposited on this structure, forming an insulator-gated (IG) FET device. Drain current I_d was measured as a function of drain-source voltage V_{ds}. n-doped Si worked as a gate, where the application of gate voltage V_g could regulate the carrier concentration in the channel formed near the interface between C_{60} and SiO_2 layers, through which I_d flew. According to a general theory for IG-FET devices (Sze, 1969), I_d is expressed as

$$I_d = \mu \left(\frac{CW}{L} \right) \left[\left(V_g - V_t \right) V_{ds} - \frac{V_{ds}^2}{2} \right] \tag{5}$$

for small V_{ds}, where μ is the mobility, C the capacitance, W the channel width, L the channel length, and V_t the threshold voltage for V_{ds} above which I_d emerges. As V_{ds} becomes larger, saturation of I_d is reached and the saturation current is expressed as

$$I_{dsat} = \mu \left(\frac{CW}{2L} \right) \left(V_g - V_t \right)^2 \tag{6}$$

An important feature in the IG-FET configuration is that the carrier concentration contributing to I_d can be controlled by V_g without altering the carrier mobility; the latter is assumed to be constant throughout the channel. Thus by plotting $(I_{dsat})^{0.5}$ versus $(V_g - V_t)$ and by using known values of the device parameters such as W, L, and C, one can obtain the field-effect mobility μ.

This technique was used to evaluate the carrier mobility in vacuum-deposited C_{60} thin films (Haddon et al., 1995). Figure 9 shows the dependence of I_d on V_{ds} for different V_g's as measured for a C_{60} thin film FET, giving I_{dsat} for each V_g. As shown in Figure 10, by plotting $(I_{dsat})^{0.5}$ as a function of V_g, the field-effect mobility of the C_{60} thin film was obtained by the use of Eq. (6). For an 80 nm thick film deposited and measured in situ in ultra high vacuum (10^{-8} Torr), a field-effect mobility as high as 0.08 $cm^2 \cdot V^{-1} \cdot s^{-1}$ was obtained. This is among the highest values of carrier mobility ever reported for molecular thin films, being comparable to those of sexithiophene (Horowitz et al., 1996). Still, it is somewhat smaller than those of C_{60} single crystals determined by the TOF measurements as described in the previous section. The discrepancy can be attributed to the difference in the sample morphology, since such polycrystalline films are normally made up of small grains, causing considerable carrier scattering at their boundaries, while the effect of metal/C_{60} interface was suggested to be an important factor (Haddon et al., 1995). Even smaller field-effect mobilities of C_{60} thin films were reported by other groups (10^{-4}–10^{-6} $cm^2 \cdot V^{-1} \cdot s^{-1}$) (Kudo et al.,

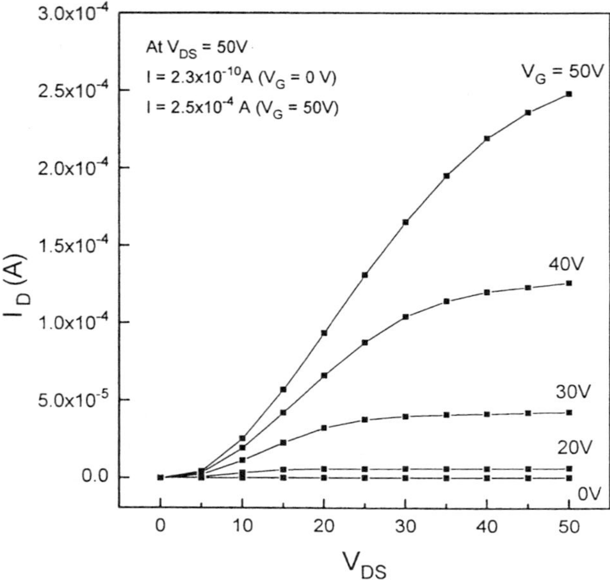

Figure 9   Drain current (I_D) versus drain-source voltage (V_{DS}) for various gate voltages (V_G), for a C_{60} thin film transistor. (*After Haddon et al., 1995.*)

1995; Kaneto et al., 1996), necessitating more elaborate experiments (such as correlation with film crystallinity, morphology, experimental conditions, etc.) for the relatively high mobility to become an established value. Nonetheless, the isotropic shape of C_{60} should be an advantageous factor for a molecular electronic material, since most other π-conjugated organic molecules have inherently aniso-tropic characters (planar or linear), complicating device performance and its eval-uation.

3.3   Carrier Dynamics by Time-Resolved Photoconductivity

In a different experimental configuration, Lee et al. (1993) worked on the fast relaxation process of photogenerated carriers by the measurement of transient photoconductivity in a picosecond time regime. They used surface type cells of C_{60} thin films instead of the sandwich cells employed in the TOF technique. It is often referred to as an Auston switch method, the details of which are described in the literature (Moses and Heeger, 1992). The experimental procedure was that short laser pulses (50 ps) excite the gap area (an order of 0.1 mm width) between coplanar strip electrodes, creating fast transient photoconductivity, reflecting the

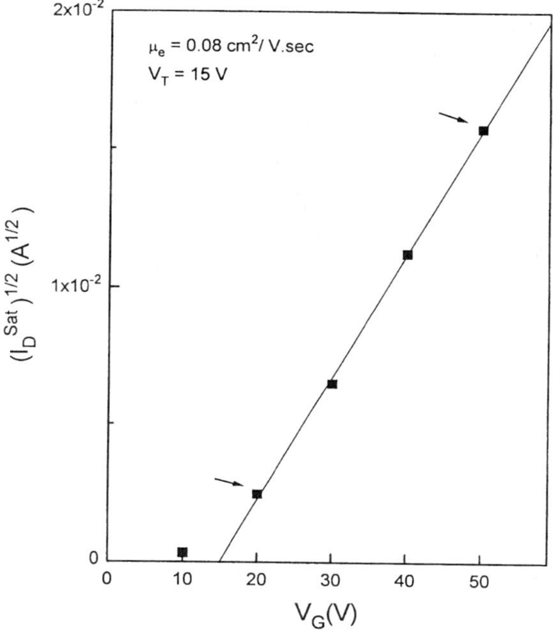

Figure 10 Plot of the square root of the saturation current at the drain versus gate voltage (data from Figure 9), together with the derived field effect electron mobility (μ_c) and threshold voltage (V_T). (*After Haddon et al., 1995.*)

generation, recombination, and trapping of charge carriers. For the detection of photoresponse, a boxcar averager is normally used in combination with a fast sampling head. By an appropriate choice of circuit elements (impedance etc.), the technique is claimed to be capable of detecting very fast transient components of photoconductivity (as short as 50 ps) that has been inaccessible by traditional methods. Generally, carrier recombination, carrier trapping, and variation in carrier mobility during response time of the experimental system affect the magnitude and transient of the fast photocurrent.

Figure 11 shows fast transient photoconductivity signals from a C_{60} thin film at various temperatures (Lee et al., 1995), for which sample oxygen contamination was strictly excluded considering its severe effects on the electrical properties of fullerene solids (see Section 4). The rise curve is largely determined by the rise of the laser pulse itself, while the decay transient varies depending on the temperature. At 300 K the photocurrent shows nonexponential decay, consisting of an initial component followed by a long-lived one. As the temperature

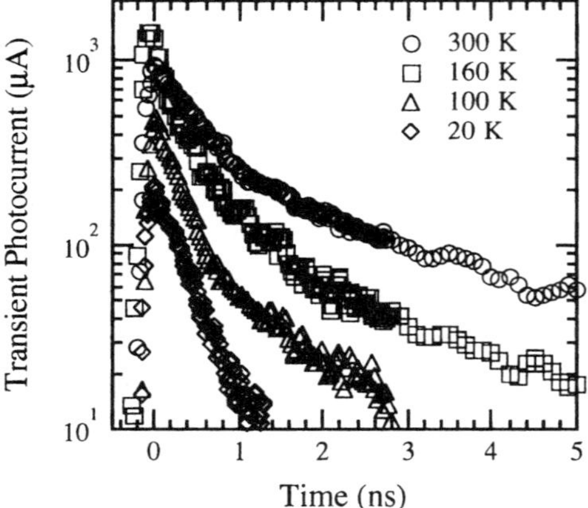

Figure 11   The transient photocurrent decay measured in an oxygen-free C_{60} film after photoexcitation at $h\nu = 2.0$ eV at various temperatures. (*After Lee at al.*, 1995.)

goes down, the long-lived component is greatly reduced and the photocurrent decay approaches an exponential one. The thermally activated component of photocurrent was interpreted in terms of carrier relaxation by multiple trapping, where the capturing of carriers occurs many times during their lifetime, while the initial component was thought to stem from photocarriers occupying extended band states (Moses et al., 1995). It was also found that the nonexponential decay became faster as the excitation intensity increased. This should be another manifestation of the predominance of the bimolecular recombination of photogenerated carriers as discussed in the Section 2.2. The same group also analyzed the evolution of photocurrent decay curves caused by the successive introduction of oxygen into the sample chamber, which will be described in Section 4 (Lee et al., 1994).

4.   EFFECTS OF OXYGEN

At this stage of the discussion, it is appropriate to refer to the effects of oxygen on electrical properties of fullerene solids including photo- and dark conductivity. In earlier works on electronic properties of fullerene, no particular precaution was taken to assess the effect of atmosphere. However, in view of the fact that the conductivity of common organic semiconductors such as anthracene and phthalocyanine is extremely sensitive to the presence of air (dark- and photocurrent increase by orders of magnitude by the introduction of oxygen), it should have been easy to suspect the necessity for more "atmosphere-alert" experiments. The first information about the effect of oxygen on electrical properties of solid C_{60} was brought about by three independent works. Minami and Sato (1993) performed in situ electrical measurements on C_{60} thin films vapor deposited in a high vacuum chamber strictly excluding the exposure to air. They showed that the introduction of oxygen into the chamber caused an orders of magnitude drop in both the photo- and dark conductivity, which could not be restored by the subsequent reevacuation of the chamber (Figure 12). From these results, oxygen molecules once absorbed in solid C_{60} did not seem to be easily removed. However, it was later shown that heating samples in vacuum at elevated temperatures considerably restored the original level of photo- and dark conductivity, suggesting that oxygen can be deintercalated almost reversibly (Kazaoui et al., 1994). Arai et al. (1992) reported that the dark conductivity of a C_{60} single crystal reversibly changed by the repeated cycle of vacuum heating and oxygen exposure. When the oxygen-exposed sample was heated in vacuum, its conductivity increased by 4 orders of magnitude and then drastically decreased again upon exposure to oxygen. Similar effects of oxygen on electrical properties were reported by Hamed et al. (1993) for C_{60} thin films.

The oxygen-induced drastic decrease of photocurrent can be understood by noting that the majority charge carriers in C_{60} solids are electrons. This is opposite to most commonly studied π-conjugated molecular solids such as phthalocyanine, in which cases, holes being the majority carriers, photoconductivity is greatly enhanced by oxygen exposure, because oxygen facilitates charge pair separation

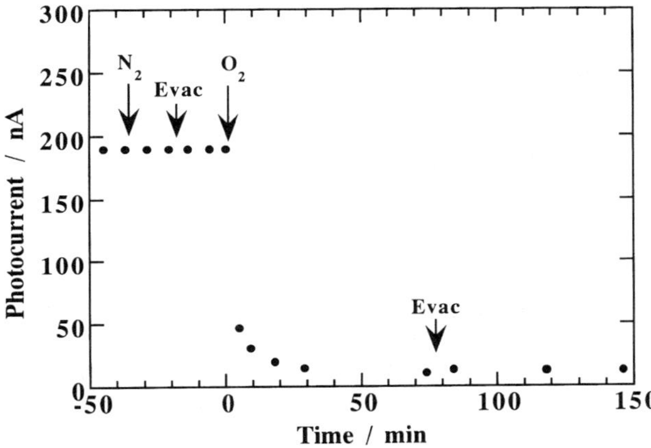

Figure 12 Effects of gas introduction and evacuation (Evac) on photocurrent in a C_{60} thin film.

by capturing electrons (Simon and André, 1984). In an n-type semiconductor like C_{60}, oxygen with strong affinity with electrons can work as a carrier trap or recombination center, thus greatly decreasing the carrier density or lifetime, resulting in the drastic drop in dark- and photoconductivity. Alternatively, in terms of a semiconductor language, it can be explained by a carrier compensation effect, where oxygen accepts electrons from existing donors in C_{60} solid, thus inactivating them, resulting in the reduction in conductivity (Arai et al., 1992).

The dynamic aspects of the oxygen effects on electronic transport processes in C_{60} thin film were revealed by time-resolved photoconductivity measurements (Lee et al., 1994). By successively exposing the samples to oxygen, they studied the evolution of the decay curves of transient photoconductivity in the time scale of nanosecond. As discussed in the Section 3.3, the way it decays represents how

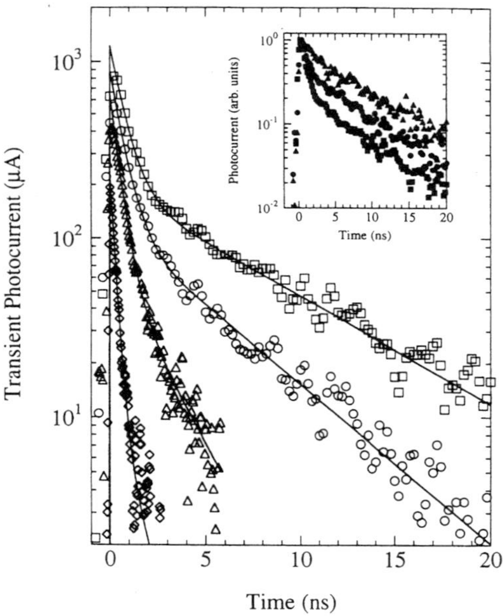

Figure 13   The time-resolved transient photocurrent ($T = 300$ K, $h\nu = 2.0$ eV) in pristine C_{60} film, and in the film at various levels of oxygen content. The dark current and relaxation time of each curve are indicated: $I_d = 3.6$ nA ($\square$; oxygen-free C_{60}, $\tau_1 = 693$ ps and $\tau_2 = 7.2$ ns), $I_d = 0.00567$ nA ($\bigcirc$; $\tau_1 = 640$ ps and $\tau_2 = 5.0$ ns), $I_d = 0.00024$ nA ($\triangle$; $\tau_1 = 463$ ps and $\tau_2 = 2.2$ ns), and $I_d < 0.00001$ nA ($\diamondsuit$; $\tau_1 = 238$ ps and $\tau_2 = 1.0$ ns). The inset shows the normalized transient photocurrent of pristine C_{60} at different laser intensities: 2.7×10^{14} ($\blacktriangle$), 1.1×10^{15} ($\bullet$), and 5.4×10^{15} photons/cm^2 ($\blacksquare$). (*After Lee et al., 1994.*)

photogenerated carriers disappear. It was found that the decay became faster as the concentration of oxygen increased (Figure 13). From these results, it was concluded that the photocarriers were lost by trapping or recombination involving oxygen absorbed in the samples.

It was known by NMR measurements that oxygen is easily intercalated into octahedral interstitial site in the face-centered cubic lattice of C_{60} (Assink et al., 1992). The oxygen diffusion was shown to be enhanced by light irradiation (Rao et al., 1993). It was even found that intense light irradiation in the presence of oxygen brought about irreversible destruction of the C_{60} cage, causing a large drop in the conductivity (Hamed et al., 1993). Furthermore, it was shown that oxygen drastically quenches not only photoconductivity but also fluorescence of C_{60} thin film (Minami et al., 1995a). Triplet excited states in C_{60} solid was also quenched by oxygen, generating singlet oxygen (Howells et al., 1994) showing infrared emission at 1281 nm. Although the detailed mechanisms of such oxygen effects on the electronic properties remain to be well established, the experimental results demonstrated so far are strong indications that extreme care must be taken while making measurements of any electronic processes in solid fullerene.

5. APPLICATIONS

5.1 Xerographic Photoreceptors

Although it seems to need a little more time before industrial uses are realized, there have been a number of works that were oriented toward practical applications of photoconductive properties of fullerene solids. At the present, the most important ones seem to be those for xerographic photoreceptors.

Wang (1992) studied the xerographic discharge behavior of C_{60}/C_{70}-doped films of polyvinylcarbazole (PVK), a well-known photoconductive polymer. Samples were made by spin casting mixed solution of C_{60}/C_{70} (85:15) and PVK with a doping level of 2.6 wt %. In this technique, sample films are positively or negatively charged by corona discharge, building electrostatic potential on the surface. The irradiation of samples generates photocarriers, which start to move by the effect of strong electric field formed there, causing a decay in the surface potential. The photoconduction process can be studied by monitoring and analyzing the decay. At low light intensity, charge carrier generation efficiency ϕ can be obtained by the time derivative of surface voltage V at time $t = 0$:

$$\phi = -\left(\frac{\varepsilon}{4\pi eLI}\right)\left(\frac{dV}{dt}\right)_{t=0} \tag{7}$$

where ε is the dielectric constant, e the electronic charge, L the film thickness, and I the absorbed photon flux. It was shown that by doping with fullerene the photoinduced discharge rate was markedly enhanced as compared with pure PVK under a tungsten lamp irradiation (Figure 14). Under monochromatic excitation at $\lambda = 500$ nm, the doping brought about a factor of 50 increase in the charge generation efficiency. Moreover, it was found that fullerene-doped PVK films had high dark resistance contributing to a very small voltage decay in dark. These are the important features to be satisfied for practical xerographic photoreceptor applications. The performance on the whole was claimed to be comparable to that of polycarbonate film doped with thiapyrylium dye aggregate, one of the best organic photoconductors used for electrophotography.

A similar system employing polymethylphenylsilane (PMPS) instead of PVK as a matrix polymer was later reported (Wang et al., 1993). Polysilanes including PMPS are known to possess high hole mobility along their σ-conjugated main chains and thus are considered as good hole conductors, while the charge photogeneration efficiency is quite low in their pristine state. Therefore, techniques to improve the efficiency had been crucially needed. By doping PMPS film with C_{60}, a strong electron acceptor, the system was expected to combine enhanced charge generation efficiency with its inherently good hole transport property. It was demonstrated that fullerene-doped PMPS possessed better photosensitivity than did fullerene-doped PVK; when compared at low field, the charge generation efficiency was shown to be improved by an order of magnitude.

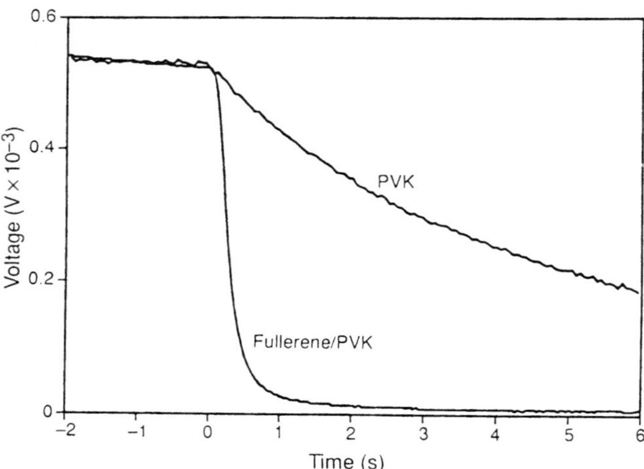

Figure 14   A qualitative comparison of the photoinduced discharge curves for pure PVK and fullerene-doped PVK under the same experimental conditions. A tungsten lamp (50 mW cm^{-2}) is used as the light source. (*After Wang, 1992.*)

Analysis of the field dependence of the charge generation efficiency gave fairly good fit to Onsager's geminate recombination model, yielding the Onsager radius of 1.9 and 2.7 nm for fullerene-doped films of PVK and PMPS, respectively (Wang et al., 1993). The larger radius indicates that charge generation at low field is higher, which was actually observed. As the mechanism of the improvement in charge generation efficiency by doping, photosensitization by fullerene or electron transfer from the polymer's excited states were considered depending on whether the excitation wavelength was >400 nm or <350 nm, respectively. In the former case, only fullerene absorbs light and its vacant ground states accept electrons from ground states of the polymer, while in the latter ground-state fullerene attracts electrons from the excited states of polymers, in either case resulting in charge carrier generation. For the case of fullerene-doped PVK, the direct excitation of a fullerene/carbazole charge transfer complex was also suggested. It is interesting to note that, in the case of polysilanes with no aromatic side groups (such as cyclohexylmethylpolysilane), fullerene doping brought about no significant improvement in charge generation efficiency. Such a result was later corroborated by the measurement of luminescence quenching of polysilane by C_{60} doping (Yoshino et al, 1995). These results suggest the importance of the phenyl side-group in PMPS and its electronic interaction with doped fullerene, for the charge generation efficiency to be improved.

While the above works used the fullerene-doped polymer films both as charge generation and as hole transport layers, Hirao et al. (1995) fabricated dual-layered xerographic photoreceptors, where fullerene thin film was used only as a charge generation layer and a polymer film doped with triphenyldiamine (TPD) as a hole transport layer. Such dual-layer configuration is the most commonly used one in commercial xerography. By separating the two essential functions of xerographic photoreceptors, this configuration can make the best use of the charge-generation capability of fullerene thin films. Four different kinds of charge generation layers were fabricated; polystyrene films doped with C_{60} and C_{70}, vacuum deposited C_{60} and C_{70} films. Among them, the photoreceptor made of C_{70} vacuum–deposited films in combination with a TPD-doped polystyrene charge transport layer showed the highest xerographic sensitivity. Figure 15 shows the sensitivity spectrum of the C_{70}/TPD dual layer photoreceptor. It was claimed to be the first fullerene-derived photoconductive device that has ever achieved a practically applicable sensitivity. C_{70} was also known to possess a higher photoconductive efficiency than did C_{60} in a sandwich cell configuration (Hosoya et al., 1994). The reason for the difference between C_{60} and C_{70} remains a subject of further studies. One factor might be the difference in the band gaps of solid C_{60} and C_{70}, 2.73 eV and 2.57 eV, respectively, as estimated from the energies of their CT excitons and polarization energies (Kazaoui et al., 1998).

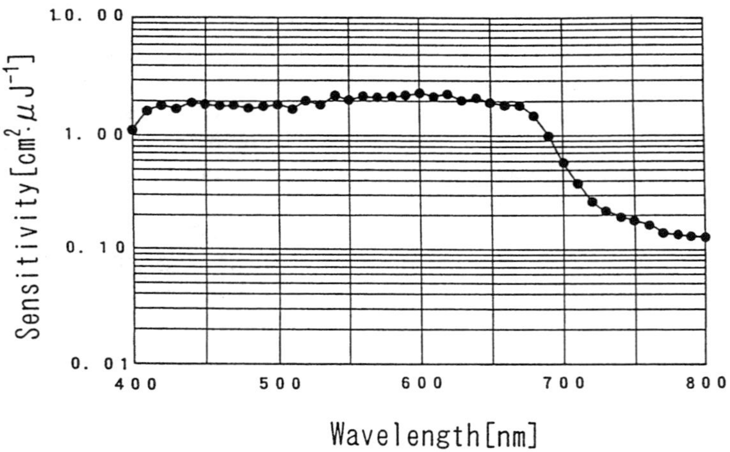

Figure 15   Sensitivity spectrum of a dual layer C_{70} photoreceptor. (*After Hirao et al.,* *1995.*)

5.2   Photovoltaic Devices

When electron accepting and donating layers are in contact with each other and molecules near the interface are photoexcited, positive and negative charges are separated from the excited states and driven in the opposite directions, generating photovoltage in the outer circuit. By analogy with inorganic devices, this type of configuration is often called organic *pn* junction. Such organic devices had been studied using phthalocyanine as electron-donating (*p* type) and perylene derivatives as electron-accepting (*n* type) compounds, giving relatively high energy conversion efficiencies of 0.5–1.0% under simulated solar irradiation (Tang, 1996). Since fullerenes were known to be relatively strong electron acceptors, attempts were made to fabricate fullerene-based photovoltaic cells. Examples of such cells with phthalocyanine compounds used as electron donors are mentioned here. A number of such works were also reported about composite systems consisting of fullerene and π-conjugated polymers (Sariciftci et al., 1993).

Xia et al. (1995) fabricated a sandwich-type cell ITO/InPcCl/C_{60}/Au, where InPcCl stands for a chloroindium phthalocyanine layer. Upon irradiation of monochromatic light (514.5 nm, 10 mW · cm^{-2}), the device developed an open-circuit voltage (V_{oc}) of 100 mV and a short-circuit current (J_{sc}) of 1 μA · cm^{-2}. The dark current–voltage curve showed rectification with the forward direction corresponding to the positive bias at the ITO electrode. These results could be understood in terms of *pn* junction formed at the InPcCl/C_{60} interface. Yonehara and Pac (1996) worked on a device with a structure ITO/TiOPc/C_{60}/

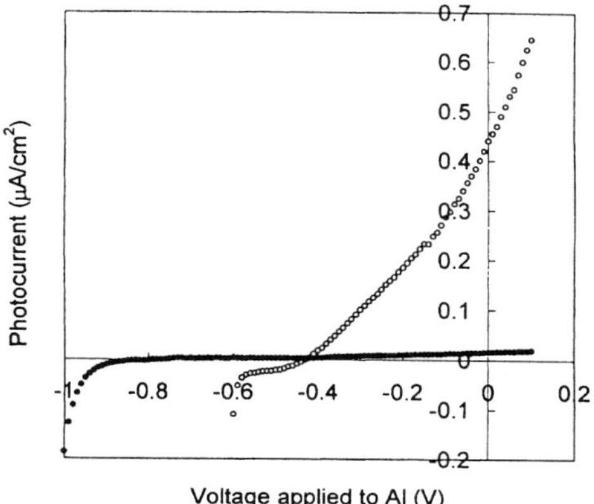

Figure 16 Photocurrent-voltage characteristics of ITO/α-TiOPc/C_{60}/Al cell ($\bigcirc$) and ITO/α-TiOPc/Al cell ($\bullet$) illuminated at 720 nm through the Al electrode (light intensity, 10 µW cm^{-2}); α-TiOPc was obtained by $CH_2 Cl_2$ vapor treatment of as-deposited TiOPc. (*After Yonehara and Pac, 1996.*)

Al, where TiOPc stands for titanyl phthalocyanine, giving V_{oc} of 0.42 V, J_{sc} of 4×10^{-7} A $\cdot$ cm^{-2}, and energy conversion efficiency (η) of 0.347% under monochromatic irradiation (720 nm, 10 µW $\cdot$ cm^{-2}) (Figure 16). The performance of the cell, however, degraded considerably under strong white light. The stability was greatly improved for a structure ITO/C_{60}/TiOPc/Cr $\cdot$ Au, in which case $V_{oc} = 0.05$ V, $J_{sc} = 1.82 \times 10^{-6}$ A $\cdot$ cm^{-2}, and $\eta = 1.37 \times 10^{-3}$% (720 nm, 1.86 mW $\cdot$ cm^{-2}). Moreover, by the insertion of a SiO layer between TiOPc and Cr $\cdot$ Au, the improvement of V_{oc} was effected. Murata et al. (1996) reported a cell whose structure was ITO/C_{60}/H_2Pc/Au, where H_2Pc stands for metal-free phthalocyanine; under white light irradiation (12.5 mW $\cdot$ cm^{-2}) this device developed $V_{oc} = 0.18$V, $J_{sc} = 89$ µA $\cdot$ cm^{-2}, and $\eta = 0.03$%. While these works showed that C_{60} thin films worked effectively as electron accepting layers in organic *pn* junctions, some breakthrough should be necessary for them to become materials of any promise for practical photovoltaic applications.

6. PHENOMENA OF INTEREST

In relation to photoconductivity in fullerene solids, several interesting phenomena were reported. While their complete establishment and mechanistic interpretation should await further studies, it appears appropriate to refer to some of them at this point.

For vacuum-deposited C_{60} thin films, Hamed et al. (1993a) observed a photoinduced increase in dark conductivity that persisted after the termination of photoexcitation. It took a long time (up to 10^6 s) for the conductivity to return to the original dark level, the length of which depended on the period of light irradiation. This kind of phenomenon had been known as persistent photoconductivity (PPC) for a number of inorganic semiconductors. However, C_{60} was claimed to be the first material that retained PPC at record high 260°C. It could not be understood in terms of thermal excitation of carriers from localized states or of macroscopic potential barriers that were proposed to explain PPC in inorganic semiconductors. The effect was tentatively associated with the strong distortions occurring on C_{60} molecules in the presence of localized charges. Later, on the basis of measurements on C_{60} single crystals, it was found that PPC behavior was strongly affected by the order-disorder phase transition at 260 K (Chiu et al., 1996). The PPC relaxation time took the maximum value at 260 K and decreased whether the temperature increased or decreased from it. From the results, it was proposed that the underlying mechanism of PPC in C_{60} solids involved the motion of C_{60} in the solid lattice.

While the majority of photoconductivity experiments on fullerenes were done using thin film samples, there have been several works aiming at the understanding of the photoconductivity in C_{60} single crystals. This can be rationalized because they generally suffer fewer disorders or grain boundaries inevitably encountered in thin films. Especially for carrier transport processes, it is highly probable that these defects work as traps or recombination centers, obscuring intrinsic properties of solid fullerenes. For that reason, crystals were used in the evaluation of charge carrier mobilities in C_{60} solid (Section 3.1). Even carrier generation from excitons may be subject to the effects of such defects if the structural coherence in fullerene solids becomes shorter than the exciton diffusion length.

Reflecting the structure-sensitive nature of single crystals, several intriguing photoconductive phenomena were reported, especially in connection with the phase transitions in C_{60} solids. Yamaguchi et al. (1995) reported three anomalies in the temperature dependence of pulsed photoconductivity in C_{60} single crystals. They took place at 85, 155, and 260 K and were proposed to be connected with the glass transition, freezing of rotational axis of C_{60} molecules, and the order-disorder transition, respectively. In another work, the phase transition near 260 K appeared to be involved in the emergence of negative photoconductivity in C_{60} single crystals observed upon excitation at 1.65 eV (Ishijima et al., 1996). The photoresponse at this energy was originally suggested to come from carrier transport through the t_{1u} state in molecular C_{60} (Matsuura et al., 1995). While more works seems necessary for these phenomena to be understood with better consistency, these results might represent some new aspects of electronic processes in C_{60} solids.

7. CONCLUSIONS

The outline of the photoconductive behavior of fullerene solids (thin films and crystals) can probably be described largely in the basic framework developed for π-conjugated organic molecular solids. The photocarrier generation proceeds by the Onsager formalism, where photocreated charge pairs bound by Coulomb force are eventually separated by the assistance of electric field and thermal energy. The charged pairs are none other than CT excitons working as precursors for photocarriers, locating above the specific energy ($\sim$2.3 eV) in C_{60} solids. Their characteristic natures are found to be reflected in various spectroscopic properties such as absorption, electroabsorption, luminescence excitation, and electric-field-induced luminescence quenching. Much more work, however, remains to be done for the establishment of their detailed features, such as a relationship between the manifold of CT states and the crystal lattice structure. In principle, it should be possible to assign each CT transition to a specific lattice point, as was done for an anthracene crystal (Sebastian et al., 1983). Such analyses are expected to contribute to the elucidation of the unique features derived from this unprecedented architecture of closed-packed, π-conjugated balls (Kazaoui et al., 1998).

About the carrier transport process in solid fullerene, the understanding is much more limited mainly because of the scarcity of the experimental data. In view of the fact that oxygen has the tremendous effects on the carrier dynamics in fullerene thin films and crystals, the direct measurements of the carrier mobility by the TOF method strictly excluding its disturbance would be desired. In addition, the effect of the crystal quality on carrier transport processes should be explored, because improvements in crystal growth techniques may contribute to the realization of much higher carrier mobility, as was demonstrated for ultrapure naphthalene crystals (Warta and Karl, 1985). Theoretical approaches are also needed for the understanding of the interrelationship between the carrier mobility and such parameters as temperature and the phase transition.

The effects of oxygen should also be important from the viewpoint of practical applications. Their extent and how they can be prevented or controlled should be widely explored especially for applications in which carrier transport constitutes a crucial step. Fabrication of structurally controlled fullerene thin films is another important factor to investigate, because the field-effect mobility seems to be limited by disorders or impurities inevitably contained in them. For example, increase of the structural coherent length in thin films (the size of crystal domains) to as long as 10–100 μm may contribute greatly to the improvement of the device performance, as was reported recently for a pentacene thin film (Gundlach et al., 1997).

Finally, it should be stressed that despite all the efforts devoted to the study of the photocarrier generation and transport processes in solid fullerene, their comprehensive understanding is still in its infant stage. Explorations of such fields as exemplified in the preceding paragraphs are sure to contribute to the revelation of still hidden aspects of this new type of π-conjugated systems. In this context, the study of carbon nanotube, another novel form of solid carbon, is extremely intriguing, in which case the carrier transport should proceed in a highly anisotropic way. The spectral and transient analysis of its photoconductivity that presumably exist in its semiconducting state must provide information essential for understanding their unique electronic properties. Research efforts in that direction have just started (Tans et al., 1997; Bockrath et al., 1997).

REFERENCES

Achiba, Y., Nakagawa, T., Matsui, Y., Suzuki, S., Shiromaru, H., Yamauchi, K., Nishiyama, K., Kainosho, M., Hoshi, H., Maruyama, Y., and Mitani, T., *Chem. Lett.* 1233 (1991).

Alers, G. B., Golding, B., Kortan, A. R., Haddon, R. C., and Theil F. A., *Science*, **257**, 511 (1992).

Arai, T., Murakami, Y., Suematsu, H., Kikuchi, K., Achiba, Y., and Ikemoto, I., *Solid State Commun.* **84**, 827 (1992).

Assink, R. A., Schirber, J. E., Loy, D. A., Morosin, B., and Carlson, G. A., *J. Mater. Res.* **7**, 2136 (1992).

Borsenberger, P. M., and Weiss, D. S., *Organic Photoreceptors for Imaging Systems*, Marcel Dekker, New York, 1993, p. 82.

Bockrath, M., Cobden, D. H., McEuen, P. L., Chopra, N. G., Zettl, A., Thess, A., and Smalley, R. E., *Science* **275**, 1922 (1997).

Chiu, K. C., Wang, J. S., Dai, Y. T., and Chen, Y. F., *Appl. Phys. Lette.* **69**, 2665 (1996).

Eilmes, A., Munn, R, W., Pac, B., and Petelenz, P., *Chem. Phys.* **214**, 341 (1997).

Frankevich, E., Maruyama, Y., and Ogata, H., *Chem. Phys. Lett.* **214**, 39 (1993).

Frankevich, E., Maruyama, Y., and Ogata, H., *J. Phys. Chem. Solids* **57**, 483 (1996).

Giro, G., Kalinowski, J., Di Marco, P., Fattori, V., and Marconi, G., *Chem. Phys. Lett.* **211**, 580 (1993).

Gundlach, D. J., Lin, Y. Y., Jackson, T. N., Nelson, S. F., and Schlom, D. G., *IEEE Electron Device Lett.* **18**, 87 (1997).

Haddon, R. C., Perel, A. S., Morris, R. C., Palstra, T. T. M., Hebard, A. F., and Fleming, R. M., *Appl. Phys. Lett.* **67**, 121 (1995).

Hamed, A., Sun, Y. Y., Tao, Y. K., Meng, R. L., and Hor, P. H., *Phys. Rev.* **B47**, 10873 (1993).

Hamed, A., Rasmussen, H., and Hor, P. H., *Phys. Rev.* **B48**, 14760 (1993a).

Harigaya, K., and Abe, S., *Phys. Rev.* **B49**, 16746 (1994).

Heiney, P. A., *J. Phys. Chem. Solids* **53**, 1333 (1992).

Hirao, A., Nishizawa, H., Miyamoto, H., Sugiuchi, M., and Hosoya, M., *Proc. SPIE*, **2526**, 71 (1995).

Horowitz, G., Garnier, F., Yassar, A., Hajlaoui, R., and Kouki, F., *Adv. Mater.* **8**, 52 (1996).

Hosoya, M., Ichimura, K., Wang, Z. H., Dresselhaus, G., and Dresselhaus, M. S., *Phys. Rev.* **B49**, 4981 (1994).

Howells, S. C., Black, G., and Schlie, L. A., *Synth. Met.* **62**, 1 (1994).

Ishijima, Y., and Ishiguro, T., *J. Phys. Soc. Jpn.* **65**, 1574 (1996).

Jeglinski, S., Vardeny, Z. V., Moses, D., Srdanov, V. I., and Wudl, F., *Synth. Met.*, **49–50**, 557 (1992).

Kaiser, M., Reichenbach, J., Byrne, H. J., Anders, J., Maser, W., Roth, S., Zahab, A., and Bernier, P., *Solid State Commun.* **81**, 261 (1992).

Kalinowski, J., Stampor, W., and Di Marco, P. G., *J. Chem. Phys.* **96**, 4136 (1992).

Kalinowski, J., Giro, G., Camaioni, N., Fattori, V., and Di Marco, P., *Synth. Met.* **77**, 181 (1996).

Kaneto, K., Yamanaka, K., Rikitake, K., Akiyama, T., and Takashima, W., *Jpn. J. Appl. Phys.* **35**, 1802 (1996).

Kazaoui, S., Ross, R., and Minami, N., *Solid State Commun.* **90**, 623 (1994).

Kazaoui, S., Ross, R., and Minami, N., *Synth. Met.* **70**, 1403 (1995).

Kazaoui, S., Ross, R., and Minami, N., *Phys. Rev.* **B52**, R11665 (1995a).

Kazaoui, S., Ross, R., and Minami, N., *Proc. SPIE.* **2530**, 41 (1995b).

Kazaoui, S., and Minami, N., *Synth. Met.* **86**, 2345 (1997).

Kazaoui, S., Minami, N., Tanabe, Y., Byrne, H. J., Eilmes, A., and Petelenz, P., *Phys. Rev.*, **B58**, 7689 (1998).

Könenkamp, R., Engelhardt, R., and Henninger, R., *Solid State Commun.* **97**, 285 (1996).

Kudo, K., Saraya, T., Kuniyoshi, S., and Tanaka, K., *Mol. Cryst. Liq. Cryst.* **267**, 423 (1995).

Lee, C. H., Yu, G., and Moses, D., *Phys. Rev.* **B48**, 8506 (1993).

Lee, C. H., Yu, G., Kraabel, B., and Moses, D., *Phys. Rev.* **B49**, 10572 (1994).

Lee, C. H., Yu, G., Moses, D., and Srdanov, V. I., *Synth. Met.* **70**, 1413 (1995).

Matsuo, T., Suga, H., David, W. I. F., Ibberson, R. M., Bernier, P., Zahab, A., Fabre, C., Rassat, A., and Dworkin, A., *Solid State Commun.* **83**, 711 (1992).

Matsuura, S., Ishiguro, T., Kikuchi, K., and Achiba, Y., *Phys. Rev.* **B51**, 10217 (1995).

Minami, N., *Chem. Lett.* 1791 (1991).

Minami, N., and Sato, M., *Synth. Met.*, **55–57**, 3092 (1993).

Minami, N., Kazaoui, S., and Ross, R., *Synth. Met.* **70**, 1397 (1995).

Minami, N., Kazaoui, S., and Ross, R., in *Physics and Chemistry of Fullerenes and Derivatives* (H. Kuzmany, J. Fink, M. Mehring, and S. Roth, eds.), World Scientific, Singapore, 1995a, p. 238.

Mort, J., Okumura, K., Machonkin, M., Ziolo, R., Huffman, D. R., and Ferguson, M. I., *Chem. Phys. Lett.* **186**, 281 (1991).

Mort, J., Machonkin, M., Ziolo, R., and Chen, I., *Appl. Phys. Lett.* **61**, 1829 (1992).

Mort, J., Machonkin, M., Ziolo, R., Huffman, D, R., and Ferguson, M, I., *Appl. Phys. Lett.* **60**, 1735 (1992a).

Moses, D., and Heeger, A. J., in *Relaxation in Polymers* (T. Kobayashi, ed.), World Scientific, Singapore, 1992, p. 134.

Moses, D., Lee, C. H., Kraabel, B., Yu, G., and Srdanov, V., I., *Synth. Met.* **70**, 1419 (1995).

Murata, K., Ito, S., Takahashi, K., and Hoffman, B. M., *Appl. Phys. Lett.* **68**, 427 (1996).

Negri, F., Orlandi, G., and Zerbetto, F., *J. Chem. Phys.* **97**, 6496 (1992).

Petelenz, P., Slawik, M., and Pac, B., *Synth. Met.* **64**, 335 (1994).

Rao, A. M., Wang, K-A., Holden, J. M., Wang, Y., Zhou, P., Eklund, P. C., Eloi, C. C., and Robertson, J. D., *J. Mater. Res.* **8**, 2277 (1993).

Sariciftci, N. S., Braun, D., Zhang, C., Srdanov, V. I., Heeger, A. J., Stucky, G., and Wudl, F., *Appl. Phys. Lett.* **6**, 585 (1993).

Schlaich, H., Muccini, M., Feldmann, J., Bässler, H., Göbel, E. O., Zamboni, R., Taliani, C., Erxmeyer, and J., Weidinger, A., *Chem. Phys. Lett.* **236**, 135 (1995).

Sebastian, L., Weiser, G., Peter, G., and Bässler, H., *Chem. Phys.* **75**, 103 (1983).

Silinsh, E. A., and Cápek, V., *Organic Molecular Crystals*, American Institute of Physics, New York, 1994.

Simon, J., and Andre, J.-J., *Molecular Semiconductors*, Springer Verlag, Berlin, 1984, p. 122.

Sumi, H., *Solid State Commun.* **28**, 309 (1978).

Sze, S. M., *Physics of Semiconductor Devices*, Wiley-Interscience, New York, 1969, p. 505.

Tang, C, W., *Appl. Phys. Lett.* **48**, 183 (1986).

Tans, S. J., Devoret, M. H., Dai, H., Thess, A., Smalley, R. E., Geerligs, L. J., and Dekker, C., *Nature* **386**, 474 (1997).

Tsubo, T., and Nasu, K., *Solid State Commun.* **91**, 907 (1994).

van den Heuvel, D. J., Chan, I. Y., Groenen, E. J. J., Schmidt, J., and Meijer, G., *Chem. Phys. Lett.* **231**, 111 (1994).

Wang, Y., *Nature* **356**, 585 (1992).

Wang, Y., West, R., and Yuan, C.-H., *J. Am. Chem. Soc.* **115**, 3844 (1993).

Warta, W., and Karl, N., *Phys. Rev.* **B 32**, 1172 (1985).

Wen, C., Li, J., Kitazawa, K., Aida, T., Honma, I., Komiyama, H., and Yamada, K., *Appl. Phys. Lett.* **61**, 2162 (1992).

Williams, R. M., and Verhoeven, J. W., *Spectrochim. Acta* **50A**, 251 (1994).

Wright, J. D., *Molecular Crystals*, 2nd ed., Cambridge University Press, Cambridge, 1995.

Xia, A. D., Fu, S. J., Pan, H. B., Zhang, X. Y., Xu, Z., Liu, Q., and Yuan, R. K., *Solid State Commun.* **95**, 713 (1995).

Yamaguchi, H., Yamaguchi, T., Kagoshima, S., Masumi, T., Li, J., and Kishio, K., *J. Phys. Soc. Jpn.* **64**, 527 (1995).

Yonehara, H., and Pac, C., *Appl. Phys. Lett.* **61**, 575 (1992).

Yonehara, H., and Pac, C., *Thin Solid Films* **278**, 108 (1996).

Yoshino, K., Yoshimoto, K., Hamaguchi, M., Kawai, T., Zakhidov, A. A., Ueno, H., Kakimoto, M., and Kojima, H., *Jpn. J. Appl. Phys.* **34**, L141 (1995).

10
Optical and Electronic Properties of Polymeric Fullerenes

H. Kuzmany, B. Burger, and J. Kürti
University of Vienna
Vienna, Austria

1. INTRODUCTION

Fullerenes have been known to undergo chemical reactions to a polymeric state. This is particularly so for C_{60} which exists in various polymeric phases. These phases can be crystalline or amorphous, they can be insulating or metallic; and the geometry of the polymer can be one-dimensional, two-dimensional or even three-dimensional. In addition, the fullerene molecules can be connected by individual single bonds or by paired single bonds between the carbon atoms. Which phase is finally obtained depends strongly on the details of how the chemical reaction for the polymerization was triggered.

The polymerization process can be observed in many ways. Standard procedures are structrual analyses with X-rays or neutrons. This chapter will focus on the electronic structure and optical properties of the polymers which means on optical and electronic methods, including light scattering spectroscopy, IR absorption, photoemission, and electron energy-loss spectroscopy.

2. OPTICAL AND VIBRATIONAL PROPERTIES OF PRISTINE C_{60}

Optical and electronic properties of C_{60} are reviewed in detail in the other chapters of this book. For the discussions in this chapter it is convenient to recall these properties and to include the activites of vibrational modes with respect to Raman scattering and IR spectroscopy.

The 60 p_z electrons per molecule are accommodated in 30 orbitals derived from the irreducible representations of the I_h point group. In the solid the highest occupied and the lowest unoccupied bands (valence band and conduction band) are derived from the h_u and the t_{1u} molecular orbitals, respectively. The widths of both bands is only of the order of 0.5 eV. Since dipole transitions are not allowed between bands of equal parity, the optical gap is determined by the energy difference between the valence band and the band above the conduction band. This band is derived from the t_{1g} molecular orbitals.

The isolated C_{60} cage has 10 Raman active vibrations: two nondegenerated modes with A_g symmetry and eight fivefold degenerated modes with H_g symmetry. At room temperature the two A_g modes are the radial breathing mode at 497 cm^{-1} and the pentagonal pinch mode at 1469 cm^{-1}. The H_g modes are observed at 270, 433, 709, 773, 1103, 1253, 1424 and 1576 cm^{-1}.

The infrared active modes of the free C_{60} molecule are of T_{1u} symmetry and located at 527, 576, 1183, and 1429 cm^{-1} for pristine solid C_{60}.

3. POLYMERIC STATES, BONDING, AND CRYSTALLINITY

To create the polymeric state one or more double bonds on the cage must be opened and reclosed to carbon atoms on a neighboring cage. This means in all polymeric phases the carbons connecting the cages are in sp^3-like states rather than in the usual sp^2 hybridization and the continuous conjugation across the cage is interrupted between the cages. π-electron overlap is only possible between carbon atoms next to the bonding atoms on both sides. This is a rather large distance and of the same magnitude as the distance between overlapping orbitals perpendicular to the polymeric bonds. As a consequence the polymers are always three-dimensional with respect to their electronic structure even though topologically or mechanically the chemical structure may be one-dimensional or two-dimensional.

3.1 The 2 + 2 Cycloaddition Reaction

The classical reaction which leads to the polymeric state is the 2 + 2 cycloaddition between two neighboring cages. As shown in Figure 1, a cyclobutane ring is established. To enable this reaction the two cages must be arranged at least statistically in time in a way that two double bonds are oriented parallel with a distance less than a critical distance d_c. The value of d_c depends on the reaction process. Since for purely sp^2-hybridized carbons the 2 + 2 cycloaddition is forbidden by symmetry, the reaction can only proceed if either one of the π electrons

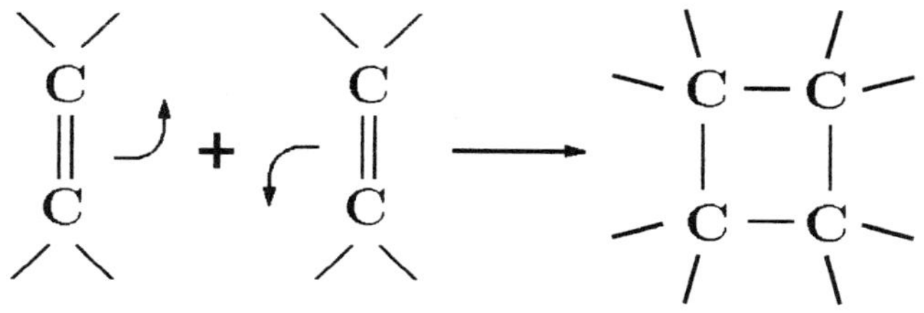

Figure 1 $2 + 2$ cycloaddition reaction between two parallel-oriented double bonds of a fullerene cage.

is excited into a higher state (photopolymerization [1]), a charge is added to the lowest unoccupied triply degenerated molecular orbital (polymerization by charge transfer [2]), or high pressure is applied at elevated temperature (polymerization by pressure [3,4]). In the first two cases the critical distance d_c is larger than the nearest carbon-carbon distance in the fcc lattice so that the polymerization reaction proceeds spontaneously. For the photochemical reaction d_c is about 4.2 Å [5]. In the third case pressure is used to squeeze the lattice and thus reduce the carbon-carbon distance below the critical value. The reaction may proceed in this case by tunneling to the lower-energy polymeric state or it may just be enabled because the selection rules are released due to the deviation of the state of hybridization of the carbons from sp^2. As it is described in the following for the polymers obtained from high-pressure experiments thermal activation of the π electrons is required.

3.2 Single-Bonded Polymers

In contrast to the $2 + 2$ cycloaddition reaction where always two adjacent carbon atoms on the one cage are bonded to two equivalent adjacent carbons on the neighboring cage, recently polymerization by bonding between single carbon atoms was reported. Prassides et al. have proved that the ground state of Na_2RbC_{60} is a single-bonded linear polymer [6, 7], whereas Oszlányi et al. have identified the Na_4C_{60} phase as a single-bonded two-dimensional polymer [8].

3.3 Dimeric Phases

Interesting phases have been observed after quenching AC_{60} (A = K, Rb, Cs) from the high-temperature fcc phase to 80 K [9, 10]. In this case C_{60} dimers have been observed where the two cages are connected by a single bond. This structure is very similar to the recently prepared $(C_{59}N)_2$ where the two intrinsically doped $C_{59}N$ molecules are also connected by a single bond [11]. The C_{60} dimers are unstable vs. an increase of the temperature [12, 13]. The formation of dimers during photopolymerization was also proposed from Raman measurements [14]. Recently Komatsu et al. reported the dimerization even for uncharged C_{60} via a pure solid-state chemical reaction [15].

3.4 Crystalline and Amorphous Polymers

Under certain conditions the high crystallinity of the original C_{60} material is retained after the polymerization process. This is the case for polymerization under mild conditions with respect to pressure and temperature but also for the dimerization and polymerization by charge transfer. On the other hand, photopolymerization and polymerization at high pressure and high temperature results in highly noncrystalline material.

4. POLYMERIZATION AT HIGH PRESSURE

Applying hydrostatic or quasi-hydrostatic pressure to C_{60} leads to structurally different materials depending on the pressure and temperature used. For pressures up to 25 GPa and temperatures of 300 K and below, the pressurized material is unstable and returns to pristine C_{60} upon release of pressure [16]. Heating the samples to several 100°C and applying pressure of several GPa yields metastable, polymeric species. Finally, pressurizing the material at even higher temperature, it becomes polymeric in three dimensions and stable, but the C_{60} cages are at least partially fragmentated.

The first characterization of the pressure polymerized phases was reported by Iwasa et al. [3]. Depending on the reaction temperature, either a rhombohedral or a fcc material was obtained. The observed small value of 9.2 Å for the nearest-neighbor distance was taken as evidence for a covalent bonding between the cages. A more detailed analysis was performed by Nuñez-Requeiro et al. [4], who observed three different polymeric structures. Polymerization was suggested to be two-dimensional for the rhombohedral and the tetragonal phase and one-dimensional for an orthorhombic phase as it is shown in Figure 2. The orthorhombic structure was generated at 8 GPa and 300°C. At lower pressures and higher temperatures (e.g., 3 GPa and 600°C) the samples were mixtures of the rhombohedral and tetragonal phases. At 4 GPa and 700°C the sample consisted of an almost pure rhombohedral phase.

Both the orthorhombic and the tetragonal phases have the space group *Immm*. The space group for the rhombohedral phase is $R\,\overline{3}m$. For the orthorhombic phase the one-dimensional linear chains are formed along the <110> direction of the original fcc phase. The C_{60} molecules are linked by a 2 + 2 cycloaddition along this chain with the four-membered ring in the (001) plane of the

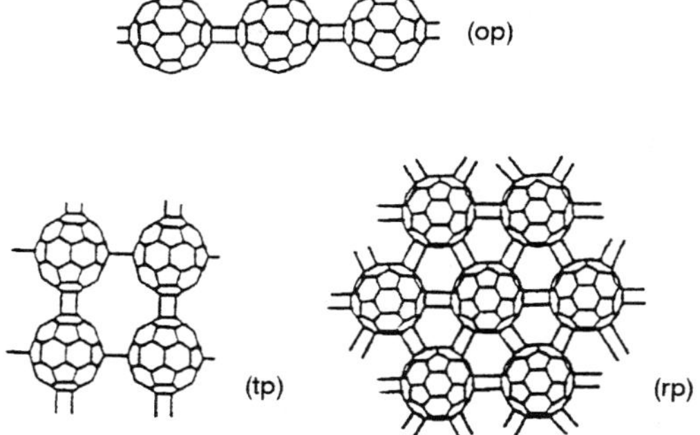

Figure 2 Schematic structural arrangement of C_{60} molecules in the one-dimensional orthorhombic (op), the two-dimensional tetragonal (tp), and the two-dimensional rhombohedral (rp) polymeric phases.

original phase. The new crystal structure is similar to the structure shown in Figure 5 for the AC_{60} compounds except that the octahedral interstitial sites are not occupied. The point group of the unit cell is D_{2h}. The rhombohedral structure originates from a shrinking of the (111) plane due to a bonding of the C_{60} molecule by a cyclobutane ring to all 6 nearest neighbors. The point group of the unit cell is D_{3d}. Calculations of Xu et al. [17] on the tight-binding level showed that the rhombohedral and tetragonal phases exhibit elongated C_{60} cages in the direction of the covalent interfullerene bonds.

A polymeric structure was also obtained for much milder conditions with respect to pressure and temperature like 1 GPa and 550–585 K [18]. X-ray analysis and Raman spectra [19] indicated an orhtorhombic unit cell with linear chains along <110> as for the orthorhombic polymer described above.

4.1 Electronic Structure

Calculations concerning the bonding between C_{60} molecules were carried out for various connections and geometries, including infinite long chains. The bonding by a cycloaddition from two parallel oriented hexagon-hexagon bonds (66/66 bonding) revealed the lowest energy. For the bonding energy of the dimer different results can be found in the literature: 1.4 eV from AM1 calculations [20], 1.2 eV with SCF-LDA [21], 0.3 eV with DF-TB and also with gradient corrected LDA [21], 0.47 eV with first-principles quantum molecular dynamics [22].

Adams et al. have calculated the electronic structure of the neutral orthorhombic polymer chain using the local-density approximation [22]. They obtained an energy gap at the Fermi level slightly smaller than the gap for the

Table 1   Correlation Table of Symmetry Groups I_h and D_{2h}, D_{3d}

I_h	D_{2h}	D_{3d}
$2\,A_g$	$2A_g$	$2A_g$
$3\,T_{1g}$	$3B_{1g} + 3B_{2g} + 3B_{3g}$	$3E_g + 3A_{2g}$
$4\,T_{2g}$	$4B_{1g} + 4B_{2g} + 4B_{3g}$	$4E_g + 4A_{2g}$
$6\,G_g$	$6A_g + 6B_{1g} + 6B_{2g} + 6B_{3g}$	$6E_g + 6A_{1g} + 6A_{2g}$
$8\,H_g$	$16A_g + 8B_{1g} + 8B_{2g} + 8B_{3g}$	$16E_g + 8A_{1g}$
$1\,A_u$	$1A_u$	$1A_u$
$4\,T_{1u}$	$4B_{1u} + 4B_{2u} + 4B_{3u}$	$4E_u + 4A_{2u}$
$5\,T_{2u}$	$5B_{1u} + 5B_{2u} + 5B_{3u}$	$5E_u + 5A_{2u}$
$6\,G_u$	$6A_u + 6B_{1u} + 6B_{2u} + 6B_{3u}$	$6E_u + 6A_{1u} + 6A_{2u}$
$7\,H_u$	$14A_u + 7B_{1u} + 7B_{2u} + 7B_{3u}$	$14E_u + 7A_{1u}$

pristine material. The Fermi level itself was found to be about 0.3 eV higher for the polymer than for the pure C_{60}. No experimental results are as yet available regarding optical properties.

4.2   Raman and Infrared Spectra

Raman spectra are most directly related to structural and electronic properties because of the high symmetry of the C_{60} molecule and the electronic nature of their excitation. Any reduction from the I_h symmetry of the cage will lead to a

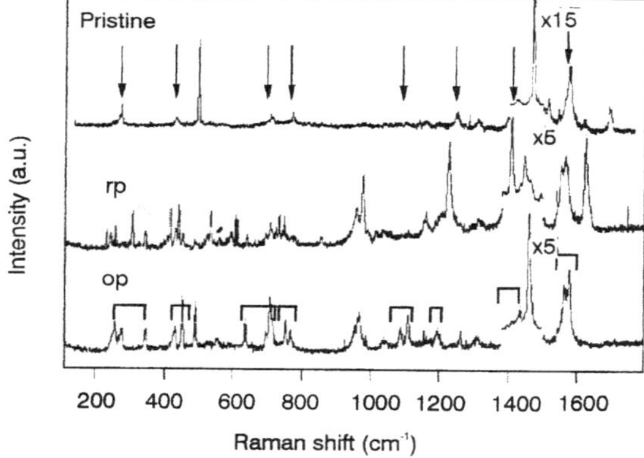

Figure 3   Overall Raman spectra of C_{60} (pristine) and of C_{60} polymerized to the rhombohedral (rp) and to the orthorhombic (op) phase. The arrows and the square brackets assign the Raman active H_g modes in their degenerate and splitted form, respectively.

splitting of the highly degenerated modes and any excitation of the spectra between two allowed transitions will lead to resonance effects. In the special case
of polymeric C_{60} the D_{2h} symmetry allows only for one-dimensional modes and
D_{3d} allows only for at most two-dimensional modes. This means all H_g modes
in the Raman spectra will be splitted. Additional modes may appear as well since
in D_{2h} and in D_{3d} all *gerade* modes are Raman active. The correlation between
the symmetry groups I_h and D_{2h} and D_{3d} is shown in Table 1.

Figure 3 shows Raman spectra for the orthorhombic and for the rhombo-
edric phase obtained at high pressure. The spectrum for pristine C_{60} is included
for comparison. Most of the new lines observed in the spectra for the polymer
can be assigned to the splitting of H_g modes. Similar spectra were reported in
Refs. 23 and 24.

The lowering of the symmetry in the polymeric phase leads to a similar

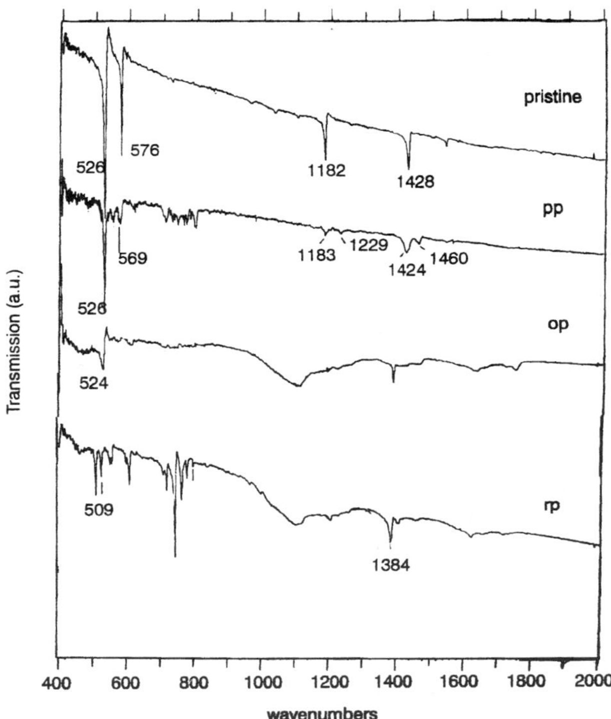

Figure 4 FTIR spectra of C_{60} (pristine) and of C_{60} polymerized to the rhombohedral
(rp) and to the orthorhombic (op) phase, together with the spectrum of the photopolymerized sample (pp). (*After Ref. 23.*)

effect in the infrared spectrum: the IR active T_{1u} bands of the pristine material split and new bands appear as it is demonstrated in Figure 4.

5. POLYMERIC PHASES AC_{60}, A = K, RB, CS

One of the many exciting properties of C_{60} fullerenes is the possibility to obtain doped compounds by intercalation of strong electron donors into the pristine C_{60} material. Two structural features of the fullerenes are in favor for such processes. The valence electrons are accommodated in π orbitals which allows for an easy addition (or subtraction) of electrons like in most of the π-electron systems, and the large size of the molecules provides ample interstitial lattice space to accommodate the donor atoms. Since C_{60} crystalizes in a fcc structure, one rather large interstitial lattice site with octahedral configuration and two smaller sites with tetrahedral configuration are available per C_{60} molecule. Filling only the octahedral site with K, Rb, or Cs leads to the famous family of AC_{60} compounds which were identified for the first time only in 1992 [25]. The delayed observation of these compounds as compared to the other members of the family $A_x C_{60}$ may be due to the lack of access to them by intercalation at ambient or moderately elevated temperatures. The compounds can only be prepared by doping at high temperatures and subsequent cooling.

5.1 The Structure of AC_{60}

The high-temperature phase of AC_{60} was identified as a fcc rock-salt structure with the alkali metals only at octahedral lattice sites and lattice constants of 14.07, 14.08, 14.13 Å at 473 K for K, Rb, and Cs, respectively [26]. All these lattice constants are smaller than the lattice constant of 14.23 Å for pristine C_{60} because of Coulomb attraction. (See also Table 2.)

On cooling the compound to room temperature a phase transition is observed. The X-ray analysis of the low-temperature structure in nearly stoichiometric AC_{60} with A = Rb was reported by Chauvet et al. [27] and in more detail by Stephens et al. [28]. It revealed an orthorhombic unit cell with lattice parameters $a = 9.13$ Å $b = 10.11$ Å $c = 14.23$ Å. This lattice can be derived from a bc tetragonal representation of the high-temperature fcc phase with lattice constants $a = b = 9.95$ Å and $c = 14.08$ Å as shown in Figure 5a, as in the orthorhombic phase obtained from polymerization by pressure. The observed lattice constant in the new a direction is only 9.13 Å, which is unusually short as compared to the center-to-center distances in the other C_{60} compounds. This finding was the reason to suggest a polymeric chain structure for the C_{60} molecules in this phase, as shown in Figure 5b.

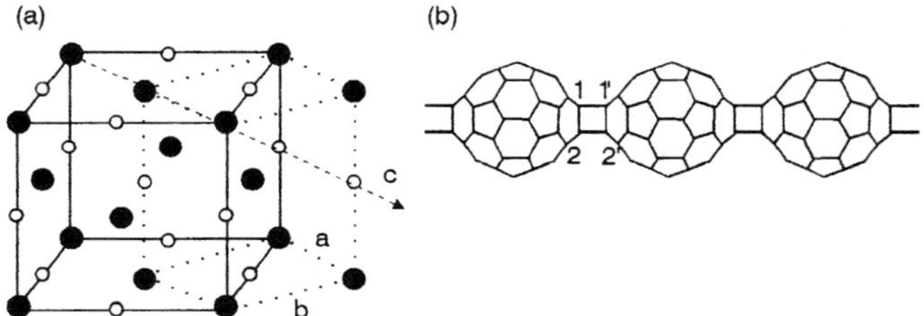

Figure 5 (a) Fcc rock salt structure (full lines) and derived bco structure (dotted lines) of AC_{60} and (b) the polymerized C_{60} chain as obtained from a 2 + 2 cycloaddition. The dashed line in (a) is the chain direction. The numbers on the chain in (b) assign the carbon atoms used to describe the bonds in Table 2.

Since the polymerization can start in six different [110] directions, a multidomain structure with nanosize clusters can be expected which makes it difficult to observe the symmetry of the bonding structure by macroscopic techniques such as optical spectroscopy. The nanoscale domains have been observed so far only by electron microscopy [29].

AC_{60} with A = Rb, Cs behaves differently from KC_{60} in several aspects. Intermediate phases during the cooling process and disproportionation into the phases C_{60} and A_3C_{60} have been reported for the potassium compound. Only if the cooling is fast enough (which means of the order of K/min) the same orthorhombic structure and the same polymeric structure is observed for KC_{60} and for RbC_{60}. Whether there is a fundamental difference between the ground state of the two compounds or simply that any cooling rate is too fast for the larger diameter Rb or Cs atoms is an open question. Activation energies for the two competing processes of phase separation and polymerization were reported recently to be 1.1 and 0.17 eV, respectively [30].

The lattice parameters and bond lengths obtained from several authors are summarized in Table 2.

The bond lengths between the fullerene cages are in agreement with typical σ bonds between carbon atoms. The intrafullerene C1—C1 bond is unusually large and indicates a very weak σ bonding. It is also important to note that the interfullerene distance C1′—C1′ is considerably shorter for RbC_{60} as compared to KC_{60}. The changes of the bond lengths on the molecule reduces the I_h symmetry to D_{2h}.

The most plausible orthorhombic space group obtained from the diffraction pattern is *Pnnm*. This means a simple cubic Bravais lattice with two C_{60} molecules per crystallographic base. In other words, two chains but only one molecule per

Table 2　Lattice Parameters and Bond Lengths for Polymeric AC_{60}, A = K, Rb, Cs*

KC_{60}, r_{ion} = 1.33 Å			Ref.
Lattice parameters	$a \times b \times c$	9.109 × 9.953 × 14.321	28
	fcc cell, 473 K	14.07	26
Bond lengths	C1—C1′	1.65	28
	C1—C1	1.74	
	C1—C2	1.50	

RbC_{60}, r_{ion} = 1.48			Ref.
Lattice parameters	$a \times b \times c$	9.138 × 10.107 × 14.233	28
	fcc cell, 473 K	14.08	26
Bond lengths	C1—C1′	1.44	28
	C1—C1	1.90	
	C1—C2	1.51	

CsC_{60}, r_{ion} = 1.69			Ref.
Lattice parameters	$a \times b \times c$	9.10 × 10.22 × 14.17	12
	fcc cell, 473 K	14.13	26

High temperature, reference cells		
Lattice parameters for C_{60}	$a \times b \times c$	10.06 × 10.06 × 14.23
for RbC_{60}	$a \times b \times c$	9.95 × 9.95 × 14.08

* All distances are in Angstroms.

chain belong to the primitive unit cell and the factor group of the crystal is the same as the point group of the molecule. This is the same point symmetry as for the C_{60} molecules on the chain.

NMR can give additional information on the structures of the polymeric phases. This information comes from line position, line intensity, line width, and relaxation rates. Both the ^{13}C and the alkali metal nuclear spins have been used as a probe [31].

A detailed analysis of the KC_{60} system with respect to temperature, cooling, and heating conditions performed by Kälber et al. [32] confirmed the results from Raman spectroscopy with respect to the lack of stability of this compound vs. a phase separation on slow cooling. By using an 18 KHz magic angle spinning NMR, all nine different carbon sites predicted by the D_{2h} symmetry and the bonding sp^3 carbons could be resolved for RbC_{60} [33, 34].

5.2 Electronic Structure and Electronic Transitions

Since the experimental results, for example, the line shift in the Raman experiments, confirm a full charge transfer between alkali metal and polymeric chain, the threefold degenerated t_{1u} derived conduction band will be occupied with one electron and render the system in a metallic state in an one electron picture. This simple model seems to work for AC_{60} as the low-temperature orthorhombic and probably also the high-temperature fcc phase are indeed metallic.

The well-defined geometry for the orthorhombic phase allowed for a more detailed evaluation of the electronic structure especially applicable to the AC_{60} family. A simple topological Hamiltonian with a dimerized unit cell on the chain revealed a gap energy of 0.13 eV [35]. More elaborate calculations using the local density approximation in a density functional theory [36] revealed a metallic structure as shown in Figure 6a. The calculation was performed for a simplified unit-cell geometry with *Immm* space group instead of *Pnnm*. This means the translation lattice is bco. The figure gives evidence for a three-dimensional electronic structure. There is no characteristic difference between the bandwidth along the chains and perpendicular to the chains. It also demonstrates the dramatic lifting of the degeneracy in the t_{1u}-derived band. The orthorhombic symmetry allows only one-dimensional representations as it was explicitly demonstrated in Section 4 for the vibrations. The splitting of the band is as large as the bandwidth itself. The lower band is about half filled as expected from the one electron transferred to the cage. The density of states at the Fermi level for the paramagnetic

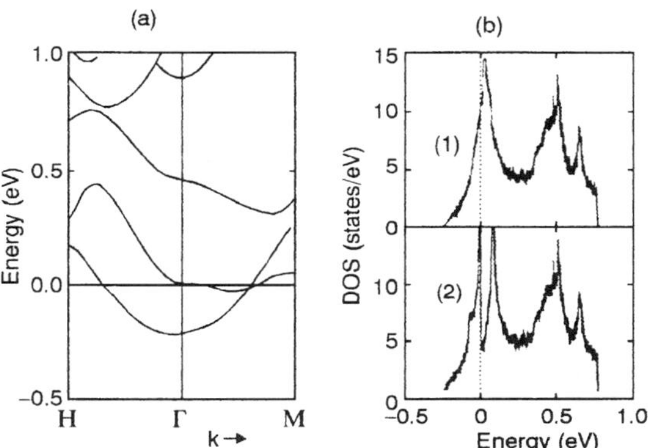

Figure 6 (a) Band structure for polymeric AC_{60} along the chains ($\Gamma - H$) and perpendicular to the chains ($\Gamma - M$); and (b) density of states for the (1) paramagnetic state and for a (2) magnetically ordered ground state at T = 0. (*After Ref. 36.*)

state is about 13 eV^{-1}, which is nearly a factor 2 larger as compared to the results from NMR [31]. Similar results were reported recently by Schulte et al. [37] for the *Pnnm* geometry.

The stability for the paramagnetic state was checked by allowing for fluctuation of magnetic ordering. A magnetic ordering temperature T_{MO} was found to depend on the electron-electron interaction potential U. The zero-temperature critical value was found to be $U = 250$ meV. For $U = 270$ meV the magnetic ordering temperature was evaluated to be 100 K. In the magnetic ground state a three-dimensional ordering is obtained. The spin orientation is ferromagnetic along the chains and antiferromagnetic between next nearest chains. The resulting density of states for the paramagnetic ground state and for the magnetically ordered ground state is shown in Figure 6b as graphs 1 and 2, respectively. In the magnetically ordered ground state a pseudogap has developed at the Fermi level leading to an electrically insulating phase. For U as high as 470 meV the pseudogap opens to a full gap.

Only preliminary data are available for the optical properties of the polymeric chains. Both early work and more recent work on photoemission reveal densities of states for the lower components of the π band similar to undoped or metallic A_3C_{60} [38, 39]. Figure 7 shows recent results from UPS for the high-temperature fcc and the low-temperature orthorhombic phase. In both cases the density of states at the Fermi level is finite.

The result supports the metallic character for the orthorhombic phase and

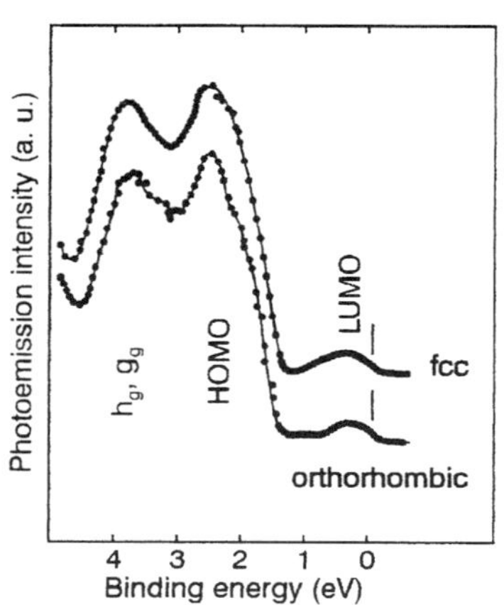

Figure 7   Photoemission for fcc RbC$_{60}$ at 525 K, and orthorhombic RbC$_{60}$ at 150 K. (*Adapted from Ref. 39.*)

the suggestion that even the high-temperature phase is metallic. This phase was originally claimed to be a semiconductor due to a Curie-like behavior of its susceptibility. An alternative interpretation of this behavior was recently given by Kirova et al. [40]. The Curie-like spin susceptibility can be a consequence of the low Fermi energy and the high temperature. The electron gas looses its degeneracy and Boltzmann statistic applies. Under these conditions the free carrier spin susceptibility is not any more temperature independent. Figure 8 shows a comparison between observed spin susceptibility as a function of temperature and the susceptibility calculated for a nondegenerate delocalized electron gas. The calculated results fit the experimental data even better than the $1/T$ Curie law.

Electron energy-loss spectroscopy is an appropriate technique to study the higher electronic transitions in the polymers. Experimental results were recently reported for RbC_{60} [41]. Figure 9 presents the loss function of C_{60} and op-RbC_{60} as well as the dielectric function between 0 and 7 eV derived by a Kramers-Kronig transformation. The loss function is determined by the allowed transitions in the system and can be directly compared to optical measurements. The loss function of the polymer shows two strong maxima at 6.4 and 23 eV, which are characteristic for sp^2 carbon systems. The maxima can be assigned to the π and the $\pi + \sigma$ plasmons, respectively. Focusing on the $\pi \rightarrow \pi^*$ features a strong broadening as compared to C_{60}. The nearly unresolved structure of the spectrum can be fitted with broad oscillators with maxima at 2.2, 3.6, and 4.8 eV. The broadening is mainly due to the lifting of the degeneracy by the distortion of the carbon cage and by the charge transfer of the Rb $5s$ electron into the molecular orbitals of the carbon cage. The π^* density is additionally reduced by the four sp^3-hybridized carbon atoms.

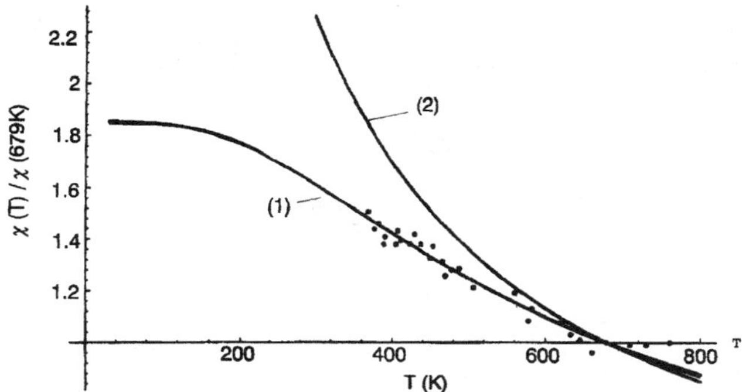

Figure 8 Electronic susceptibility for RbC_{60} in the fcc phase as measured (dots) and as calculated for a (1) nondegenerate electron gas and (2) Curie law. (*After Ref. 40.*)

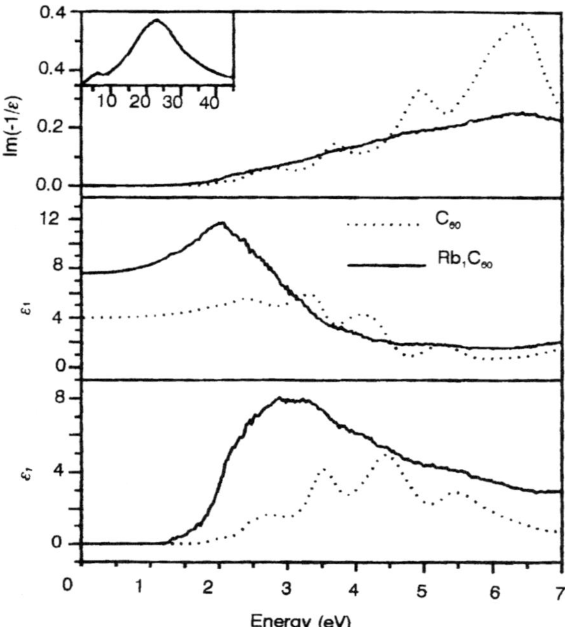

Figure 9   Electron energy-loss spectra and dielectric function for C_{60} (....) and for RbC_{60} in the orthorhombic phase (-). The insert is the high-energy response of the loss function. (*After Ref. 41.*)

No optical anisotropy could be observed for the orthorhombic phase [42]. This can be considered as evidence for the distribution of microdomains in the material.

5.3   Raman and IR Spectra

Raman scattering has proven to be of considerable importance for the understanding of the polymeric AC_{60} phases. This refers not only to the evidence for the phase and to the amount of charge transfer between the alkali metal and the cage but also to the symmetry of the resulting chain and to the coupling between the conduction electrons and the vibrational modes.

For slow cooling the Raman response for the transition from the high-temperature state of RbC_{60} is characteristically different to the one for KC_{60}. An

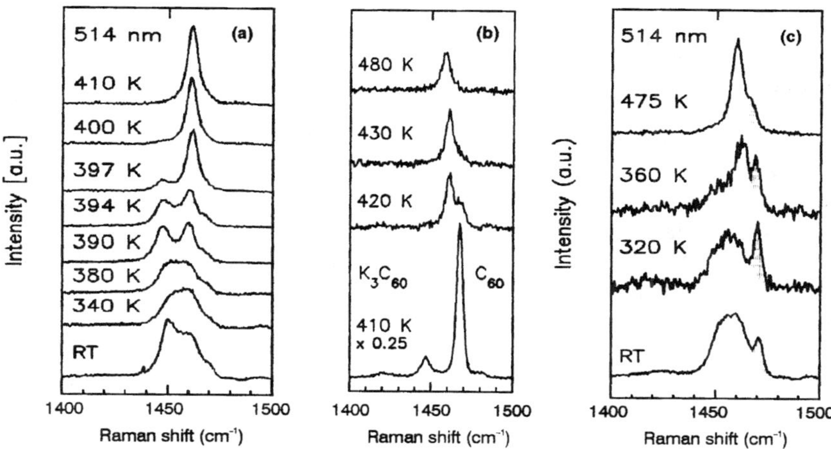

Figure 10 Temperature dependence for the Raman response of the pinch mode on cooling from the high-temperature phase for (a) RbC_{60}, (b) for KC_{60} during slow cooling, and (c) for the same system during moderate fast cooling. The hatched line in (c) is from a small fraction of undoped phase. [*Part (a) after Ref. 43.*]

example is given in Figure 10a,b for a single crystal and the spectral range of the pinch mode. In part (a) the response of this mode during cooling from above the phase transition to room temperature is shown for RbC_{60}. In the high-temperature phase it is observed at 1461 cm^{-1} consistent with the 7 cm^{-1} downshift due to the transfer of 1 electron from Rb to C_{60}. Below 400 K new features appear which finally at room temperature end up as a two-component line characteristic for the orthorhombic phase. The component at 1453 cm^{-1} corresponds to the former A_g (1) mode whereas the component at 1459 cm^{-1} originates probably from a G_g mode which became Raman active due to symmetry breaking. A similar behavior is observed for the other Raman lines [43]. In contrast to this result, part (b) of the figure shows the behavior of the modes in the same spectral range for KC_{60}. At the critical temperature of 410 K two new lines appear which are characteristic for the phase K_3C_{60} and pristine C_{60}. Only if the cooling process was performed with a moderate rate [44], a response similar to the result of the spectrum in (a) is obtained as shown in Figure 10c.

The comparison clearly demonstrates the influence of the cooling speed on the final product. The two spectra in Figure 10c taken for 360 and 320 K are more noisy because they were recorded with a reduced scanning time.

The splitting of the highly degenerated modes due to the reduction of symmetry was already demonstrated in the section on the pressure polymerized phases. It is instructive to compare the splitting of the Raman lines obtained from polymerization by pressure with the results from polymerization by charge transfer.

The equivalency of the two phases becomes evident from an inspection of the whole range of internal vibrational modes. The correlation between the two types of spectra is so close that the op-AC_{60} material can be considered as the doped version of the orthorhombic polymer (op-C_{60}) [19]. The relation between the spectra and their correlation to the high-temperature phase of AC_{60} is demonstrated in Figure 11. Spectrum (1) is the response from the high-temperature fcc phase of RbC_{60}. Spectra (2) and (3) are for the orthorhombic phases of RbC_{60} and KC_{60}. The two spectra are identical even with respect to small details which is a good proof for the equivalency of the two materials. As compared to spectrum (1) the H_g modes for both spectra are highly splitted, most of them into five components as it is expected for the reduction of the symmetry from I_h to D_{2h}. The splitting can be as high as 95 cm^{-1} as, for example, for the mode H_g (1). This is evidence for the dramatic distortion of the cages as it can be expected from a covalent bonding between them. The Raman response for the orthorhombic pressure polymerized C_{60} is shown as graph (4) in the figure for comparison. The correlation between the response from the undoped and the doped phases is evident.

The difference in linewidth between modes of the neutral and of the charged chain deserves attention. The linewidth is expected to be broader for the charged chain, since electron-phonon coupling broadens the lines in a system with free

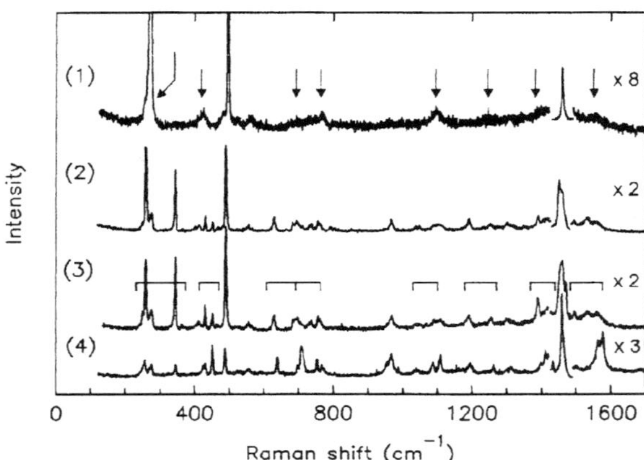

Figure 11 Raman spectra for various phases of AC_{60}. (1): fcc-RbC_{60}, (2): op-RbC_{60}, (3): op-KC_{60}, (4): pressure polymerized op-C_{60}. Arrows and brackets indicate the H_g-derived modes. (*After Ref. 43.*)

carriers. This effect was explicitly demonstrated for the system A_3C_{60} in several papers. From the nearly identical linewidths between op-RbC$_{60}$ and op-C$_{60}$ the coupling to the free carriers is small in general. The mode H_g (8) is an exception as it shows a very strong lineshift and line broadening upon doping. Line shapes for this mode are explicitly shown in Figure 12. The broadening of the line is consistent with a metallic state of the op-RbC$_{60}$ phase. From Allen's formula $\lambda = g_i \gamma_i / 2\pi \, N \, (\epsilon_F) \omega_i^2$ the dimensionless electron-phonon coupling constant λ is determined as 0.001 for a density of states at the Fermi level $N \, (\epsilon_F)$ as derived from the NMR experiments of Tycko. This is a factor of 3 smaller than in the case of Rb_3C_{60}. Since the other modes exhibit an even smaller electron-phonon coupling the lack of superconductivity for the metallic AC_{60} compounds is understandable. The weak coupling is a consequence of the reduction of the symmetry from I_h to D_{2h}. In the lower symmetry the very effective Jahn-Teller electron-phonon coupling is not active any more.

Infrared spectroscopy can give similar results as Raman scattering with respect to phase separation and phase transitions from line shift and line splitting.

From early experiments on thin films the four T_{1u} lines were observed for RbC$_{60}$ at room temperature at 525, 574, 1182, and 1392 cm^{-1} [45]. For a fit of the reflectivity spectra a Drude term was required with a 2 times smaller plasma frequency as compared to the plasma frequency for K_3C_{60}. This indicated already metallic behavior of the polymer and the reduced carrier density. For KC$_{60}$ phase separation on slow cooling from the high temperature fcc phase was reported [46] similarly to the results from Raman scattering.

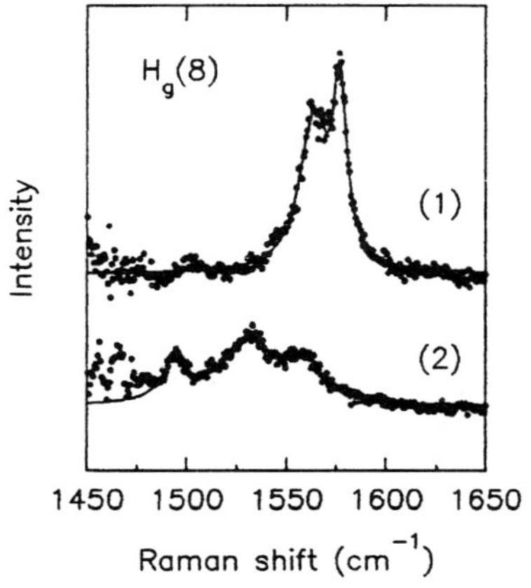

Figure 12 Raman line shape for the H_g (8) derived mode of pressure polymerized orthorhombic (1) C$_{60}$ and (2) op-RbC$_{60}$. (*After Ref. 19.*)

Table 3 Comparison of Model Parameters Given in cm^{-1} for Fitting the Optical Conductivity of AC_{60} with a Drude-Lorentz Dielectric Function

	KC_{60}		RbC_{60}	
	300 K	10 K	300 K	10 K
ω_p	736	681	493	
Γ_D	71	50	61	
$\omega_{p,1}$	1180	1180	515	671
Γ_1	490	490	281	229
ω_1	173	173	143	116
$\omega_{p,2}$	2860	2860	8230	8230
Γ_2	2360	2360	24100	24100
ω_2	2180	2180	2180	2180
$\omega_{p,3}$	14800	14800	16100	16100
Γ_3	12600	12600	14900	14900
ω_3	9680	9680	9680	9680

Source: After Ref. 47.

The FIR response can be modeled with the classical dispersion theory of Drude and Lorentz. The optical conductivity originating from the electronic transitions is given by a Drude term and three harmonic oscillators:

$$\sigma_1(\omega) = \omega_p^2 \frac{\Gamma_D}{(\omega^2 + \Gamma_D^2)} + \sum_{i=1}^{3} \omega_{p,i}^2 \frac{\omega^2 \Gamma_i}{(\omega_i^2 - \omega^2)^2 + \Gamma_i^2 \omega^2} \tag{1}$$

where ω_p and Γ_D are the plasma frequencies and the damping of the free carriers, respectively, while ω_i, $\omega_{p,i}$, and Γ_i are the resonance frequencies, the oscillator strengths, and the damping constants of the harmonic oscillators. The values for the parameters as obtained for KC_{60} and RbC_{60} are summarized in Table 3. From this table the plasma frequency originating from the free carriers within the t_{1u} band is larger for KC_{60} than for RbC_{60}. This may be interpreted as being due to a larger carrier concentration in the former material. The difference in carrier concentration may be crucial for the low-temperature behavior of these materials. On the other hand, the plasma frequency from the bonded electrons represented by $\omega_{p,2}$ is much larger for RbC_{60}. The oscillator ω_3 originates from the doping induced $t_{1u}-t_{1g}$ transition.

5.4 Electronic Structure at Low Temperature

Interesting new features in the electronic structure were obtained from magnetic resonance and from IR conductivity at low temperatures. Both techniques confirmed the metallic state for all three members of the AC_{60} family. For the ESR experiment this was evident from a nearly temperature independent spin susceptibility and a linewidth decreasing with decreasing T down to 50 K [27]. Results are shown in Figure 13 for the spin susceptibility, the linewidth, and the g factor as a function of temperature.

The small width of only 2 to 6 G for the observed resonance line was taken as evidence for a quasi-one-dimensional character of the free electrons. The reduced dimensionality was derived from the chain structure in the orthorhombic phase. Even though this conclusion is not straightforward, as discussed above, it is consistent with several details of the experimental results. In conventional metallic systems the width of the ESR line is determined by the Elliott mechanism based on a static or dynamic disorder of the lattice coupled to the free electrons by spin-orbit interaction. This mechanism is not effective in one-dimensional systems and the line remains narrow. In contrast, the linewidth is much broader in the three-dimensional metal $Rb_3 C_{60}$ where the Elliott mechanism is fully active.

The observed drop of the spin susceptibility below 50 K and the simultaneous increase of the linewidth with decreasing temperature is also consistent with a quasi-one-dimensional character of the electronic system. For strong electron-electron interaction such systems are known to undergo a transition to a spin-density wave state [49].

For the spin-density wave a gap opens at the Fermi energy (without any

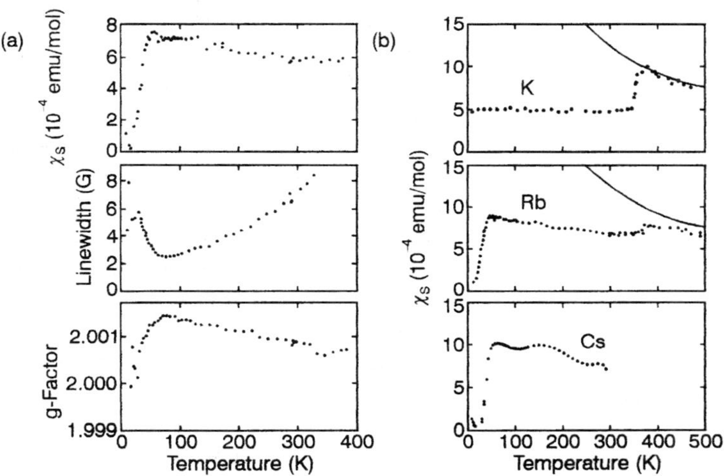

Figure 13 Spin susceptibility, linewidth and g-factor for polymeric RbC_{60} versus temperature, After [27] (a) and spin susceptibility for three different polymeric C_{60} compounds versus temperature, after [48] (b). The result for Cs is from [12].

lattice distortion) and the system becomes an insulator. Interestingly, the K polymer behaves differently at this point as shown in Figure 13b. The figure compares the spin susceptibilities as a function of temperature for KC_{60}, RbC_{60}, and CsC_{60}. In the high-temperature phase for $T > T_c$ the susceptibilities are Curie type. Below this the susceptibilities are more or less independent of temperature. Around 50 K the Rb and the Cs compounds are subjected to the metal-insulator transition whereas the K compound stays metallic even at very low temperatures. This suggests that the electronic system in KC_{60} remains three-dimensional in the paramagnetic op-phase. Since the width of the ESR absorption is still only of the order of several G in this compound, other reasons must be found why the Elliott mechanism is not effective. As the atomic number is known to have a considerable influence on the effectivity of the latter the smaller atomic number of K as compared to Rb or Cs was suggested to be the reason for the lack of line broadening in the K polymer.

The response for KC_{60} in NMR is related to the one for RbC_{60} but different in detail. Since the spin susceptibility for the K polymer is larger in the metallic phase as compared to the Rb polymer, the observed shift of the resonance lines in the spectra from magic angle spinning experiments is larger. On the contrary, the spin-lattice relaxation rate T_1^{-1} is as much as a factor 4 larger in RbC_{60} as compared to KC_{60}. Moreover, whereas KC_{60} exhibits a Korringa behavior as expected for three-dimensional metals T_1 is nearly independent of the temperature in RbC_{60}. Thus, $(T_1T)^{-1}$ increases strongly with decreasing temperature. The situation is demonstrated in Figure 14.

The fact that T_1 and χ_{spin} are both independent of temperature is again reminiscent for low dimensional electronic systems with strong correlations as it is observed in orgainc charge transfer metals and oxidic superconductors. Accordingly, antiferromagnetic fluctuations on the chain were suggested to be responsible for the unusual behavior of the Rb and the Cs polymer [50]. At low enough temperatures such fluctuation can condense to a magnetically ordered state.

Additional information on the conductivity and on the low-temperature metal-insulator transition has been obtained from far infrared and microwave reflectivity [48, 51]. Again, the behavior is different for the K polymer and for the Rb polymer. In the case of the former the reflectivity approaches 1 for reducing the frequency to zero for all temperatures below the fcc-o transition. This means a metallic behavior throughout. For the Rb-polymer a similar behavior is observed only for temperatures above 100 K. For temperatures as low as 10 K the reflectivity for $\omega \rightarrow 0$ is reduced to below 60%. From a Kramers-Kronig transformation the optical conductivity is obtained as shown in Figure 15. For K it increases for all temperatures and reaches 150 Scm^{-1} for the lowest measured frequency of 10 cm^{-1}. In contrast, RbC_{60} shows the developing of an energy gap for temperatures lower than about 100 K. Since the mean field energy gap for a spin-density wave transition is related to the transition temperature T_{SDW} by a

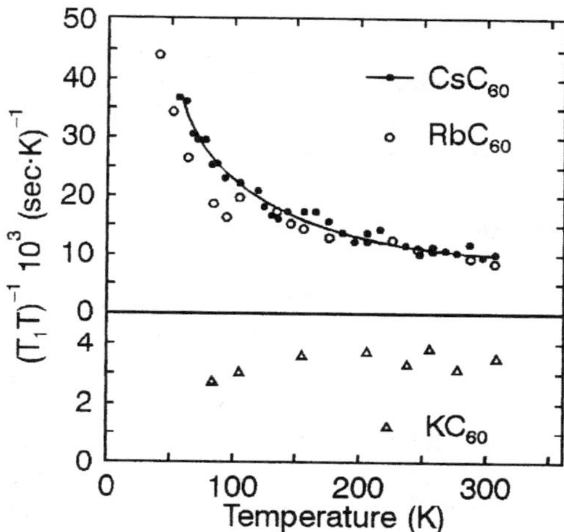

Figure 14 Temperature dependence of $(T_1T)^{-1}$ for ^{13}C NMR in AC_{60} systems. The full drawn line is a fit for $(T_1T)^{-1} = a + bT^{-\kappa}$ with $\kappa \approx 1$. (*After Ref. 50.*)

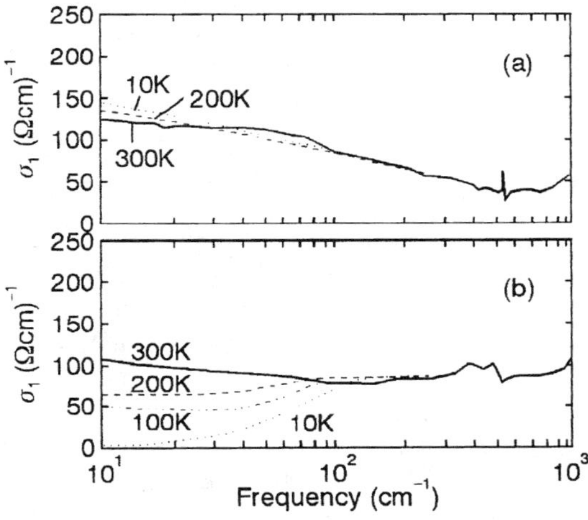

Figure 15 Optical conductivity for (a) KC_{60} and (b) RbC_{60} versus frequency for the temperatures indicated. (*After Ref. 50.*)

BCS relation as $2\Delta = 3.52k_BT_{\text{SDW}}$ the gap energy is expected at 122 cm^{-1} for $T_{\text{SDW}} = 50$ K, in reasonable good agreement with the beginning of the decrease of the optical conductivity for very low temperatures in Figure 15.

Early experiments on electrical transport were appropriate to reveal the fcc to orthorhombic phase transition as a discontinuity in the resistance [52]. They failed, however, to determine the electrical transport properties because of internal strains and granularity of the material. Recently more convincing results were obtained at least for KC_{60} from free-standing single crystal films [53]. The films were obtained from single crystals doped to a certain penetration depth. After this process the superficially doped crystals were exposed to toluene in which the undoped part of the crystal was dissolved away. The resulting films were well suited for four probe resistivity measurements in the temperature range from 2 to 450 K. In the intermediate temperature range from 50 to 350 K, $\rho(T)$ showed metallic character for KC_{60} and could be fitted with a power law in T where the linear term played an important role. Above 400 K the resistivity increased rapidly indicating the proposed transition to a nonmetallic phase. Below 50 K, $\rho(T)$ increased consistent with the observed magnetic ordering transition in the ESR and optical experiments [27, 51], even though these transitions were only observed for the Rb and for the Cs compounds. Whether or not the transition occurs seems to be slightly dependent on the sample quality. By applying a hydrostatic pressure the low-temperature increase in $\rho(T)$ can be suppressed.

Absolute values for the conductivities for the KC_{60} compounds are about 50 Scm^{-1} at 300 K and about 100 Scm^{-1} at low temperatures [54]. This compares to 2500 Scm^{-1} for K_3C_{60} and 1000 Scm^{-1} for Rb_3C_{60} at 300 K.

RbC_{60} and CsC_{60} do not show an explicit metallic dependence of ρ on T. ρ is rather strongly increasing with decreasing T [55]. This semiconducting behavior is transformed to a metallic temperature dependence under hydrostatic pressure.

5.5 Fermi-Surface Instability

The argumentation for the different behavior of KC_{60} on the one side and RbC_{60} and CsC_{60} on the other side came from a different dimensionality of the electronic system in the three compounds. This difference has not been confirmed from theoretical studies. Some evidence for it may be drawn from the structural parameters in Table 2. Relevant data are plotted in Figure 16 which shows the intermolecular (center-to-center) distance along the chains (1), the nearest interchain carbon-carbon distance in the *ab* plane (2), and the next nearest intermolecular carbon-carbon distance (3) vs. the ionic radius r for the compounds. Whereas for increasing r the distance between the carbons on two different chains increases the distance along the chains decreases. This means with increasing ionic radius the transfer integral along the chains will increase relative to the transfer integral

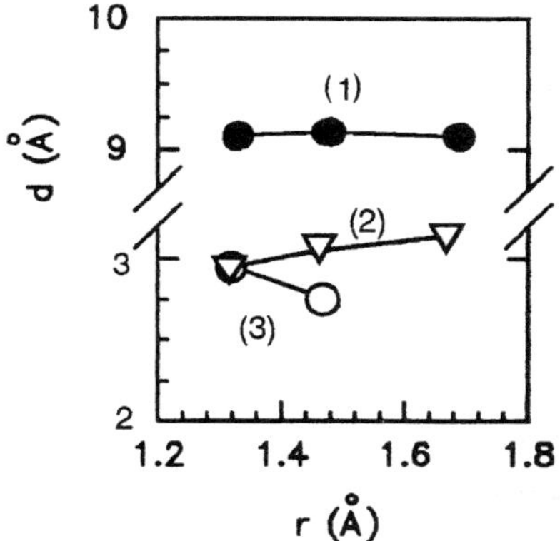

Figure 16 Intermolecular distance along a (1), smallest interchain carbon-carbon distance in the ab plane (2), and second nearest carbon-carbon distance along the chain (3) versus ionic radius r for AC_{60} compounds. The values for (3) were obtained from the experimental results for the C_1—C'_1 distance and then scaled with a geometry optimization.

between the chains and the electronic structure may become anisotropic. However, since the evaluated shifts are small it will certainly not become one-dimensional.

A better understanding for the problem of the dimensionality may be obtained from a recent paper by Fally et al. where the possibility of a spin-wave instability was demonstrated even for a three-dimensional electronic system with a half-filled tight-binding band [56]. For such a system perfect nesting of the Fermi surface exists for a special vector q_c. Starting from a linearized dispersion relation close to the Fermi surface and using the static density-density response function $\chi_0\,(q,\,T)$ for noninteracting particles, Stoner's criterion

$$\chi(q_c,\,T_c) = \frac{1}{\lambda} \tag{2}$$

was used to evaluate the temperature for the phase transition if an interaction λ between the particles was introduced. The transition temperature approaches rapidly to zero for deviation from half filling. In Figure 17 the calculated transition temperature is plotted as a function of band filling and anisotropy in a three-dimensional phase diagram. Only small differences in anisotropy or band filling for the two polymers RbC_{60} and CsC_{60} on the one side and KC_{60} on the other side may be sufficient to drive the latter compound across the boundary to a configuration where no phase transition will occur.

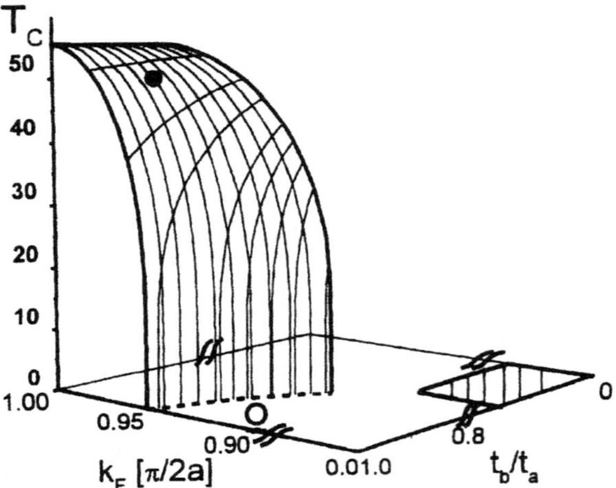

Figure 17 Three-dimensional phase diagram for AC_{60}. The calculated transition temperature T_c is plotted vs. band filling and anisotropy, described by k_F and t_b/t_a, respectively. The full circle represents possible parameters k_F and t_b/t_a for Rb whereas the open circle holds for K. The hatched area for low values of k_F and t_b/t_a refers to conventional low-dimensional material. (*After Ref. 57.*)

5.6 Single-Bonded Polymeric Phases

The stoichiometric phase of Na_4C_{60} has been identified recently by Oszlányi et al. as a two-dimensional single-bonded polymer [8]. The existence of single bonds between the C_{60} molecules was confirmed also by Raman experiments [58] and by ^{13}C magic angle spinning NMR measurements by Rachdi et al. [59]. The polymer was synthesized at ambient pressure by the usual solid-state reaction of C_{60} and Na metal. The X-ray pattern obtained could be indexed with a body-centered monoclinic unit cell with lattice parameters $a = 11.24$ Å $b = 11.72$ Å, $c = 10.28$ Å, and $\beta = 96.2°$. The C_{60} molecules arrange in planes where each molecule forms four single bonds within the plane as depicted in Figure 18. The nearest interfullerene distance is 9.28 Å within the plane while it is 9.93 Å out of the plane. The spin susceptibility measured by ESR has a weak temperature dependence and cannot be fitted simply as a sum of Curie and Pauli contributions. It rather resembles the susceptibility of a strongly correlated metal, like the RbC_{60} and CsC_{60} polymers. The result on Na_4C_{60} prompted a reinvestigation of Na_2AC_{60} (A = Rb, Cs) high-pressure linear polymers [60]. The structure was analyzed on the basis of a $2 + 2$ cycloaddition but considering the reported lattice parameters single bonds turned out to be more likely [6, 7].

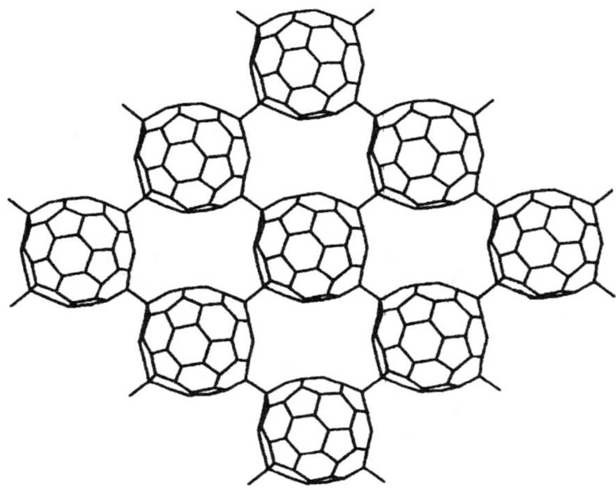

Figure 18 Geometric structure of the polymeric phase Na_4C_{60}. The bond connecting the fullerene molecules is 9.28 Å. (*After Ref. 8.*)

6. DISORDERED POLYMERS FROM PHOTOCHEMICAL REACTIONS

The sensitivity of C_{60} to visible light was first observed in Raman experiments by Duclos et al. [61]. Illumination with the laser used for the excitation of the spectra caused a dramatic intensity loss of the most prominent mode of the Raman spectrum at 1469 cm^{-1} and a new mode at 1458 cm^{-1} appeared at the same time. This effect is shown in Figure 19. Furthermore, an overall intensity loss of the Raman lines, a splitting of the fivefold degenerated H_g modes and a new mode at 118 cm^{-1} were observed. These results were explained by the formation of cyclobutane rings between neighboring C_{60} molecules in photochemical 2 + 2 cycloaddition reactions [1].

The level scheme and the energetics of the photochemical cycloaddition process are elucidated in Figure 20. Light absorption transfers a C_{60} molecule into its first excited singlet state S_1. An efficient intersystem crossing leads to a highly reactive triplet state T_1. In this state the molecules are able to break up their double bonds and to form a four-membered ring between neighboring C_{60} molecules. Since the triplet state is long-living, this process is very efficient. From the triplet state the molecules can either be excited into higher states or relax to the ground state S_0. The former process is very efficient as so that nonlinear optical processes such as light limiting can be observed [62].

The resuting polymeric structures are not crystalline so that X-ray diffrac-

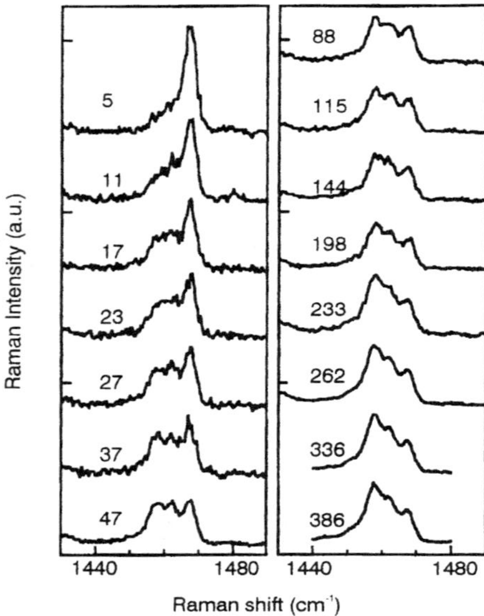

Figure 19 Changes of the Raman spectrum in the spectral range of the pinch mode of C_{60} with laser irradiation during a Raman experiment. Numbers are irradiation times in minutes. (*After Ref. 63.*)

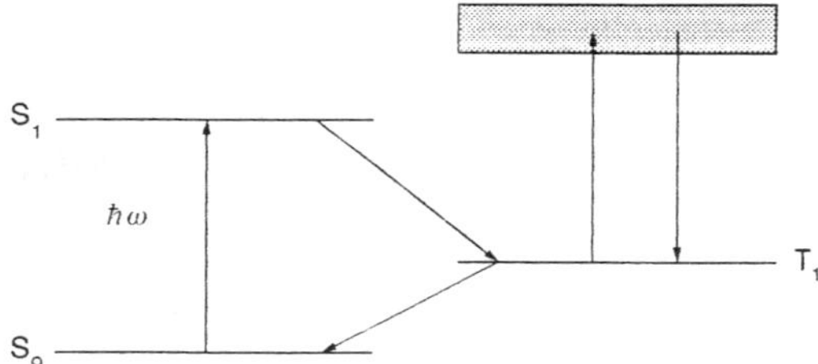

Figure 20 Energy levels and reaction processes for the 2 + 2 cycloaddition after photoexcitation of C_{60} to a triplet exciton. The transition $S_0 \to S_1$ gives the excitation to the first excited singlet state. The system relaxes into the first triplet state T_1 via an intersystem crossing. From there it can relax to the ground state or it can be excited to higher states.

tion is not an appropriate tool to analyze them. Rather, Raman scattering and IR spectroscopy have turned out to be the best analytical tools to apply.

6.1 The Polymerization Process

The photopolymerization of C_{60} depends crucially on sample treatment and experimental conditions. Since oxygen is a triplet quencher, oxygen-free samples are required and the photoreaction must be carried out in an oxygen-free atmosphere. Any gases in the environment where the samples are transformed can change the polymerization mechanism dramatically [64]. In the presence of nitric oxide (NO), for example, C_{60} polymerizes via a radical assisted bridging by NO radicals rather than via the 2 + 2 cycloaddition. The transformation rate for radical assisted process is a factor of 100 higher as compared to the 2 + 2 cycloaddition.

Since one of the essential boundary conditions for the 2 + 2 cycloaddition process is the parallel orientation of the double bonds, and since on the other hand the stability of all fullerene polymers discussed so far is limited to about 450 K, the polymerization of C_{60} takes place only within a very narrow temperature window as it is shown in Figure 21. The window is limited on the low-temperature side by the ordering transition at 260 K and on the high-temperature

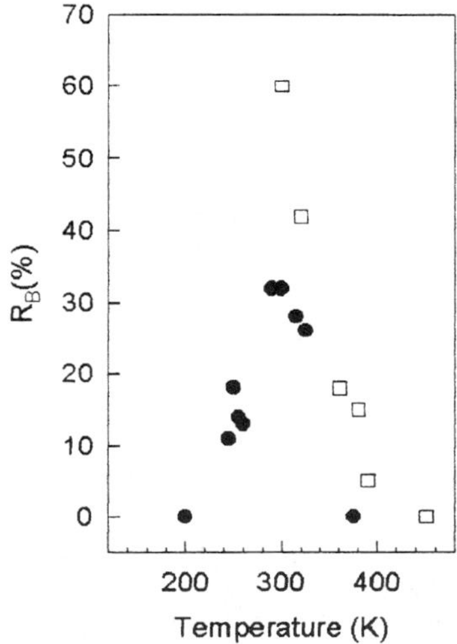

Figure 21 Temperature window for the photopolymerization. Shown is the intensity loss of the pinch mode in C_{60} after the first 20 s of irradiation vs. the temperature of reaction. The full circles represent thin films, the open squares represent single crystals. (*After Ref. 14.*)

side by the instability of the polymer. In the figure the transformation rate of C_{60} samples as derived from the percental intensity loss within the first 20 seconds of illumination is given for the pinch mode as a function of temperature. Below 260 K C_{60} forms a simple cubic lattice and the free rotation of the molecules is restricted to a hopping between certain positions around the [111] axes. Thus, no parallel arrangement of double bonds can be established between neighboring molecules.

An analysis of Raman spectra shows in addition that different polymeric structures are formed upon variation of the temperature at which the photoreaction proceeds [14]. Different species are produced in particular for irradiation at room temperature and at 380 K. The difference in the two materials can be seen from the low frequency part of the Raman spectra depicted in Figure 22, together with spectra for the pristine material and also for the *op* and *rp* pressure polymerized materials. This part covers the spectral region of the external stretching and librational modes, the modes derived from H_g (1), and the modes derived from the radial breathing vibration. The splitting of the H_g lines in the polymeric phases

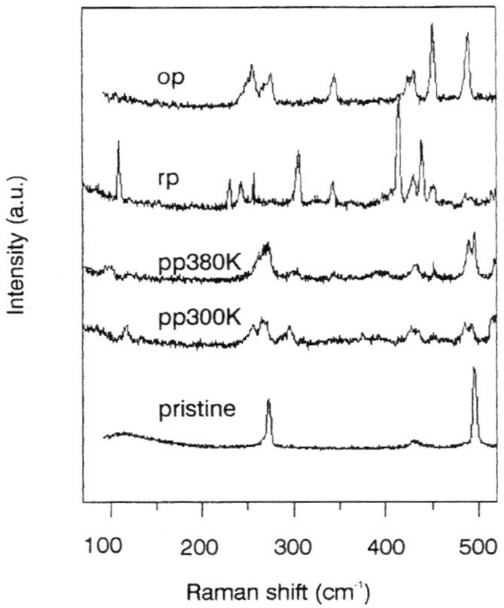

Figure 22   Raman response of C_{60} and polymeric C_{60} in the low-frequency spectral range extending from the external modes to the A_g (1) derived modes. The spectra op, rp, pp380, and pp300 are for orthorhombic pressure polymerized material, rhombohedral pressure polymerized material, material photopolymerized at 380 K, and material photopolymerized at 300 K, respectively. The spectrum at the bottom is for pristine C_{60} for comparison.

is again evident. However, the spectra from the photopolymers differ characteristically from the spectra obtained for the other polymeric species. The most significant changes are below the H_g (1) mode and in the region of the mode H_g (2). For example, the mode observed at 345 cm^{-1} for all orthorhombic polymers and for the rhombohedral polymer is missing in the photopolymers. The different nature for the crystals photopolymerized at 300 K and at 380 K is also seen from these spectra. For the former the new low-frequency mode is found at 118 cm^{-1}, whereas it is found at 96 cm^{-1} for the latter. Characteristic differences are also observed for the mode A_g (2).

From a quantum-mechanical calculation using the local density approximation in the density functional theory [21] the spectra observed for the material polymerized at elevated temperature gives best correlation with model spectra of dimerized C_{60} molecules. A remaining mismatch between theory and experiment implies that the observed product is not a simple dimer. The material prepared at room temperature corresponds to a more disordered polymer, probably with clustered covalent bonding in different directions or a herring bone structure for the arrangement of the C_{60} molecules.

These interpretations are supported by AFM measurements [65] and recent investigations of chemically produced and purified C_{60} dimer samples [66].

6.2   The IR Spectrum of Photopolymerized C_{60}

The IR spectrum of the photochemically produced polymer is only known for the material transformed at room temperature. It is shown in Figure 4 in comparison with the response from pristine C_{60} and C_{60} polymerized at high pressure. The four principal modes of T_{1u} symmetry expire a strong decrease of intensity. The T_{1u} (1) mode at 526 cm^{-1} becomes the strongest feature and the T_{1u} (2) mode at 576 cm^{-1} vanishes completely. The highest T_{1u} mode at 1428 cm^{-1} downshifts by 4 cm^{-1} to 1424 cm^{-1}. Due to the loss of symmetry many new lines appear in the spectrum, mostly in the region between 500 and 800 cm^{-1}.

6.3   Photoluminescence and Optical Absorption

Like for the IR, photoluminescence and optical absorption studies are only available for the material transformed at room temperature.

In Figure 23 the optical absorption spectrum of a photopolymerized C_{60} film is shown in comparison to the spectrum for pristine C_{60}. The transitions at 2.7, 3.6, 4.7 and 5.6 eV show a weak blueshift of 0.05 eV, a broadening by 0.3 eV, and a decrease in intensity. The assignment for the transitions is the same as for the pristine material. The peak at 2.7 eV comes from the first allowed transition between the bands derived from the HOMO and LUMO + 1 level and from the HOMO − 1 and LUMO level. The broadening can be associated to the

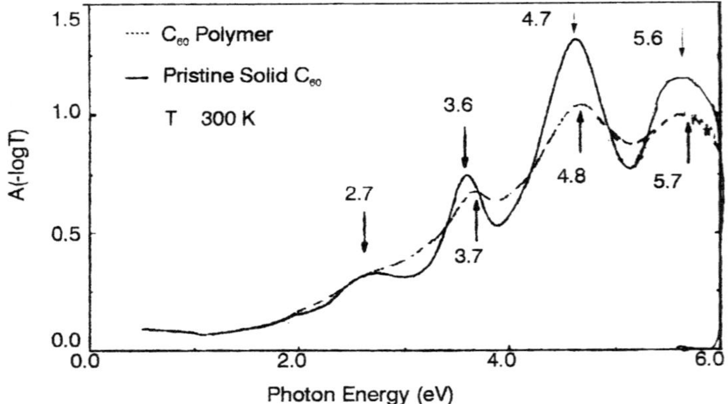

Figure 23   Optical absorption of pristine C$_{60}$ ($-$) and photopolymerized C$_{60}$ (....) plotted as minus log of the transmission through a thin film. (*After Ref. 67.*)

random nature of the cross-linking of molecules. In addition, the degeneracy of electronic energy levels is removed by the loss of symmetry upon transformation, which contributes to the broadening as well.

Spectra of photoluminescence (PL) for pure and photopolymerized C$_{60}$ are shown in Figure 24. Upon transformation the transitions at 1.69 and 1.52 eV are

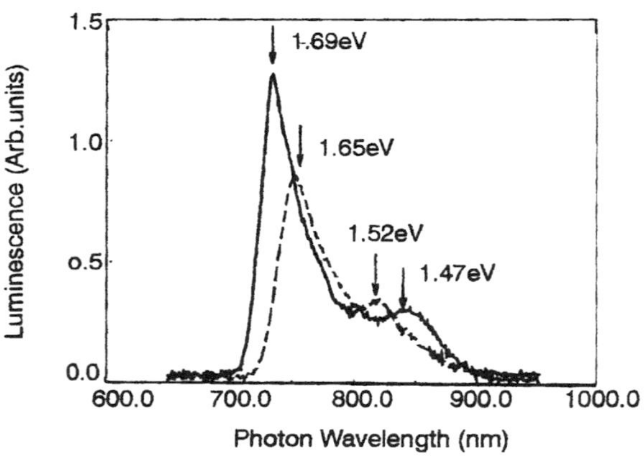

Figure 24   Photoluminescence of pristine C$_{60}$ ($-$) and photopolymerized C$_{60}$ (....) observed at room temperature. (*After Ref. 67.*)

red-shifted to 1.65 and 1.47 eV. The red-shift of the PL is consistent with the red shift observed for the luminescence of oligomers as compared to monomers, in general. The widths of the peaks is due to inhomogeneous broadening and lifting of the degeneracy of the vibrational modes, as discussed in connection with the optical absorption.

6.4 Oxygen Bridged and other Disordered Polymers from Photochemical Reactions

Oxygen-bridged dimers have been prepared from chemical reactions between C_{60} and $C_{60}O$ [68]. These compounds are soluble in $C_6H_4Cl_2$ and have been investigated by IR and Raman spectroscopy [69].

If the photochemical transformation is carried out under ambient conditions the interaction with oxygen becomes important. This interaction leads not only to the quenching of the triplet state as already mentioned but also to a reaction of oxygen with the C_{60} molecules. These compounds are seriously linked together, since they are neither soluble nor sublimable, not even at rather high temperatures. In the case of single crystals they were shown to grow as a transparent skin around the bulk of the crystal [70, 71]. Because of their stability at elevated temperatures such materials have been suggested to be appropriate for lithography and as seeds for the growth of diamond in CVD processes [72].

Photopolymerization was also demonstrated for C_{70}. In thin films the cycloaddition was found to be active and leads to a polymeric material in the same sense as for photopolymerized C_{60} [5]. Alternately, photoreactions were observed in solution and an insoluble and unsublimable material was obtained [73]. In both cases Raman scattering was the appropriate tool to analyze the material.

7. DISORDERED POLYMERS FROM VERY HIGH PRESSURE AND HIGH TEMPERATURE TREATMENT

Blank et al. applied high nonhydrostatic pressure (9.5 and 13 GPa) at high temperature (620–1830 K) to bulk C_{60} samples [74]. They obtained superhard and ultrahard polymeric phases with a hardness exceeding that of diamond.

X-ray studies showed that the fcc structure of fullerite was retained in both superhard and ultrahard materials, but long-range order was gradually lost with increasing temperature at which the material was processed. The lattice constants decreased with increasing temperature. The hard phases of the fullerite quenched from $T > 620$ K did not return to pristine C_{60} at annealing. From Raman measurements it was concluded that neighboring molecules are still connected by

Table 4 Hardness H, Cubic Lattice Parameter a, Density ρ and ρ_{exp} as Calculated from X-ray Data and as Measured (Both in g/cm^3), Energy Gap ϵ_g, and Resistivity R of C_{60} After Treatment at High Pressure and High Temperature

	No.	T (K)	H^*	a (Å)	ρ	ρ_{exp}	ϵ_g (eV)	R (Ωcm)
9.5 GPa	1	300	—	14.17	1.68	—	1.5	—
	2	520	soft	13.75	1.82	—	—	—
	3	570	soft-hard	13.6	1.87	—	—	—
	4	620	superhard	13.3	2.01	—	—	—
	5	650	superhard	12.98	2.07	2.10	—	—
	6	700	ultrahard	12.6	2.20	2.25	0.015	1–10
	7	520+SD	ultrahard	hex	2.80	—	—	—
	8	770	ultrahard	al	—	2.90	—	—
	9	900	ultrahard	hex+al	—	2.95	0.3	1–10
	10	1770	—	turbostrate	graphite			
13GPa	11	900	ultrahard	12.4+a2	—	2.95	—	—
	12	1270	ultrahard	12.4+a2	—	3.3	0.2	10^5
	13	1830	ultrahard	—	—	3.3	—	—

* "Hard" stands for phases harder than sapphire, "superhard" for phases harder than cubic BN, and "ultrahard" for phases harder than diamond; SD means shear deformation, and "hex" refers to the hexagonal phase; a1 and a2 are disordered phases.
Source: After Ref. 74.

4-membered rings. The most important parameters of each material are summarized in Table 4.

The superhard and ultrahard materials are semiconductors, some of them with a rather low bandgap and a low resistivity, as evident from the last two columns in Table 4.

8. DIMERIC PHASES AND ANNEALING AFTER QUENCHING OF AC$_{60}$

A large amount of work has been dedicated to rapid cooling, quenching, and annealing after various thermal treatments. The original idea for such work was to retain the high temperature fcc phase at low temperatures and to check on special magnetic properties as they are known for some monoanionic C_{60} compounds like TDAE-C_{60}. Since cooling rates appear to be a critical experimental detail, three categories will be discriminated in the following: quenching, rapid cooling, and slow cooling. As already discussed for slow cooling, phase separation is obtained in KC$_{60}$. On the other hand, as proven from Raman experiments, a great part of the high-temperature phase can be retained after quenching KC$_{60}$

from high temperatures to liquid nitrogen [75]. With annealing several intermediate states were recorded until finally phase separation was obtained. Similarly Martin et al. [76] found a dramatic increase of the infrared transmission after quenching RbC_{60} from the high temperature phase. The phase obtained was therefore assigned to a nonconducting compound.

More detailed information on the structure of the intermediate phases obtained after quenching and annealing was derived from X-ray, calorimetric, and ESR analysis [9, 12, 10].

A refined x-ray analysis for KC_{60} after quenching was reported by *Zhu* et al. [9]. Quenching from 573 K to ice water and subsequent cooling to 19 K revealed a new phase. The X-ray pattern of this phase showed a superlattice structure which required a doubling of the unit cell and a distortion from a cubic or even from the orthorhombic symmetry. The final assignment was to a monoclinic cell of space group $C2/m$ with lattice constants $a = 17.092$ Å, $b = 9.771$ Å, $c = 19.209$ Å, and $\beta = 124.06°$. The C_{60} molecules were arranged in a dimeric structure with a center-to-center distance of 9.34 Å. This distance is slightly larger than in op-C_{60} but still much shorter than in all other known C_{60} compounds. Similar results were obtained by Oszlányi et al. for KC_{60} and RbC_{60} [77].

The structure was observed to be unstable on heating. At 270 K the superlattice peaks disappeared and finally, at room temperature, the polymeric phase of KC_{60} (and not the phase-separated compound) was obtained. These experiments triggered a new series of quenching and annealing experiments for the AC_{60} phases. The dimerized phase was expected to be a Peierls insulator as a consequence of the doubling of the unit cell. This idea was confirmed by photoemission spectra which revealed no density of states at the Fermi level. If the samples are quenched from high temperatures to liquid nitrogen a metallic phase with structure $Pa\bar{3}$ is obtained [78].

The variety of new metastable phases obtained during an annealing process after quenching was demonstrated in experiments of differential thermal analysis (DTA) by Kosaka et al. [12]. To identify the various phases, ESR and X-ray structural analysis were used simultaneously. Figure 25 shows two DTA diagrams, one for annealing after quenching and one for annealing of a slowly cooled sample of CsC_{60}. The figure displays five phase transitions for the quenched sample at the indicated temperatures. Two of the transitions are exothermic and three are endothermic. For the slowly cooled compound only one transition is observed. Similar results with a quantitative thermal analysis have been reported by Gránásy et al. [13, 79, 80]. As supported from the simultaneous measurements of the ESR, the phase transitions exhibit the following scenario: for $T < 160$ K the phase is three-dimensional metallic; for 160 K $< T < 220$ K the phase is dimeric and semiconducting; for 220 K $< T < 260$ K the phase is sc, three-dimensional metallic and orientationally ordered; for 260 K $< T < 270$ K the phase is fcc (I), three-dimensional metallic and disordered; for 270 K $< T < 370$ the phase is polymeric and metallic; for $T > 370$ K the phase is fcc and semiconducting

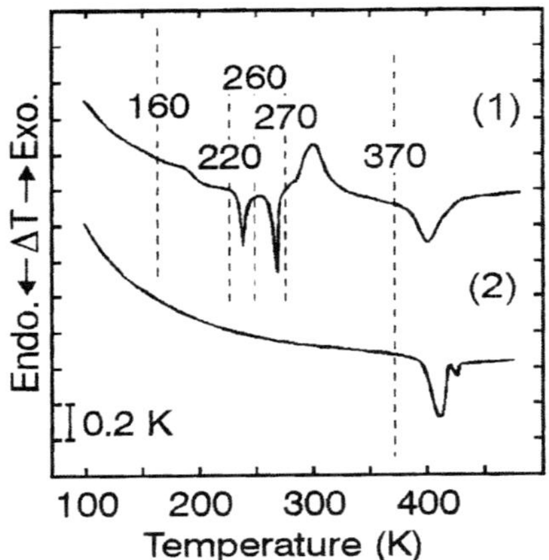

Figure 25   Differential thermal analysis of (1) quenched CsC_{60} and (2) slowly cooled CsC_{60}. The numbers label the temperatures of transition. (*After Ref. 12.*)

(or metallic with the degeneracy of the electron gas lifted). The temperatures as given above are for the Cs compound. The same scenario with similar temperatures holds for the Rb and the K compound, except that for the latter no discrimination between a sc and a fcc(I) phase was observed. Also, for KC_{60} another intermediate phase between the polymeric phase and the high temperature fcc phase was reported by Faigel et al. [13,81].

The heat of transition for the change from the polymeric phase to the high temperature phase was observed to be 35 J/g which is much higher than the value of 12 J/g reported earlier [26].

The electronic structure and optical transitions for the charged dimer as a representation for the charged polymeric chain were calculated by Fagerström et al. [82]. From this a substantially smaller barrier toward the formation of the polymer was found as compared to the neutral dimer. However, these calculations were carried out with the assumption that the charged dimer has the same 2 + 2 cycloaddition structure as the polymer.

Very recently the structure of the dimer phase was optimized by determining the mutual orientation of the two molecules in the dimer [83]. Several possible geometries were taken into consideration. The double-bonded interconnection is obtained from the 2 + 2 cycloaddition and known from quantum-chemical calculations as the lowest energy state for the neutral compound [22, 21, 20]. The larger center-to-center distance observed for the dimers and its instability at rather low temperatures suggests the possibility of a single bond.

The best fit of the X-ray pattern was indeed obtained for a single-bonded dimer in a *trans*-configuration with C_{2h} geometry and a space group $P2_1/a$. The intermolecular C—C distance was determined to be 1.54 Å This result explains why the dimer is dissolved on heating before the polymer is generated. The two compounds have a completely different bonding structure. As a matter of fact the dissolution of the dimer is a necessary condition for the generation of the polymer since the latter relies on freely rotating molecules.

Infrared measurements of Kamarás et al. on various RbC_{60} phases are consistent with the picture of stronger interball bonds in the polymer than in the dimer [84].

Ground-state optimized geometries and total energies have been reported by Kürti et al. [20] for the neutral and for the double negatively charged dimer of C_{60} for various single-bonded and double-bonded structures. The semiempirical calculations were carried out using AM1 and PM3 parametrization. The calculation revealed the 66/66 D_{2h} configuration (2 + 2 cycloaddition product) and the single-bond *trans*-configuration with geometry C_{2h} as the lowest energy states for the neutral pairs and for the dianion $(C_{60})_2^{2-}$, respectively, both in full agreement with the experiment. The reason for the difference in the ground-state structure of the charged dimer and that of the AC_{60} polymer was investigated by carrying out semiempirical calculations [85]. For the case of one extra electron per buckyball it was shown that the single-bonded configuration becomes less stable with increasing number of C_{60} units. The tetramer is already unstable and dissociates into two dimers, whereas the stability of the cycloadduct product increases with increasing chain length. However, the stability of the single-bonded polymers increases with increasing the negative charge of the C_{60} molecule as it was pointed out recently by Pekker et al. using semiempirical model calculations [86], in agreement with the observation of the linear single-bonded polymer phase of Na_2RbC_{60} [6, 7].

Very recently, theoretical and experimental investigations have been carried out on the $(C_{59}N)_2$ azafullerene dimer which is isoelectronic to the $(C_{60})_2^{2-}$ dianion [11, 87, 88]. The intrinsically doped heterofullerene dimer has the same single-bonded structure as the charged buckyball dimer discussed above. From electron energy-loss spectroscopy photoemission spectroscopy, and density functional based calculations it was concluded that the extra electron is predominantly localized on the N atom. This results in only a weak perturbation of both the occupied and unoccupied C-derived states [88].

9. DISCUSSION AND OPEN PROBLEMS

Polymeric phases of fullerenes, in particular those of C_{60}, represent a colorful new class of materials with interesting optical and electronic properties. The basic techniques to prepare them are applying high pressure at elevated temperatures, doping the cage to -1 with a rather small counterion, or to illuminating the monomer with light of an appropriate energy. Some of the obtained polymers are highly crystalline which is rather surprising because of the dramatic shrinking of the lattice in the direction of the chains or within the polymeric planes. Also, the metallicity of the AC_{60} phases is not so trivial as it might be anticipated form the one-electron picture. Electron correlation or Jahn-Teller distortions are expected to be important in these materials. For example, the compounds A_4C_{60} (A = K, Rb, Cs) are not metallic, even though the t_{1u} band is partially filled. There is also a claim that A_3C_{60} is not a metal if the stoichiometry between alkali metal and C_{60} is really $3:1$. The metallicity of AC_{60} may have to do with its polymeric state. Due to the reduced symmetry it is not any more a Jahn-Teller system and may stay metallic.

Even though some of the basic properties of the polymerization process and of the polymeric phases are reasonably well understood, several open questions remain and provide a wide field for future research work.

Open questions are concerned, for example, with the structure of the photo-polymer or with the optical properties of the polymers obtained from high-pressure treatment. Also, the low stability vs. elevated temperatures is not immediately understandable. After all, two single bonds are a rather strong connection.

In the case of the AC_{60} phases the dimensionality of the electronic system should be rediscussed and checked experimentally to shine more light on the nature of the low-temperature phase. Doping of the two-times single-bonded planar polymers (rhombohedral and tetragonal) has not been successful so far and there is a substantial lack of knowledge about linear and nonlinear optical properties in such materials.

The nature of the ultrahard material and the requirements to obtain it are another example of a decent lack of knowledge.

Finally, more effort in the problems of selfdoping, like in $C_{59}N$, is expected to reveal new materials with interesting optical and electronic properties.

ACKNOWLEDGMENTS

This work was supported by the Hochschuljubiläumsfonds der Österreichischen Nationalbank, project 5799, the Fonds zur Förderung der Wissenschaftlichen Forschung, project P10922-PHY in Austria, and the OTKA T022980 in Hungary. One of the authors (JK) gratefully acknowledges financial support from the Soros Foundation and from the Universität Wien. Valuable discussions with M. Krause from the IWF Dresden and S. Pekker from the KFKI in Budapest are gratefully acknowledged.

REFERENCES

1. A. M. Rao et al., *Science* **259**, 955 (1993).
2. S. Pekker, L. Forró, L. Mihály, and A. Jánossy, *Solid State Communic.* **90**, 349 (1994).
3. Y. Iwasa et al., *Science* **264**, 1570 (1994).
4. M. Nuñez-Regueiro et al., *Phys. Rev. Lett.* **74**, 278 (1995).
5. P. C. Eklund et al., *Thin Solid Films* **257**, 185 (1995).
6. K. Prassides et al., *J. Am. Chem. Soc.* **119**, 834 (1997).
7. Bendele et al., Proc. Int. Winterschool on Electronic Properties of Novel Materials (H. Kuzmany, J. Fink, M. Mehring, and S. Roth, eds.), World Scientific, Singapore 1997, p. 337.
8. Oszlányi et al., *Phys. Rev. Lett.* **78**, 4438 (1997).
9. Q. Zhu, D. E. Cox, and J. E. Fischer, *Phys. Rev.* **B51**, 3966 (1995).
10. G. Oszlányi et al., *Phys. Rev.* **B51**, 12228 (1995).
11. J. C. Hummelen et al., *Science* **269**, 1554 (1995).
12. M. Kosaka et al., *Phys. Rev.* **B51**, 12018 (1995).
13. L. Gránásy, S. Pekker, and L. Forró, *Phys. Rev.* **B53** 5059 (1996).
14. B. Burger, J. Winter, and H. Kuzmany, *Z. Physik* **B101**, 227 (1996).
15. Komatsu et al., *Nature* **387**, 583 (1997).
16. V. Blank et al., *Physics Letters* **A188**, 281 (1994).
17. C. H. Xu and G. E. Scuseria, *Phys. Rev. Lett* **74**, 274 (1995).
18. Soldatov et al., Proc. Int. Winterschool on *Electronic Properties of Novel Materials*, H. Kuzmany, J. Fink, M. Mehring, and S. Roth, eds.), World Scientific, Singapore, (1996), p. 344.
19. J. Winter, H. Kuzmany, A. Soldatov, et al., *Phys. Rev.* **B54**, 1 (1997).
20. J. Kürti, K. Németh, Chem. *Phys. Lett.* **256**, 119 (1996).
21. D. Porezag, M. R. Pederson, T. Frauenheim, and T. Köhler, *Phys. Rev.* **B52**, 14963 (1995).
22. G. B. Adams et al., *Phys. Rev.* **B50**, 17471 (1994).
23. A. M. Rao et al., *Appl. Phys.* **A64**, 231 (1997).
24. A. M. Rao et al., *Phys. Rev.* **B55**, 4766 (1997).
25. J. Winter and H. Kuzmany, *Solid State Communic.* **84**, 935 (1992).
26. Q. Zhu et al., *Phys. Rev.* **B47**, 13948 (1993).
27. O. Chauvet et al., *Phys. Rev. Lett.* **72**, 2721 (1994).
28. P. Stephens et al., *Nature* **370**, 636 (1994).
29. N. G. Chopra, J. Hone, and A. Zettl, *Phys. Rev.* **B53**, 8155 (1996).
30. J. Robert, P. Petit, and J. E. Fischer, *Synthetic Metals* **77**, 119 (1996).
31. R. Tycko et al., *Phys. Rev.* **B48**, 9097 (1993).
32. T. Kälber, G. Zimmer, and M. Mehring, *Phys. Rev.* **B51**, 16471 (1995).
33. K. F. Thier, G. Zimmer, M. Mehring, and F. Rachdi, *Phys. Rev.* **B53**, R496 (1996).
34. H. Alloul et al., *Phys. Rev. Lett.* **76**, 2922 (1996).
35. P. Surján and K. Németh, *Solid State Communic.* **92**, 407 (1994).
36. S. C. Erwin, G. V. Krishna, and E. J. Mele, *Phys. Rev.* **B51**, 7345 (1995).

37. J. Schulte and M. Böhm, Chem. *Phys. Lett.* **252**, 367 (1996).
38. G. P. Lopinski, M. G. Mitch, J. R. Fox, and J. S. Lannin, *Phys. Rev.* **B50**, 16098 (1994).
39. D. M. Poirier, D. W. Owens, and H. J. Weaver, *Phys. Rev.* **B51**, 1830 (1995).
40. N. Kirova, L. Firlej, *Synthetic Met.* **77**, 63 (1996).
41. T. Pichler et al., *Applied Phys.* **A64**, 301 (1997).
42. K. Kamarás, S. Pekker, D. B. Tanner, and L. Forró, Proc. Int. Winterschool on *Electronic Properties of Novel Materials* (H. Kuzmany, J. Fink, M. Mehring, and S. Roth, eds.), World Scientific, Singapore, 1996, p. 119.
43. J. Winter and H. Kuzmany, *Phys. Rev.* **B52**, 7115 (1995).
44. J. Winter and H. Kuzmany, *J. Raman Spectroscopy* **27**, 373 (1996).
45. T. Pichler, R. Winkler, and H. Kuzmany, *Phys. Rev.* **B49**, 15879 (1994).
46. D. Koller, M. C. Martin, and L. Mihály, *Mol. Cryst. Liqu. Cryst.* **256**, 275 (1994).
47. F. Bomelli, L. Degiorgi, P. Wachter, and L. Forro, *Synth Met.* **77**, 111 (1996).
48. F. Bomelli et al., *Phys. Rev.* **B51**, 14794 (1995).
49. G. Grüner, *Rev. Mod. Phys.* **66**, 1 (1994).
50. V. Brouet et al., *Phys. Rev. Lett.* **76**, 3638 (1996).
51. F. Bomelli, L. Degiorgi, P. Wachter, and L. Forró, Proc. Int. Winterschool on *Electronic Properties of Novel Materials* (H. Kuzmany, J. Fink, M. Mehring, and S. Roth, eds.), World Scientific, Singapore, 1995, p. 351.
52. T. Pichler, R. Winkler, and H. Kuzmany, Proc. Int. Winterschool on *Electronic Properties of Novel Materials* (H. Kuzmany, J. Fink, M. Mehring, and S. Roth eds.), World Scientific, Singapore, 1994, p. 61.
53. J. Hone, M. S. Fuhrer, and K. Khazeni, A. Zettl, *Phys. Rev.* **B52**, R8700 (1995).
54. A. Zettl, private communication.
55. A. Zettl, Proc. Int. Winterschool on *Electronic Properties of Novel Materials* (H. Kuzmany, J. Fink, M. Mehring, and S. Roth eds.), World Scientific, Singapore, 1996, p. 14.
56. M. Fally, H. Kuzmany, *Phys. Rev.* **B56**, 13861 (1997).
57. H. Kuzmany et al., *Physica* **B244**, 186 (1998).
58. J. Winter, H. Kuzmany, unpublished.
59. F. Rachdi et al., Carbon, **36**, 607 (1998).
60. Q. Zhu, *Phys. Rev.* **B52**, R723 (1995).
61. Duclos et al., *Solid State Commun.* **80**, 481 (1991).
62. T. N. Thomas et al., *Int. J. Mod. Phys. B*, **6**, 3931 (1992).
63. M. Matus, H. Kuzmany, *Phys. Rev. Lett.* **68**, 2822 (1992).
64. A. F. Hebard, K. B. Lyons, R. C. Haddon, Proc. of the Symp. on Recent Advances in the Chem. and Phys. of Fullerenes and Related Materials, El. Chem. Soc., Pennington, New Jersey, 1995, p. 11.
65. A. Hassanien, Proc. Int. Winterschool on *Electronic Properties of Novel Materials* (H. Kuzmany, J. Fink, M. Mehring, and S. Roth, eds.), World Scientific, Singapore, 1995, p. 486.
66. B. Burger, unpublished.
67. M. S. Dresselhaus, G. Dresselhaus, and P. C. Eklund, *Synthetic Met.* **78**, 313 (1996).
68. S. Lebedkin et al., *Tetrahedron Letters* **36**, 4971 (1995).
69. M. Krause et al., Proc. Int. Winterschool on *Electronic Properties of Novel Materials*

(H. Kuzmany, J. Fink, M. Mehring, and S. Roth eds.), World Scientific, Singapore, 1997, p. 101.
70. M. Haluska et al., *Appl. Phys.* **A56**, 161 (1993).
71. J. Li et al., *Chem. Phys. Lett.* **227**, 572 (1994).
72. S. McGinnis et al., *Appl. Phys. Lett.* **70**, 586 (1997).
73. J. A. Nisha et al., Carbon, **B36**, 637 (1998).
74. V. D. Blank et al., *Phys. Lett.* **A205**, 208 (1995).
75. H. Kuzmany, M. Matus, T. Pichler, J. Winter, Proc. Nato Conference on *Physics and Chemistry of Fullerenes* (K. Prassides, ed.), Kluwer Acad. Pub., NL, 1994.
76. M. C. Martin et al., *Phys. Rev.* **B49**, 10818 (1994).
77. G. Oszlányi et al., *Phys. Rev.* **B54**, 11849 (1996).
78. A. Lappas et al., *J. Am. Chem. Soc.* **117**, 7560 (1995).
79. L. Gránásy et al.: Proc. Int. Winterschool on Electronic Properties of Novel Materials (H. Kuzmany, J. Fink, M. Mehring, and S. Roth eds.), World Scientific, Singapore, 1995, p. 331.
80. L. Gránásy et al., *Solid State Commun.* **97**, 573 (1996).
81. G. Faigel et al., *Phys. Rev.* **B52**, 3199 (1995).
82. J. Fagerström, S. Stafström, *Phys. Rev.* **B53**, 13150 (1996).
83. G. Oszlányi et al., Proc. Int. Winterschool on *Electronic Properties of Novel Materials* (H. Kuzmany, J. Fink, M. Mehring, and S. Roth, eds.), World Scientific, Singapore, 1996, p. 354.
84. K. Kamarás et al., *Fullerene Sci. Tech.* **5**, 465 (1997).
85. J. Kürti, P. Rajczy, *Carbon*, **B36**, 653 (1998).
86. S. Pekker et al., *Chem. Phys. Letters* **282**, 435 (1998).
87. W. Andreoni et al., *J. Am. Chem. Soc.* **118**, 11335 (1996).
88. T. Pichler et al., *Phys. Rev. Lett.* **78**, 4249 (1997).

11
Electronic Properties of Fullerene/ π-Conjugated Polymer Composites

N. Serdar Sariciftci
Johannes Kepler University of Linz
Linz, Austria

1. INTRODUCTION

Conjugated polymers in their undoped, semiconducting state are electron donors upon photoexcitation (electrons promoted to the antibonding π^* band). The idea of using this property in conjunction with an molecular electron acceptor to achieve long-living charge separation was based on the stability of the photo-induced nonlinear excitations (such as polarons) on the conjugated polymer backbone. Once the photoexcited electron is transferred to an acceptor unit, the resulting cation radical (positive polaron) species on the conjugated polymer backbone is known to be highly delocalized and stable, as shown in electrochemical and/or chemical oxidative doping studies. Analogous to the chemical doping process, we will describe the photoinduced electron transfer from the conjugated polymer donor onto an acceptor moeity as *photodoping*.

On the other hand, the mixing of buckminsterfullerene with donors like di-alkylanilines result in ultrafast photoinduced charge transfer phenomena, as reported by Hochstrasser and his group [1]. Wang and Cheng suggested the formation of a charge transfer complex in the excited state [2] rather than a whole photoinduced electron transfer. Comparison of toluene with dialkylaniline shows a strong increase in nonlinear optical response for the latter due to stronger donor properties [3]. Another example of photoinduced charge transfer phenomena with the participation of C_{60} was presented by Kamat by utilizing ZnO suspension composites with buckminsterfullerene [4]. There are many other reports on charge transfer interactions involving fullerenes which are described in detail in Ref. 5.

Independently, Santa Barbara group [6–16] and Osaka group [17–23] reported studies on the photophysics of mixtures of conjugated polymers with C_{60}. The observations clearly evidenced an ultrafast, reversible, metastable photoinduced electron transfer from conjugated polymers onto Buckminsterfullerene in solid films. Schematic description of this phenomenon is displayed in Figure 1. Using this molecular effect at the interface between bilayers consisting of semiconducting polymer [poly(2-methoxy,5-(2′-ethyl-hexoxy)-p-phenylene) vinylene, hereafter referred to as MEH-PPV] and C_{60} films, diodes were demonstrated with rectification ratios on the order of 10^4 which exhibited a photovoltaic effect [14]. Significant improvement of the relatively low collection efficiency of the D/A bilayer has been achieved by using phase-separated composite materials through control of the morphology of the phase separation into an interpenetrating network. Power conversion efficiency of solar cells made from MEH-PPV/C_{60} composites was subsequently increased by 2 orders of magnitude to approximately 3% [15].

In this chapter, we review different experimental results to determine the

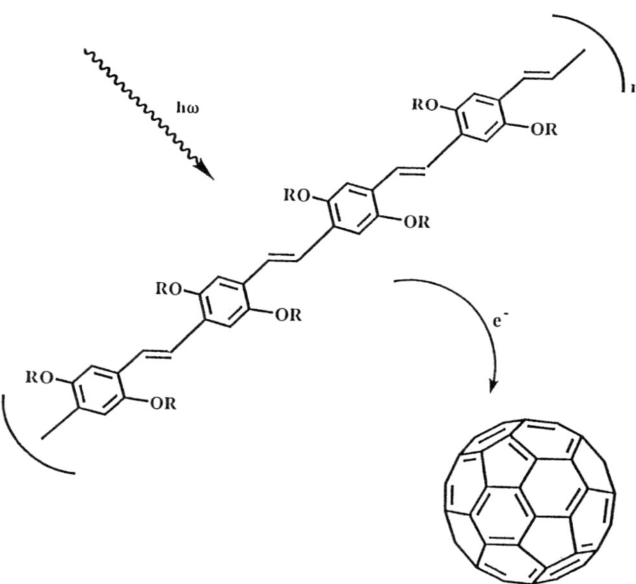

Figure 1 Schematic illustration of the photoinduced electron transfer from conjugated, semiconducting polymers onto Buckminsterfullerene, C_{60}.

photophysical properties of these conjugated polymer/C_{60} composites. Strong emphasis is given to the results which evidence the photoinduced electron transfer.

2. PHOTOPHYSICS OF CONJUGATED POLYMER/ FULLERENE COMPOSITE FILMS

Figure 2 shows the optical absorption spectrum of a MEH-PPV/C_{60} film with different C_{60} content compared to the optical absorption spectrum of the compo-

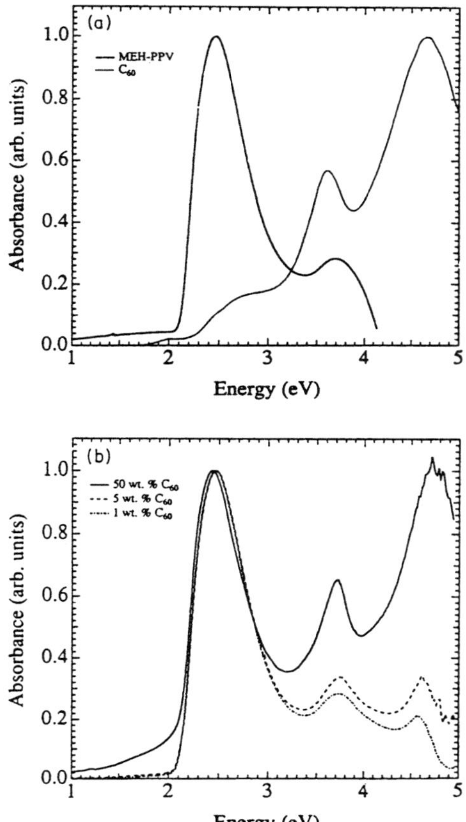

Figure 2   (a) Optical absorption spectra of MEH-PPV and C_{60}. (b) Optical absorption spectra of MEH-PPV/C_{60} films with different concentrations of C_{60}.

nents alone. The π-π^* absorption of MEH-PPV (peak at 2.5 eV) is clearly observed along with the first dipole allowed transition in C_{60} (at 3.75 eV). The spectrum is a simple superposition of the two components without any indication of states below the π-π^* gap of the conducting polymer as might arise from interaction between the two materials in the ground state.

Figure 3 shows the luminescence spectrum of MEH-PPV/C_{60} composites compared to MEH-PPV alone. The strong photoluminescence of MEH-PPV is quenched by a factor in excess of 10^3. The luminescence decay time is reduced from $\tau_0 = 550$ ps to $\tau_{rad} \ll 60$ ps (the instrumental resolution), indicating the existence of a rapid quenching process, e.g., subpicosecond electron transfer [6, 7].

The strong quenching of the luminescence of another conjugated polymer P3OT reported by Morita et al. [17] is also consistent with efficient photoinduced electron transfer. Figure 4 shows the intensity of the photoluminescence as a function of the C_{60} concentration in P3OT/C_{60} composites.

Luminescence quenching in an organic dye may have very different reasons. Photoinduced electron and/or energy transfer from the singlet excited state is a prominent quenching mechanism; however, other nonradiative relaxation channels have to be ruled out to assign the luminescence quenching to an electron transfer.

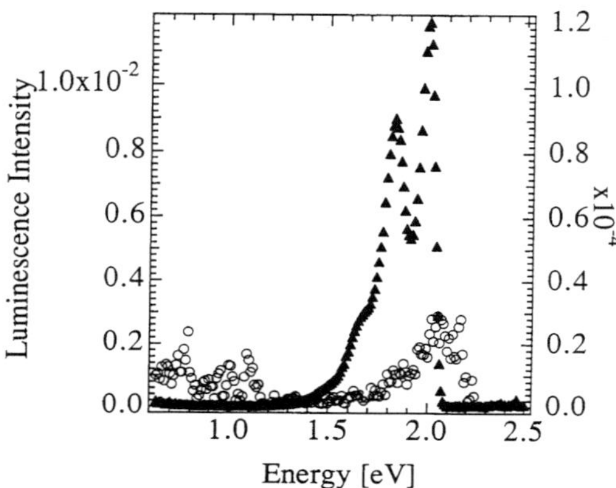

Figure 3   Luminescence of MEH-PPV (solid triangles, left-hand axis) and MEH-PPV/ C60 composite (circles, right-hand axis).

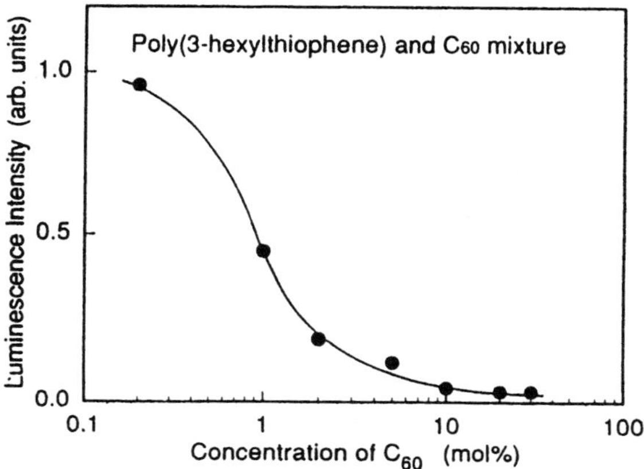

Figure 4 The intensity of the photoluminescence as a function of the C_{60} concentration in P3HT/C_{60} composites. (*From Ref. 17.*)

2.1 Quenching of the Intersystem Crossing to the Triplet State

Since the above results demonstrate that the excited states of the conjugated polymers strongly interact with C_{60}, a comparative spectroscopic study of this excited state is necessary. The first results on the excited-state spectroscopy of these composites originate from near-steady-state photoinduced absorption (PIA) studies. To clearly distinguish the spin multiplicity of the photoinduced absorption (PIA) peaks, photoinduced absorption detected magnetic resonance experiments were performed [13]. The full spectral range of photoinduced absorption spectra are displayed in Figure 5 [13]. Results show that the PIA bands centered around 0.3 and 1.1 eV are significant only in the composite material. Pure MEH-PPV shows the single PIA band from the triplet-triplet absorption, as assigned earlier [24].

The PIA band at 1.35 eV in the pure MEH-PPV shows a strong response at the (forbidden) half field resonance of spin $= 1$, indicating triplet character for this PIA band and dominance of the neutral photoexcitations in MEH-PPV (Figure 6). There is a small, residual spin $= \frac{1}{2}$ response even within the pure MEH-PPV. Upon adding C_{60}, the triplet signal for the 1.35 eV PIA band is completely quenched. Instead, a strong spin $= \frac{1}{2}$ signal dominates, indicating charged polarons as photoexcitations on the polymer donor (Figure 6 [13]). This confirms that the photoinduced electron transfer occurs on a time scale sufficiently fast to quench the intersystem crossing to the triplet state.

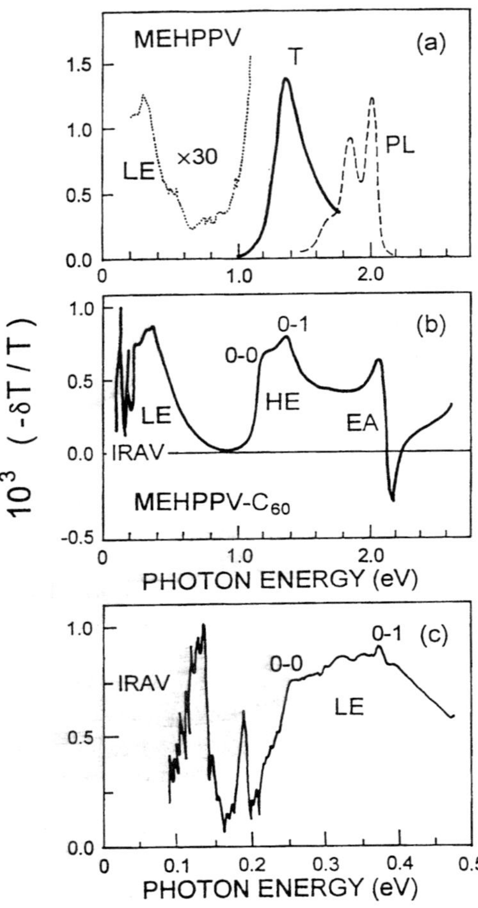

Figure 5　(a) Photoinduced absorption and photoluminescence (PL) spectra of MEH-PPV at 80 K. (b) and (c): The photoinduced absorption spectra of MEH-PPV/C$_{60}$ at 4 K with argon ion laser at 514.5 nm modulated at 200 Hz. The abbreviations T, LE, HE, EA stand for triplet-triplet absorption, low-energy band, high-energy band and electroabsorption features, respectively. IRAV stands for infrared activated modes. (*From Ref. 13.*)

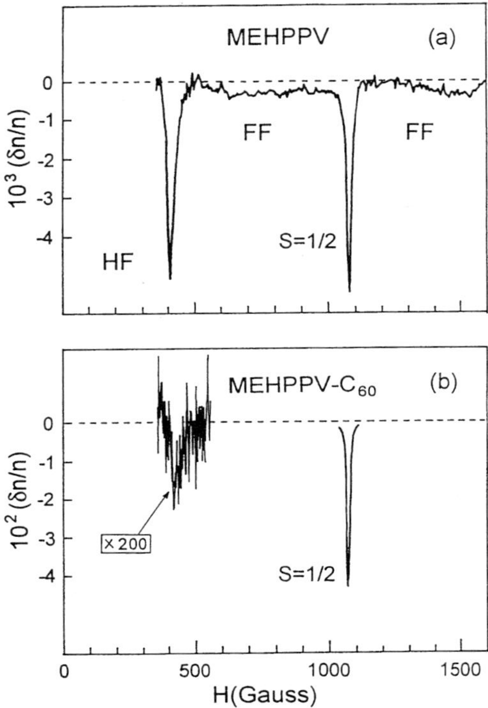

Figure 6 (a) Photoinduced absorption detected magnetic resonance (ADMR) spectrum of MEH-PPV. HF and FF signify the half-field and full-field powder pattern for the triplet ($S = 1$) resonance, respectively. (b) ADMR spectrum of MEH-PPV/C_{60} composite film. Both spectra were measured at probe energy 1.35 eV, $T = 4$ K, and 3 GHz resonant microwave frequency. (*From Ref. 13.*)

2.2 Subpicosecond Photoinduced Absorption

The time-resolved photoinduced absorption (PIA) spectra of pristine P3OT and of P3OT with 1% C_{60} are shown in Figure 7*a* and *b*, respectively [9]. The spectrum for the pristine material consists of two distinct features, one at 1.9 eV, and one at 1.2 eV, which we shall refer to as the high-energy (HE) feature and the low-energy (LE) feature, respectively. After 10 ps there is no change in the relative spectral weight of the HE and LE features; the spectrum at 250 ps can be scaled to precisely fit the spectrum at 10 ps [9]. The time evolution of the HE feature consists of two components: a fast exponential component with a lifetime of 800 fs, and a slow component which matches the dynamics of the LE feature

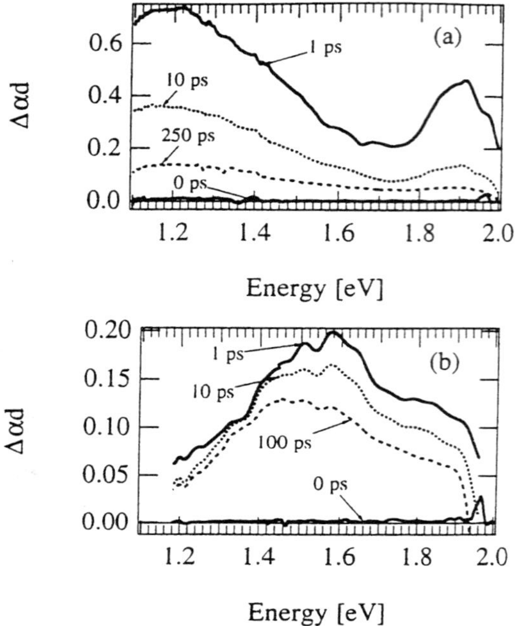

Figure 7 Photoinduced absorption spectra at 300 K for various delay times after a 2.01 eV, 100 fs pump pulse for (a) P3OT and (b) P3OT/C_{60} (1%). (*From Ref. 9.*)

[9]. The fast component of the HE feature matches that seen in poly(3-methylthiophene) [25] and polythiophene [26], and has been attributed to a self-trapped singlet exciton, or a singlet polaron-exciton. Upon adding C_{60} to P3OT, the PIA spectrum, decay kinetics, and intensity dependence all change dramatically [9].

At 1 ps after photoexcitation by a 100 fs pump pulse at 2.01 eV, the PIA spectrum for P3OT/C_{60} (1%) consists of a single broad peak at approximately 1.55 eV, Figure 7*b*, with virtually no evidence of the HE and LE features seen in pristine P3OT at 1.9 and 1.2 eV [9]. The risetime of this feature is resolution limited (250 fs). After photoexcitation, the peak shifts to the red, from 1.55 eV at 1 ps to 1.45 eV at 100 ps. The spectrum at 100 ps closely matches that seen in millisecond (ms) PIA experiments; the latter has been assigned to positively charged polaron band as discussed above (see Figure 8) [7]. The ultrafast (<1 ps) formation of the PIA band at 1.55 eV again demonstrates that the electron transfer occurs on a subpicosecond time scale, conforming earlier estimates based on luminescence quenching.

In order to compare the time decay of the PIA in pristine P3OT to that of

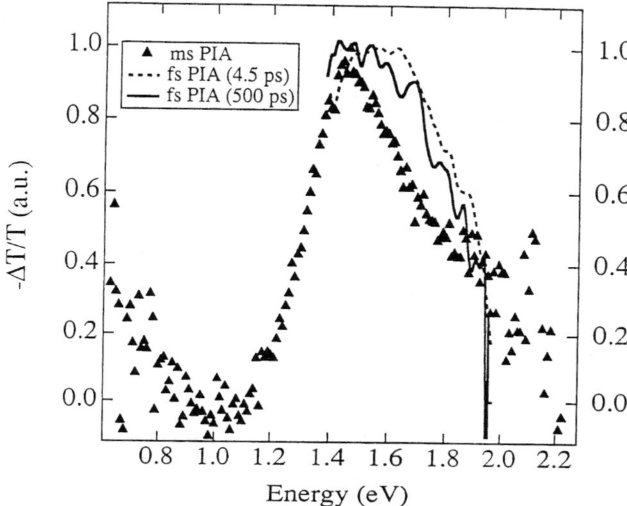

Figure 8 Comparison of the photoinduced absorption spectra for near steady-state (millisecond, Figure 15) and ultrafast (picosecond, Figure 22) time domains for P3OT/C_{60} composite films. (*From Ref. 7.*)

the peak of the PIA feature in P3OT/C_{60} (1%), we show the kinetics in P3OT at 1.45 eV (Figure 9) [9]. The fit shown is to the response function $R(\tau) = Ae^{-t/\tau_1} + Be^{-(t/\tau_2)^{1/3}}$ with $\tau_1 = 300$ fs and $\tau_2 = 20$ ps. The decay of the photoexcitations in P3OT with 1% C_{60} fits the same response function used above for pristine P3OT. The lifetimes in P3OT/C_{60} (1%) are $\tau_1 = 300$ fs and $\tau_2 = 25$ ps. The ratio of the coefficients (B/A) used in the fitting increases by over an order of magnitude upon addition of 1% C_{60}, implying that a much larger fraction of the photoexcitations contribute to the long-lived signal in the composite material than in pristine P3OT. In P3OT/C_{60} (10%), the lifetimes used to fit to the data are $\tau_1 = 300$ fs and $\tau_2 = 700$ ps. These results are consistent with those obtained earlier using P3OT/C_{60} (50%) [27] and with photoconductivity measurements made on P3OT/C_{60} (5%) composites (see below). The increase in τ_2 observed in the 10% composite implies that not only does the charge transfer occur on an subpicosecond time scale, but the reverse transfer (i.e., the electron transferring from C_{60} back to P3OT) is inhibited, yielding metastable charge-separated photoexcitations. Hence, in the 10% composite, both mechanisms discussed above are contributing to the increase in the yield of long-lived charged photoexcitations.

The addition of 1% C_{60} also affects the dependence of the PIA signal on

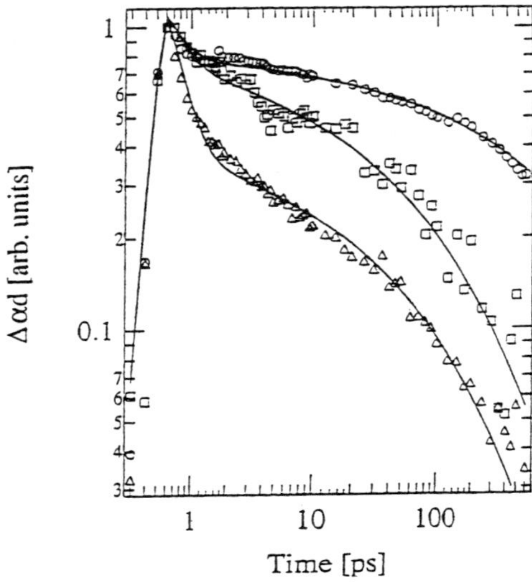

Figure 9 Decay of the photoinduced absorption at 1.45 eV in pristine P3OT (open triangles), P3OT/C$_{60}$ (1%) (open squares), and P3OT/C$_{60}$ (10%) (open circles) at 300 K. The formula and fit parameters are described in the text. (*From Ref. 9.*)

the incident pump fluence [9]. In order to compare with the behavior of the peak of the PIA feature in P3OT/C$_{60}$, we consider the intensity dependence of the same spectral region (1.45 eV) in pristine P3OT. At this energy, pristine P3OT follows an $I^{0.64}$ dependence at 1 ps delay time, which becomes $I^{0.47}$ by 20 ps. An exciton is expected to decay geminately and therefore exhibit a linear intensity dependence. However, this superimposed HE band in P3OT has a long-lived species which follows a square-root dependence on the pump fluence. This results in an intensity dependence which is intermediate between linear and square root, which persists during the lifetime of the polaron-exciton. After the polaron-exciton has decayed ($t > 10$ ps) the intensity dependence of the entire PIA spectrum follows a square-root behavior. In contrast, the P3OT/C$_{60}$ (1%) system displays a linear dependence on pump fluence from the hundred femtosecond time scale to the hundred picosecond time scale [9].

2.3 Photoinduced Dichroism

By measuring the dichroic ratio, the time resolution is improved since we detect only a 100 meV band of the probe pulse at a given energy [9]. The dichroic ratio is defined as the ratio of the photoinduced absorption with the pump and probe polarization vectors parallel, to that with the polarization vectors perpendicular [28, 9]:

$$\frac{\delta\alpha_{\parallel}}{\delta\alpha_{\perp}} = \frac{3(f_e + f_{p\parallel})\alpha_{\parallel} + f_{p\perp}\alpha_{\perp}}{(f_e + f_{p\parallel})\alpha_{\parallel} + 3f_{p\perp}\alpha_{\perp}}$$

In this expression the subscripts e, $p_{\parallel}$, and $p_{\perp}$ refer to neutral bipolaron excitons, *intra*chain and *inter*chain excitations, respectively, and f_i is defined by $f_i \equiv \eta_i \Delta\alpha_i(\omega)$, where η_i is the quantum efficiency for generating excitations of type i, and $\Delta\alpha_i(\omega)$ is the change in absorption due to these excitations. Because of the strong delocalization of the π electrons on an individual chain and the weak interchain coupling, the optical absorption matrix elements are strongest when the light is polarized along the chains. Highly oriented MEH-PPV, for example, shows $\alpha_{\parallel}/\alpha_{\perp} = 100$ [29]. Thus, since the initially excited electron-hole pairs should be predominantly on individual chains, the dichroic ratio should be 3 at the earliest times. In the pure material, the dichroic ratio will approach the isotropic value, $\delta\alpha_{\parallel}/\delta\alpha_{\perp} = 1$, after sufficient time has elapsed for interchain transfer or diffusion of the excitation over relatively long distances. If charge transfer from the polymer to C_{60} is rapid and if the absorption in the charge-transferred state results from an interchain dipole transition, the charge transfer process will quickly destroy any excitation with a memory of the direction of the original transition dipole; hence, the dichroic ratio will approach the isotropic value more rapidly.

The dichroic ratio of the HE feature in pristine P3OT (Figure 10, open circles) in the first picosecond after photoexcitation approaches the theoretical limit of 3, indicating a large contribution from intrachain excitations in this time regime [9]. By 10 ps after photoexcitation, the excitations decay and/or diffuse to polymer segments that are randomly oriented with respect to the initial segment, so that the dichroic ratio decreases to 1.6. The dichroism of the LE feature (Figure 10) begins at 2.25, then drops within several hundred femtoseconds to approximately 1.6, where it remains relatively constant for over 100 ps. The decay of the photoinduced dichroism discriminates the HE and LE PIA bands in pristine P3OT, even though their behavior is identical after 20 ps [9].

Upon adding 1% C_{60} to P3OT (Figure 10, closed squares), the dichroic ratio at 1.5 eV decreases from 2.7 to 1.5 within 300 fs [9]. The high initial value is again attributed to the initial intrachain excitations on the polymer. The dichroism decreases rapidly to 1.5, and then remains relatively constant for over 100 ps. This behavior is attributed to the ultrafast charge transfer as described for the PPV derivatives above; i.e., once the electron transfers to a C_{60} molecule, the polarization memory is destroyed. This measurement sets an upper limit of 300 fs on the photoinduced electron transfer time in P3OT/C_{60} composites.

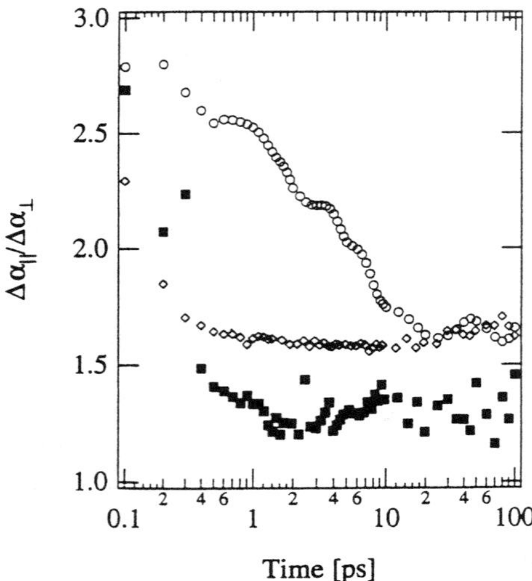

Figure 10 Dichroic ratio for pristine P3OT at 1.87 eV (open circles), and at 1.21 eV (open diamonds), and for P3OT/C_{60} (1%) at 1.5 eV (closed squares), at 300 K. (*From Ref. 9.*)

2.4 Steady-State Infrared Photoinduced Absorption

In semiconducting, conjugated polymers, the quasi-one-dimensional electronic structure is strongly coupled to the chemical (geometrical) structure. As a result, the nonlinear excitations (solitons, polarons, and bipolarons) are dressed with local structural distortions creating states with energies within the energy gap. "New" infrared bands with large intensities (IRAV modes) are induced by doping and/or photoexcitation. Solitons, polarons, and bipolarons are charged "defects" which break the local symmetry and therefore make Raman modes infrared active [30–31].

Figure 11 shows that there is no interaction between the conjugated polymer backbone and the C_{60} molecule in the ground state for the conjugated polymer/C_{60} composites [32]. The infrared spectrum, which is dominated by the vibrational absorptions of the conjugated polymer backbone, shows no change upon addition of 5% C_{60}. In the excited state, however, the photoinduced electron transfer and the long-lived charge separation in conjugated polymer/C_{60} composites greatly enhance both the concentration and the lifetime of the photoinjected charges on the polymer backbone resulting in an enhancement of the strength of the IRAV modes (Figure 12) [32]. In Figure 12 the electronic absorption bands associated with the C_{60}^{-} (at 1.15 and 1.25 eV) are also observable. Thus, conju-

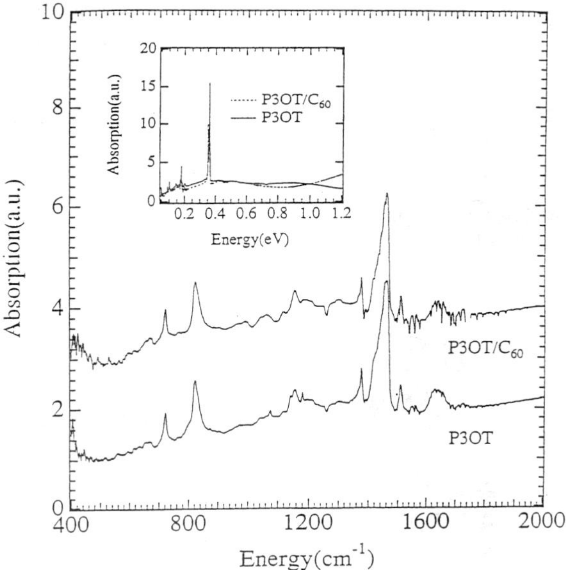

Figure 11 Infrared absorption spectra of P3OT and P3OT/C$_{60}$ (5%) films at 80 K. The inset shows the data over the whole IR range up to 1.0 eV. (*From Ref. 32.*)

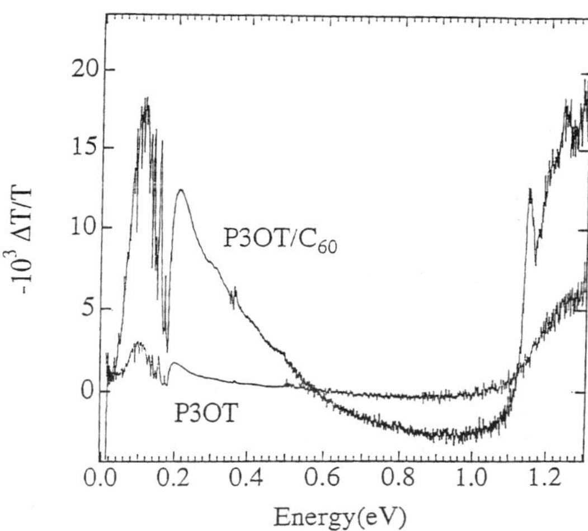

Figure 12 Photoinduced IR absorption spectra of P3OT and P3OT/C$_{60}$ (5%) at 80 K obtained by pumping with argon-ion laser at 2.41 eV with 50 mW/cm^2. (*From Ref. 32.*)

gated polymer/C_{60} composites may serve as valuable for the study of the IRAV modes of conjugated polymers by strongly enhancing the absorption signals.

2.5 Sensitization of Photoconductivity

Figure 13 shows the time-resolved transient photocurrent (PC) of a P3OT film containing 5% C_{60} and a pure P3OT film upon photoexcitation at 2.9 eV; the addition of 5% C_{60} to P3OT results in an enhancement of the photocurrent by nearly an order of magnitude [8]. The rise time is limited by the temporal resolution of the instrumental response (50 ps). The increase of the photocarrier generation efficiency results in the enhancement of the initial photocurrent. This enhancement for other nondengenerate semiconducting polymers is demonstrated in Figure 14 for the case of MEH-PPV/C_{60} composites with different C_{60} content [8]. The admixture of 1% of C_{60} results in an increase of initial photocurrent by an order of magnitude. This increase of the photocarrier generation efficiency is accompanied by the successive increase in lifetime of the photocarriers upon adding more C_{60}. Thus, the ultrafast photoinduced electron transfer from the semiconducting polymer onto C_{60} not only enhances the charge carrier generation in

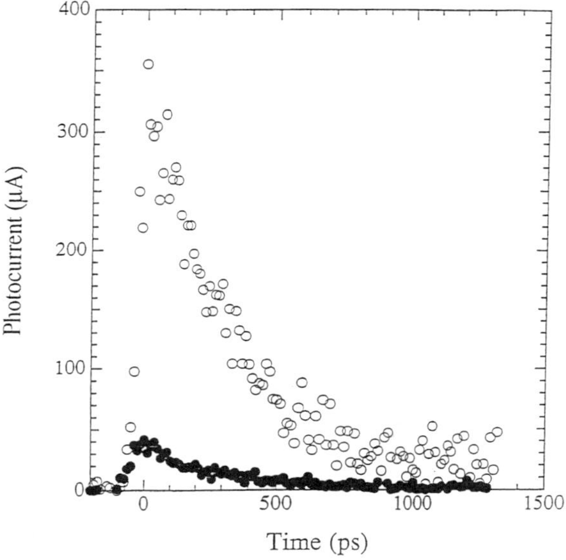

Figure 13 Transient photoconductivity in P3OT alone (full circles) and in P3OT/C_{60} (5%) (open circles). (*From Ref. 8.*)

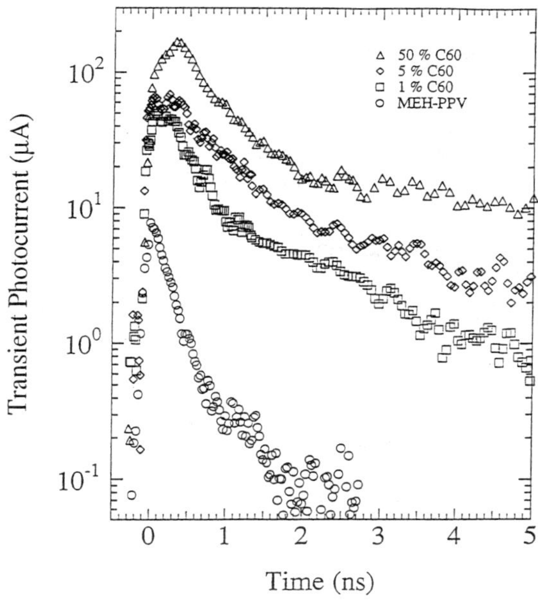

Figure 14 The transient photoconductivity of MEH-PPV and MEH-PPV/C60 for various concentrations. (*From Ref. 8.*)

the host polymer but also serves to prevent recombination by separating the charges and stabilizing them [8].

Figure 15 shows, on a semilog plot, the spectral response of the cw-PC of MEH-PPV alone and MEH-PPV/C_{60} composite for different concentrations of C_{60} [8]. These room temperature data are normalized to the constant incident photon flux of about 7.5×10^{14} photons/cm^2 s. The photoconductive spectral response of MEH-PPV shows a sharp onset at about 2 eV which coincides with the optical absorption edge across the energy gap. However, the photoconductive response of the MEH-PPV/C_{60} composites increases sharply at about 1.3 eV, which is lower than the photoconductivity onset of the individual components alone [8]. This issue has been recently addressed by Zhang et al. [33] with calculations which show an enhancement of electron affinity of C_{60} upon photoexcitation, in agreement with the experimental results that photoexcited C_{60} is a better electron acceptor compared to C_{60} in the ground state [34]. Furthermore, the composite films exhibit a remarkably enhanced photoconductivity over the broad spectral range from the near infrared to the ultraviolet. At photon energies below 2 eV, the enhancement reaches several orders of magnitude [8]. Although the effect of

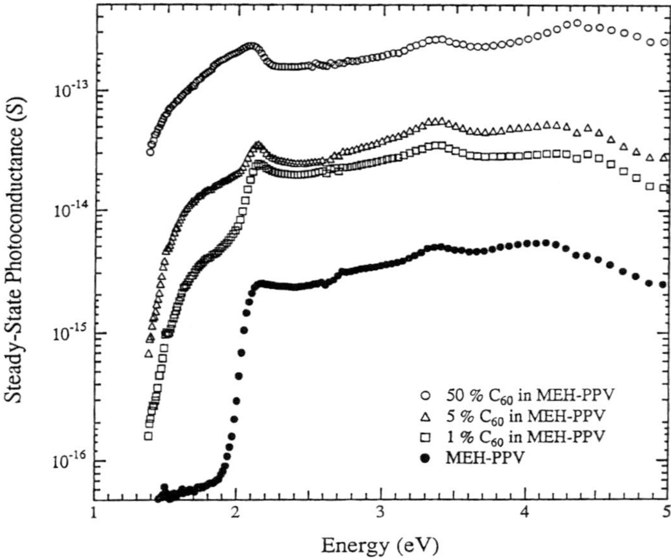

Figure 15   Spectral response of the steady state photoconductivity of MEH-PPV alone and MEH-PPV/C60 for several concentrations at 300 K and biasing field of 10^4 V/cm. (*From Ref. 8.*)

the C_{60} on the mobility of the carrier should be considered (the mobility will probably decrease as a result of the deep trap sites associated with C_{60} clusters), the significant effect of C_{60} on the charge carrier generation efficiency is evident from the observation that even 1% C_{60} in the polymer matrix enhances the cw-PC by more than an order of magnitude [8]. This observation is in full agreement with the photoinduced electron transfer phenomenon which leaves metastable positive polarons on the polymer backbone after the electron transfer, i.e., *photo-doping*.

2.6   Direct Experimental Observation of the Metastable Charge Separation: Light-Induced Electron Spin Resonance (LESR)

Definitive evidence of charge transfer and charge separation is obtained from LESR experiments [6]. Figure 16 shows the ESR signal upon illuminating the $P3OT/C_{60}$ composites with light of $h\nu = E_{\pi\text{-}\pi}{}^*$, where $E_{\pi\text{-}\pi}{}^*$ is the energy gap of the conjugated polymer (donor). Two photoinduced ESR signals can be resolved:

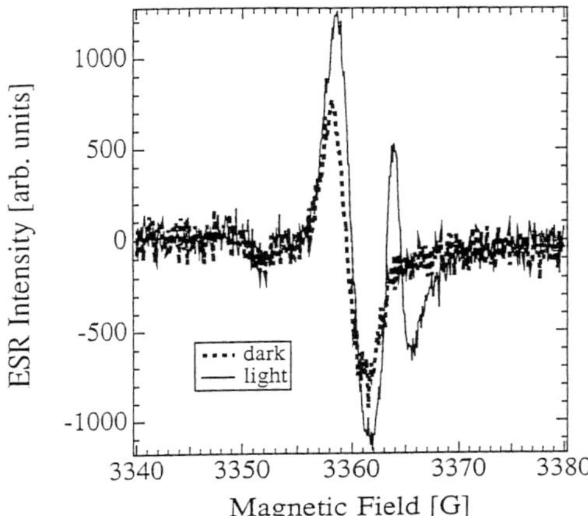

Figure 16 Light-induced electron spin resonance spectra of P3OT/C_{60} upon succesive illumination with 2.41 eV argonion laser with 100 mW. The dark ESR signal is also displayed as a dashed line. (*From Ref. 6.*)

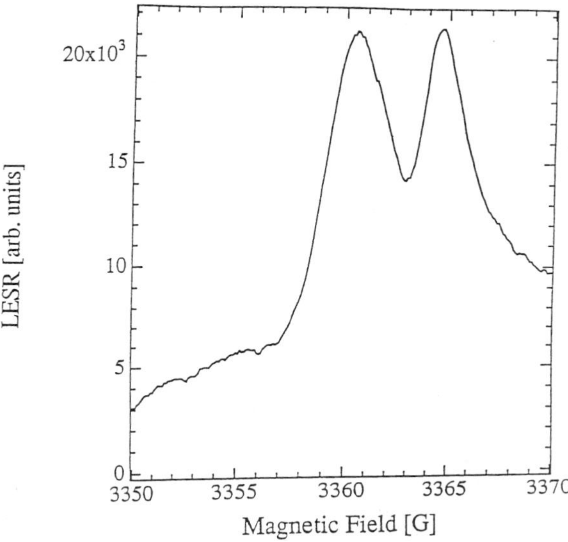

Figure 17 Integrated LESR intensity [ESR (illuminated)–ESR (dark)] as displayed in Figure 16 of P3OT/C_{60}.

one at $g = 2.00$ and the other at $g = 1.99$ [6]. The higher g value line is assigned to the conjugated polymer cation (polaron), and the lower g value line to C_{60}^- anion. The assignment of the lower g value line to C_{60}^- is unambiguous, for this low g value was measured earlier [35]; the higher g value is typical of conjugated polymers. The LESR signal vanishes above 200 K; this rules out permanent photochemical changes as the origin of the ESR signal and demonstrates the reversibility of the photoinduced electron transfer. The integrated LESR intensity shows two peaks with equivalent intensities (Figure 17).

3. PHOTOPHYSICS OF CONJUGATED POLYMER/ FULLERENE MIXTURES IN SOLUTION

The photoinduced electron transfer in conjugated polymer/C_{60} composites has been investigated in two different media using 1,2-dichlorobenzene (ODCB, ϵ = 9.93) and p-xylene ($\epsilon = 2.27$) [36]. The PIA spectrum of a 0.5 mg/ml solution of P3MBET in ODCB is displayed in Figure 18. In this more polar solvent, a strong transition is observed centered at 1.47 eV with a second one at low energy with maximum below 0.64 eV. The high-energy transition at 1.47 eV is attributed to the superposition of a PIA signal due to a triplet-triplet absorption and a charged photoexcitation [36]. The low-energy PIA band, however, is solely associated with charged excitations. Thus, in high polarity solvents the neutral photoexcitations (long-lived triplet) and charged photoexcitations coexist.

Upon mixing C_{60} (5×10^{-4} M) into the ODCB solution of P3MBET, the high-energy PIA band is quenched [36]. The resulting PIA spectrum exhibits the two peaks associated with charged excitations of the polymer only [36]. In addition to the features of the charged excitations on the P3MBET conjugated backbone, the PIA spectrum of the P3MBET/C_{60} solution in ODCB exhibits a small absorption at around 1.15 eV which is assigned to the optical absorption of C_{60}^-.

The PIA spectrum of a 0.5 mg/ml solution of P3MBET in p-xylene, however, exhibits one strong, well-defined band at 1.50 eV assigned to the triplet-triplet absorption band of the polymer backbone [36]. This clearly shows that the nature of the long-lived photoexcitations in conjugated polymers can change from neutral triplets to charged photoexcitations by changing the polarity of the surrounding medium.

Upon mixing C_{60} into the p-xylene solution of P3MBET, the triplet-triplet absorption is quenched without an increase in the charged photoexcitation density. Since the observed PIA spectrum of P3MBET/C_{60} solution in p-xylene is identified as the triplet-triplet absorption of C_{60} with significantly enhanced intensity, a triplet energy transfer from the conjugated polymer triplet state onto the C_{60} triplet state has been proposed [36]. The results account for the solvent dependent

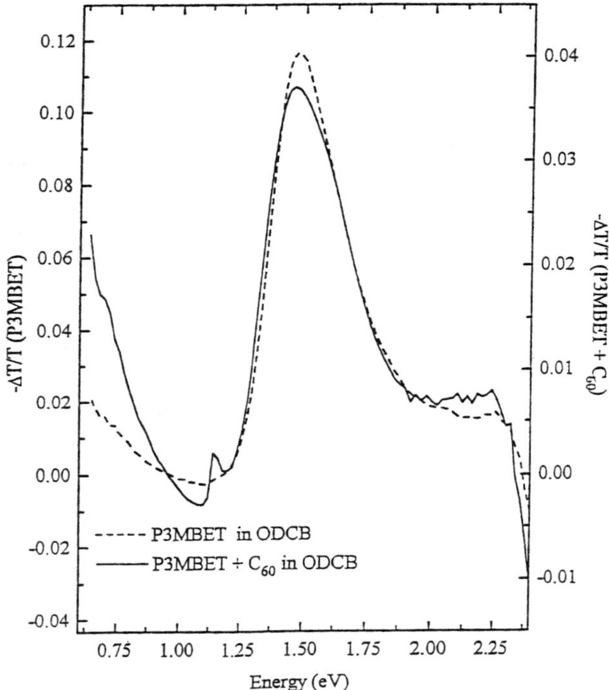

Figure 18 Photoinduced absorption spectra of 0.5 mg/ml P3MBET and P3MBET/C_{60} (5×10^{-4}M) *o*-dichlororbenzene solutions. (*From Ref. 36.*)

formation of the charged photoexcitations in conjugated polymers and for a shift between energy and/or electron transfer in cosolutions with C_{60} [36].

4. PHOTOVOLTAIC AND PHOTODETECTOR APPLICATIONS

The use of organic bilayers with photoinduced electron transfer at the interface has been investigated heavily during the last couple of decades (for a summary of the early reports see, for example, Refs. 37 and 38). Tang demonstrated photovoltaic activity in small molecular bilayers which were vacuum evaporated [39]. Extensive literature exists on the fabrication of solar cells based on small molecule dyes as well as donor-acceptor systems (see, for example, Ref. 40 and references therein). Inorganic oxide semiconductors have also been used to facilitate electron transfer from organic dyes to achieve charge separation and photovoltaic conversion (see, for example, Refs. 41 and 42 and references therein). Furthermore, conjugated polymer layers have been used in solar cells (see, for example, Refs. 43 to 45 and references therein). Yamashita and coworkers reported a bilayer photodiode based on the organic donor tetratiafulvalene (TTF) and C_{60} [46]. The Au/TTF/C_{60}/Au devices do not exhibit rectification in the dark, indicating that there is no charge transfer in the ground state. Low-charge carrier density results in negligible barriers, and the current-voltage (I-V) characteristic is nearly linear. Upon illumination, however, the I-V characteristic changes dramatically, exhibiting a large rectification ratio and photocurrent.

Because of the ultrafast photoinduced electron transfer with long-lived charge separation, the conjugated polymer/C_{60} system offers a special opportunity. Using conjugated polymers as donors with different acceptors, results in photoinduced charge separation with quantum efficiency near unity and with correspondingly enhanced device performance.

4.1 Conjugated Polymer/C_{60} Heterojunction Photodiodes

From the energy band diagram it is clear that the heterojunction formed at the interface between a semiconducting polymer and a C_{60} thin film should function as a diode with a rectifying current-voltage characteristic (analogous to a *pn* junction, but with a different mechanism based on molecular redox properties). In reverse bias, electron injection into the semiconducting polymer and electron removal from C_{60} are energetically unfavorable. This inherent polarity of the device results in very low reversed bias current densities. On the other hand, electron injection into C_{60} and electron removal from the semiconducting polymer are energetically favorable, thus resulting in relatively high current densities under forward bias. Analogous arguments for molecular diodes were first proposed by Aviram and Rattner years ago for Langmuir-Blodgett D-A structures [47].

Figure 19 shows the current-voltage characteristics of a heterojunction device consisting of successive layers of ITO/MEH-PPV/C_{60}/Au; a schematic description of the device operation [14, 48] is shown in Figure 20. Positive bias is defined as positive voltage applied to the ITO contact. Exponential turn-on up to 500 mV in forward bias is clearly observable; the rectification ratio is approximately 10^4.

In order to test if the ITO or the gold electrodes form blocking (or rectifying) contacts, the following three layer devices were prepared: gold/MEH-PPV/gold, gold/MEH-PPV/ITO, gold/C_{60}/gold, and gold/C_{60}/ITO. Since all such devices have symmetric and linear current-voltage characteristics, we conclude that gold and ITO form nonrectifying contacts both to MEH-PPV and to C_{60}, in agreement with the reported absence of rectification in Au/C_{60}/ITO devices [49]. We,

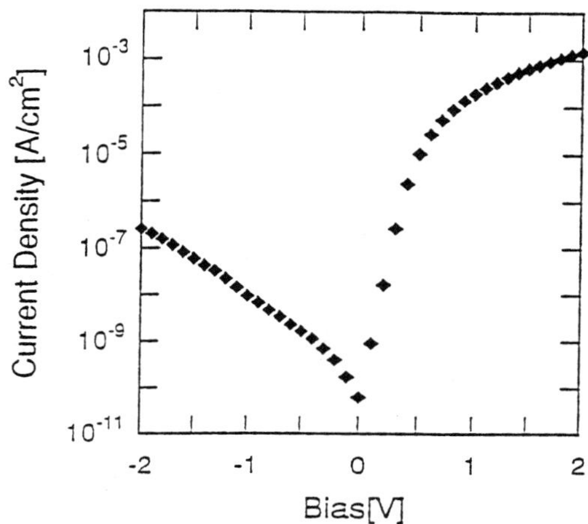

Figure 19 Dark current vs. voltage characteristics of the ITO/MEH-PPV/C_{60}/Au device at room temperature. (*From Ref. 14.*)

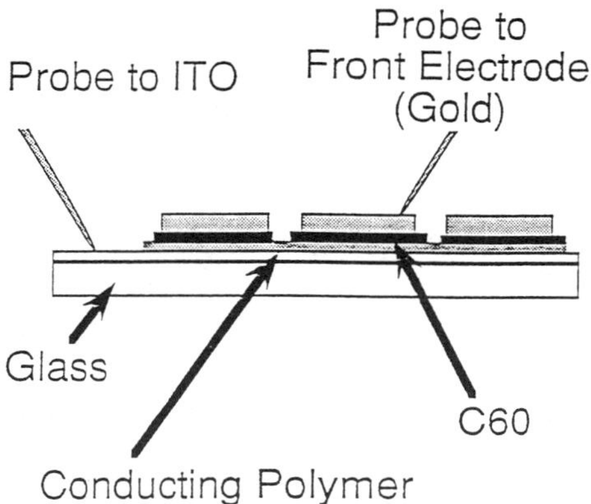

Figure 20 Schematic cross section of the heterojunction devices fabricated from MEH-PPV and C_{60}. (*From Ref. 14 and 48.*)

therefore, attribute the rectifying behavior of the four layer devices to the hetero-junction between the semiconducting polymer and C_{60}.

The current-voltage characteristic of the device changes dramatically upon illumination by visible light. Figure 21 shows the current-voltage data, with the ITO/MEH-PPV/C_{60}/Au device in the dark and with the device illuminated by 514.5 nm light with intensity (P_{in}) of 1 mW/cm^2. The open circuit voltage (V_{oc}) is 0.44 V which saturates to around 0.53 V under stronger illumination. The short circuit current density (J_{sc}) is 2.08×10^{-6} A/cm^2, and the fill factor (FF) can be obtained from the relation

$$FF = \frac{\int_0^{V_{oc}} J \, dV}{J_{sc} V_{oc}}$$

The data of Figure 21 give a fill factor of 0.48 and a power conversion efficiency of 0.04%. An open circuit voltage of about 0.5 V appears to be characteristic of the MEH-PPV/C_{60} interface. Similar values were obtained with ITO/MEH-PPV/C_{60}/Al devices and with Al/MEH-PPV/C_{60}/ITO devices.

As expected from the observation of fast photoinduced electron transfer from MEH-PPV to C_{60} [14, 48], the major increase in both forward and reverse bias current results from photoinduced charge separation at the heterojunction

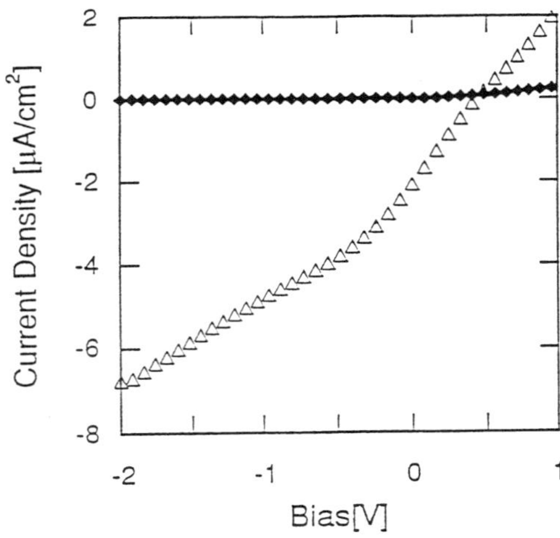

Figure 21 Current vs. voltage characteristics of the ITO/MEH-PPV/C_{60}/Au device in the dark (diamonds) and upon illumination with the 514.5 nm line of argon-ion laser of 1 mW/cm^2 (triangles). (*From Ref. 14.*)

interface. The increase in photocurrent by nearly 4 orders of magnitude (from 1 $\times$ 10^{-9} A/cm^2 to 6 $\times$ 10^{-6} A/cm^2) upon illumination at -1 V (reverse) bias demonstrates that the heterojunction serves as a relatively sensitive photodiode.

Figure 22 shows the dependence of the short curcuit current and the photocurrent at -1 V (reverse) bias as a function of the illumination intensity (514.5 nm); the data show no indication of saturation at light intensities up to approximately 1 W/cm^2, i.e., one order of magnitude greater than the terrestrial solar intensity.

The spectral dependence of the photoresponse of these bilayer heterojunction devices is displayed in Figure 23. The onset of photocurrent at $hv = 1.7$ eV follows the absorption of MEH-PPV, which initiates the photoinduced electron transfer [14, 48]; note that illumination is from the ITO/MEH-PPV side of the device. The minimum in the photocurrent at $hv = 2.5$ eV corresponds to the photon energy of maximum absorption of MEH-PPV. The MEH-PPV layer, therefore, acts as a filter which reduces the number of photons reaching the MEH-PPV/C$_{60}$ interface. This implies that diffusion of charge carriers in these devices is limited; the photoactive region is restricted to a thin layer adjacent to the interface between the MEH-PPV and C$_{60}$ layers. Assuming a charge carrier mobility of $\mu = 10^{-4}$ cm^2/V $\cdot$ s, with an applied field of 10^5 V/cm and with a photoinduced carrier lifetime approximately 1 ns, the maximum distance a carrier could travel

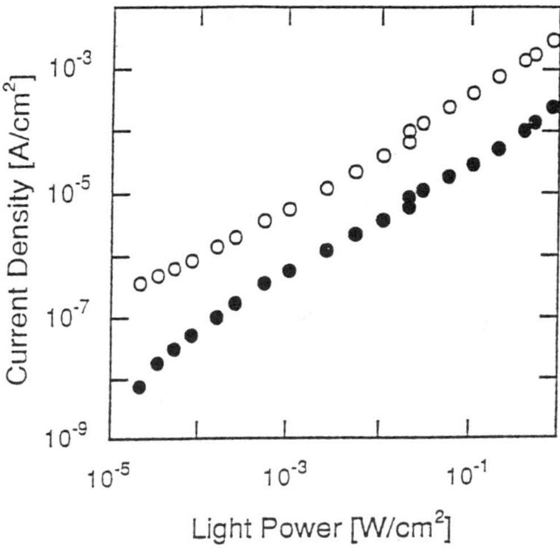

Figure 22 Short circuit current (closed circles) and photocurrent at -1 V bias (open circles) as a function of light intensity for the ITO/MEH-PPV/C$_{60}$/Au device. (*From Ref. 14.*)

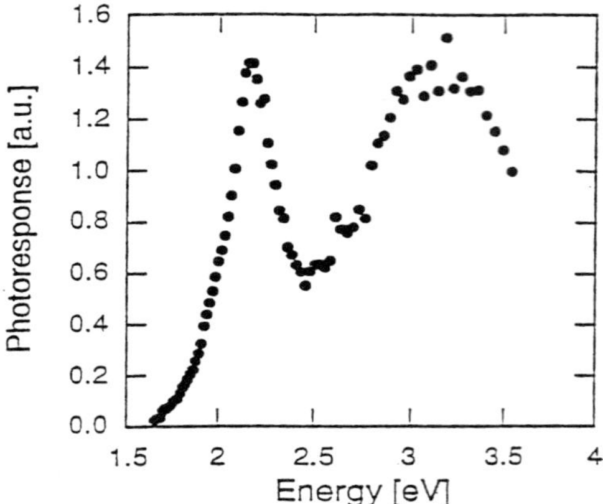

Figure 23   Spectral response of the photocurrent in ITO/MEH-PPV/C_{60}/Au photodiode at (reverse)-1 V bias. (*From Ref. 14.*)

in the MEH-PPV is estimated to be only a few Angstroms. Thus, the generation of photoexcitations which result in separated charge carriers occurs only at the heterojunction interface.

Yoshino and coworkers also reported the optical response of a heterojunction device comprising a P3OT and C_{60} bilayer [50]. The photoresponse of these devices shows a broad excitation profile ranging from 750 nm into the UV.

Tripathy and coworkers reported the enhancement of the photocurrent upon mixing 10% C_{60} into poly(3-undecylthiophene) (P3UT) in Al/P3UT + C_{60}/ITO devices [51], but did not quantify the device parameters; thus we are unable to compare device performances. Developing similar device architectures resulted in very promising efficiencies, as discussed in the next section.

4.2   Diodes Made of Conjugated Polymer/Fullerene Composites: Bulk Heterojunctions

A semiconducting polymer with asymmetric contacts (a low work function metal on one side and a high work function metal on the opposite side) functions as a "tunneling injection diode"; such devices have been described by Parker [52]. A schematic cross-sectional view of such devices is displayed in Figure 24.

In forward bias, tunneling injection diodes exhibit relatively high-efficiency electroluminescence which is promising for flat panel and/or flexible, large-area display applications. In reverse bias, on the other hand, the devices exhibit a strong photoresponse with a quantum yield >20% (el/ph at -10 V reverse bias)

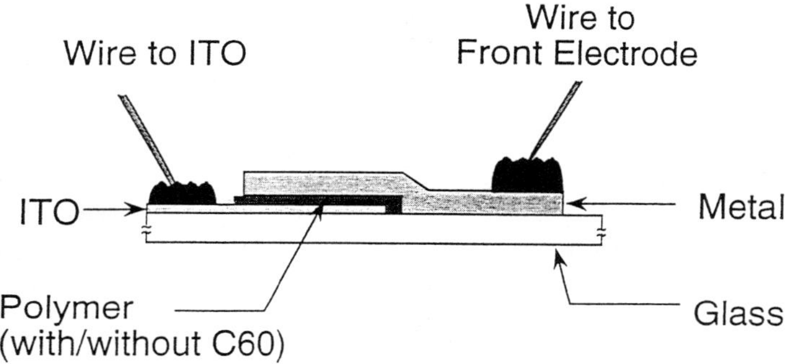

Figure 24 Schematic cross section of the tunnel diodes fabricated from conjugated polymer/C_{60} composites.

[53, 54]. Devices based on derivatives of polythiophene exhibit even better photoresponse (80% el/ph at -15 V), competetive with UV-sensitized Si photodiodes (Figure 25) [53, 54]. A photovoltaic response was observed under zero bias conditions [53–55]. The energy band structure for such thin film devices can be approximated by the rigid band model displayed in Figure 26. The charge carrier

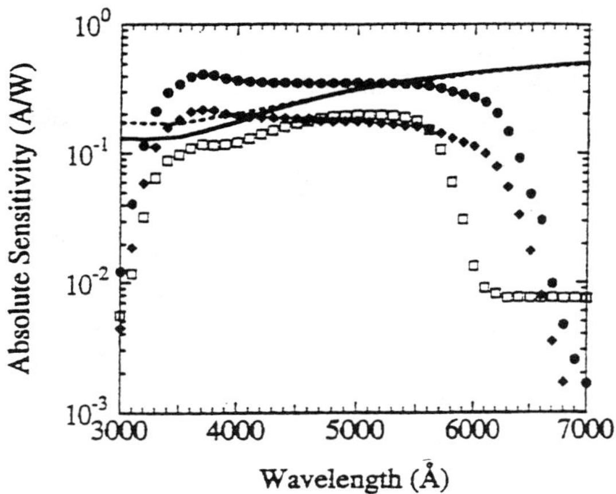

Figure 25 Absolute sensitivities of an Au/P3OT/ITO photodiode at -10 V (full diamonds) and -15 V (full circles), and a Ca/MEH-PPV + C_{60}/ITO photodiode at -10 V (empty squares). Also shown are the photosensitivities of two commercial UV-enhanced photodiodes at -15 V bias; (solid line) EG&G UV444B, serial No. 7926$-$01, and (dashed line) UDT, model No. FIL-UV20. (*From Ref. 53 and 54.*)

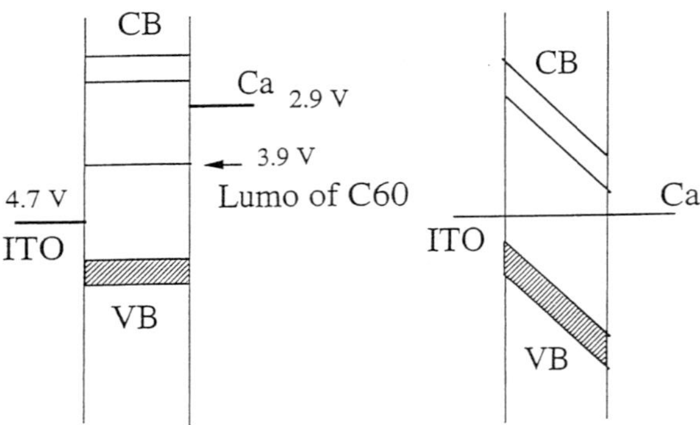

Figure 26 Flat band scheme of the Ca/Polymer + C_{60}/ITO tunnel diodes at open circuit and short circuit conditions.

concentration in these undoped, pristine materials is sufficiently low that the semiconducting layer (a few thousand angstroms) is fully depleted; there is negligible band bending at the metal-polymer interface. Note, however, that the mobility of injected charge carriers is not symmetric; in most conjugated polymers, the electron mobility is lower than the hole mobility.

For photovoltaic cells made with pure conjugated polymers, energy conversion efficiencies were typically 10^{-3}–$10^{-2}\%$, too low to be used in practical applications [53–55]. The photoinduced electron transfer in composites of semiconducting polymers (as donors) and buckminsterfullerene, C_{60}, and its derivatives (as acceptors) has provided a "molecular approach" to high-efficiency photovoltaic conversion; the initial processes are analogous to the initial steps in natural photosynthesis. Since the time scale for photoinduced charge transfer is subpicosecond, more than 10^3 times faster than the radiative or nonradiative decay of photoexcitations, the quantum efficiency for charge transfer and charge separation from donor to acceptor is close to unity. Thus, photoinduced charge transfer across a donor/acceptor (D/A) interface provides an effective method to overcome early time carrier recombination in organic systems and thus to enhance their optoelectronic response. As noted above, for example, with the addition of only 1% C_{60}, the photoconductivity of MEH-PPV + C_{60} increases by an order of magnitude over that of pure MEH-PPV.

Although the quantum efficiency for photoinduced charge separation is near unity for a D/A pair, the conversion efficiency in a bilayer heterojunction device is limited [14]:

1. Due to the molecular nature of the charge separation process, efficient charge separation occurs *only* at the D/A interface; thus, photoexcit-

ations created far from the *D/A* junction recombine prior to diffusing to the heterojunction.

2. Even if charges are separated at the *D/A* interface, the photovoltaic conversion efficiency is limited by the carrier collection efficiency; i.e., the separated charges must be collected with minimum losses.

Consequently, interpenetrating phase-separated *D/A* network composites would appear to be ideal photovoltaic materials [15]. Through control of the morphology of the phase separation into an interpenetrating network, one can achieve a high interfacial area within a bulk material. Since any point in the composite is within a few nanometers of a *D/A* interface, such a composite is a *bulk D/A heterojunction* material. If the network is bicontinuous, as shown schematically in Figure 27, the collection efficiency can be equally efficient.

Important progress has been made toward creating bulk *D/A* heterojunction

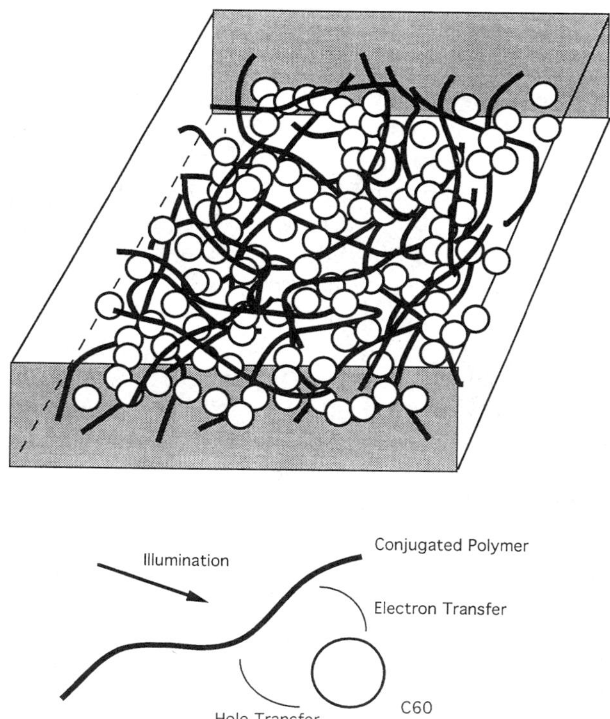

Figure 27 Schematic illustration of the interpenetrating donor/acceptor (conjugated polymer/C60) network of internal heterojunctions.

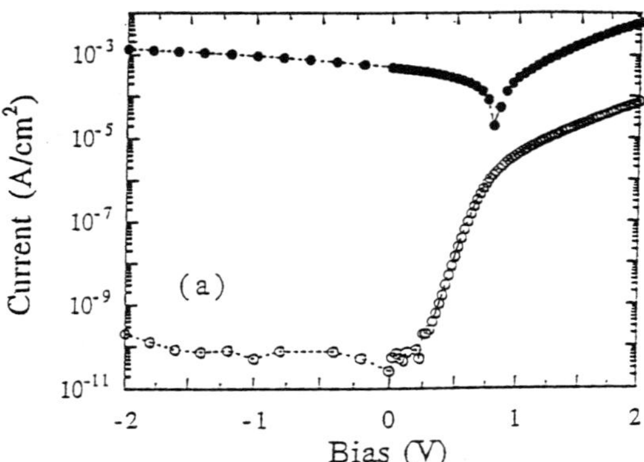

Figure 28 Current-voltage characteristics of a Ca/MEH-PPV:[6,6]PCBM/ITO device in the dark (open circles) and under 20 mW/cm² illumination at 430 nm (solid circles). (*From Ref. 15.*)

materials [15, 16]. As shown in Figure 28, the short circuit current, $I_{sc} = 0.5$ mA/cm² under 20 mW/cm² illumination, corresponding to a collection efficiency of $\eta_c = 7.4\%$ electrons per incident photon [15] approximately 2 orders of magnitude higher than that of pure MEH-PPV tunneldiodes as well as of the MEH-PPV/C_{60} heterojunction device as described in the previous section. The electroluminescence quantum efficiency of this blend device was 10^3–10^4 times *less* than in pure MEH-PPV devices, consistent with the ultrafast photoinduced charge separation which quenches the emission of the donor [15].

Uniform films with even higher concentrations of soluble C_{60} derivatives were cast from 1,2-dichlorobenzene solutions containing up to 1:4 weight ratio MEH-PPV:C_{60} derivative. For devices made from these high concentration blends, charge collection efficiencies around $\eta_e = 45\%$ (electron/incident photon) and power conversion efficiencies around $\eta_e = 3\%$ (electrical power out/incident light power) have been realized [15]. The efficiencies are nearly independent of the incident illumination intensity, as shown in Figure 29. Furthermore, the internal efficiencies are even *higher* when corrected for the small thickness of the film which absorbs only ~60% of the incident photons.

The excellent photosensitivity and relatively high-energy conversion efficiencies obtained from the network heterojunction materials are promising. Further optimization of device performance can be achieved by opimization of the device physics:

1. Optimize the choice of metallic electrodes to achieve good ohmic contacts on both sides for collection of the oppositely charged photocarriers.

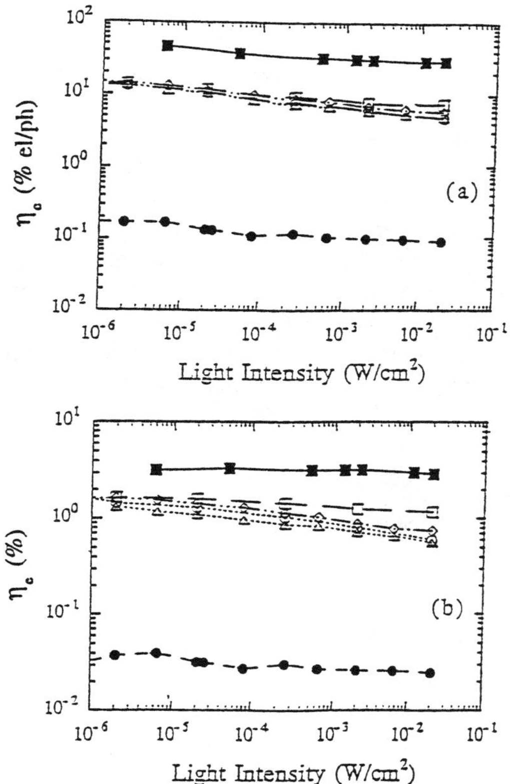

Figure 29 Carrier collection efficiency, η_c, and energy conversion efficiency, η_e, of a Ca/MEH-PPV: [6,6]PCBM (1:4)/ITO (solid squares), Ca/MEH-PPV: [6,6]PCBM (1:1)/ITO (open squares), A1/MEH-PPV: [6,6]PCBM (1:1)/ITO (open diamonds), Ca/MEH-PPV: [5,6]PCBM (1:1)/ITO (open circles), Ca/MEH-PPV: C_{60} (3:1)/ITO (open triangles), and Ca/MEH-PPV/ITO (solid circles). (*From Ref. 15.*)

2. Optimize the choice of the *D/A* pair.
3. Optimize the network morphology of the phase separated composite material. In addition, the band gap of the semiconducting polymer should be chosen for efficient harvesting of the solar spectrum.

Because of the advantages which would be realized with polymer-based photovoltaics, such as low-cost fabrication in large sizes and in desired shapes with mechanical flexibility, efficient ''plastic'' solar cells would have a major impact.

5. CONCLUSIONS

The discovery of an ultrafast, reversible, metastable photoinduced electron transfer in conjugated polymer/fullerene composites has opened a large area of scientific and technological interest. Research areas affected by this discovery for the time being are schematically displayed in Figure 30.

Growing alternating donor/acceptor multiple-layer heterostructures, using, for example, oligothiophenes as donors and C_{60} as acceptor with the vacuum deposition of thin films, can result in photoinduced quantum well phenomena.

Yoshino, Zakhidov, and coworkers [57, 58] reported that doping the conjugated polymer/C_{60} composites with alkali atoms results in granular superconductivity of the C_{60} component embedded in the polymer matrix. Photoinduced electron transfer may increase the number of superconducting domains in these composites.

Embedding the conjugated polymers and fullerenes in different matrices such as zeolites can reveal interesting effects due to decoupling of the individual *D/A* pairs in a dielectric environment which can be varied.

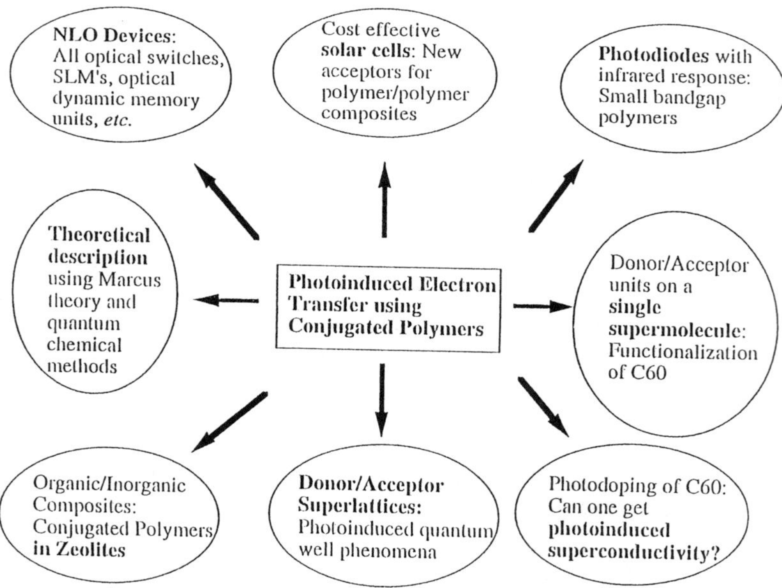

Figure 30 Some examples of current and future scientific research directions utilizing the ultrafast, reversible, photoinduced electron transfer from conjugated polymers onto fullerenes.

Nonlinear optical effects derived from the photoinduced electron transfer have potentials in areas such as all optical spatial light modulators, holographic memory units, and optical limiters.

Photodetector and photovoltaic applications have reached the stage where development work in an industrial level is commencing.

Thus, the photophysics and the associated device applications of the conjugated polymer/fullerene composites implies an important scientific and industrial opportunity.

ACKNOWLEDGMENTS

The author gratefully acknowledges Prof. Alan J. Heeger and Prof. Fred Wudl for the great scientific experience at the Institute for Polymers and Organic Solids at the University of California, Santa Barbara, for their personal support, help, and committment. Furthermore, we gratefully acknowledge all our coworkers who contributed to this body of work.

REFERENCES

1. R. J. Sension, A. Z. Szarka, G. R. Smith, and R. M. Hochstrasser, *Chem. Phys. Lett.* **185**, 179 (1991).
2. Y. Wang and L.-T. Cheng, *J. Phys. Chem.* **96**, 1530 (1992).
3. Y. Wang, *Nature* **356**, 585 (1992); Y. Wang, *J. Phys. Chem.* **96**, 764 (1992); Y. Wang, R. West, and C.-H. Yuan, *J. Am. Chem. Soc.* **115**, 3844 (1993).
4. P. V. Kamat, *J. Am. Chem. Soc.* **113**, 9705 (1991).
5. M. S. Dresselhaus, G. Dresselhaus, and P. C. Eklund, *Science of Fullerenes and Carbon Nanotubes*, Academic Press, San Diego, 1996.
6. N. S. Sariciftci, L. Smilowitz, A. J. Heeger, and F. Wudl, *Science* **258**, 1474 (1992).
7. L. Smilowitz, N. S. Sariciftci, R. Wu, C. Gettinger, A. J. Heeger, and F. Wudl, *Phys. Rev.* **B47**, 13835 (1993).
8. C. H. Lee, G. Yu, N. S. Sariciftci, D. Moses, K. Pakbaz, C. Zhang, A. J. Heeger, and F. Wudl, *Phys. Rev.* **B48**, 15425 (1993).
9. B. Kraabel, D. McBranch, N. S. Sariciftci, D. Moses, and A. J. Heeger, *Phys. Rev.* **B50**, 18 (1994).
10. N. S. Sariciftci and A. J. Heeger, *Int. J. Mod. Phys.* **B8**, 237 (1994).
11. N. S. Sariciftci, *Prog. Quant. Electr.* **19**, 131 (1995).
12. N. S. Sariciftci and A. J. Heeger, in *Handbook of Organic Conductive Molecules and Polymers*, vol. 1 (H. S. Nalwa, ed.), John Wiley, New York, 1997.
13. X. Wei, Z. V. Vardeny, N. S. Sariciftci, and A. J. Heeger, *Phys. Rev.* **B53**, 2187 (1996).

14. N. S. Sariciftci, D. Braun, C. Zhang, V. Srdanov, A. J. Heeger, G. Stucky, and F. Wudl, *Appl. Phys. Lett.* **62**, 585 (1993).

15. G. Yu, J. Gao, J. C. Hummelen, F. Wudl, and A. J. Heeger, *Science* **270**, 1789 (1995).

16. N. S. Sariciftci, B. Kraabel, C. H. Lee, K. Pakbaz, A. J. Heeger, and D. J. Sandman, *Phys. Rev.* **B50**, 12 (1994).

17. S. Morita, A. A. Zakhidov, and K. Yoshino, *Sol. State Commun.* **82**, 249 (1992).

18. K. Yoshino, X. H. Yin, S. Morita, T. Kawai, and A. A. Zakhidov, *Sol. State Commun.* **85**, 85 (1993).

19. K. Yoshino, X. H. Yin, K. Muro, S. Kiyomatsu, S. Morita, A. A. Zakhidov, T. Noguchi, and T. Ohnishi, *Jpn. J. Appl. Phys.* **32**, L357 (1993).

20. S. Morita, S. Kiyomatsu, M. Fukuda, A. A. Zakhidov, K. Yoshino, K. Kikuchi, and Y. Achiba, *Jpn. J. Appl. Phys.* **32**, L1173 (1993).

21. S. Morita, S. Kiyomatsu, X. H. Yin, A. A. Zakhidov, T. Noguchi, T. Ohnishi, and K. Yoshino, *J. Appl. Phys.* **74**, 2860 (1993).

22. S. Morita, A. A. Zakhidov, and K. Yoshino, *Jpn. J. Appl. Phys.* **32**, L873 (1993).

23. K. Yoshino, T. Akashi, K. Yoshimoto, S. Morita, R. Sugimoto, and A. A. Zakhidov, *Sol. State Commun.* **90**, 41 (1994).

24. L. Smilowitz and A. J. Heeger, *Synth. Met.* **48**, 193 (1992).

25. U. Stamm, M. Taiji, M. Yoshizawa, K. Yoshino, and T. Kobayashi, *Mol. Cryst. Liq. Cryst.* **182 A**, 147 (1990).

26. G. S. Kanner, X. Wei, B. C. Hess, L. R. Chen, and Z. V. Vardeny, *Phys. Rev. Lett.* **69**, 538 (1992).

27. B. Kraabel, C. H. Lee, D. McBranch, D. Moses, N. S. Sariciftci, and A. J. Heeger, *Chem. Phys. Lett.* **213**, 389 (1993).

28. M. Sinclair, D. Moses, and A. J. Heeger, *Phys. Rev.* **B36**, 4296 (1987).

29. T. W. Hagler, K. Pakbaz, K. F. Voss, and A. J. Heeger, *Phys. Rev.* **B44**, 8652 (1991).

30. B. Horowitz, *Solid State Commun.* **41**, 729 (1982); E. Ehrenfreund, Z. V. Vardeny, O. Brafman, and B. Horowitz, *Phys. Rev.* **B36**, 1535 (1987).

31. G. Zerbi, M. Gussoni, and C. Castiglioni, in *Conjugated Polymers* (J. L. Bredas and R. Silbey, eds.), Kluwer Academic Publ., Dordrecht, 1991, pp. 435.

32. K. H. Lee, R. A. J. Janssen, N. S. Sariciftci, and A. J. Heeger, *Phys. Rev.* **B49**, 5781 (1994).

33. G. P. Zhang, R. T. Fu, X. Sun, K. H. Lee, and T. Y. Park, *J. Phys. Chem.* **99**, 12301 (1995).

34. J. W. Arbogast, A. O. Darmanyan, C. S. Foote, Y. Rubin, F. N. Diederich, M. M. Alvarez, S. J. Anz, and R. L. Whetten, *J. Phys. Chem.* **95**, 11 (1991).

35. P. M. Allemand, G. Srdanov, A. Koch, K. Khemani, F. Wudl, Y. Rubin, F. Diederich, M. M. Alvarez, S. J. Anz, and R. L. Whetton, *J. Am. Chem. Soc.* **113**, 2780 (1991).

36. R. A. J. Janssen, N. S. Sariciftci, and A. J. Heeger, *J. Chem. Phys.* **100**, 8641 (1994).

37. H. Meier, *Organic Semiconductors*, Verlag Chemie, Weinheim, 1974.

38. J. Simon and J. J. Andre, *Molecular Semiconductors*, Springer Verlag, Berlin, 1985.

39. C. W. Tang, *Appl. Phys. Lett.* **48**, 183 (1986).

40. J. B. Whitlock, P. Panayotatos, G. D. Sharma, M. D. Cox, R. R. Sauers, and G. R. Bird, *Optical Engineering* **32**, 1921 (1993).

41. B. O'Regan and M. Gratzel, *Nature* **353**, 737 (1991).
42. W. A. Nevin and G. A. Chamberlain, *IEEE Transaction on Electron Devices* **40**, 75 (1993).
43. J. Kanicki, in *Handbook of Conducting Polymers* (T. A. Skotheim, ed.), vol. 1, pp. 543, Marcel Dekker, New York, 1986, pp. 543.
44. G. Horowitz, *Adv. Mater.* **2**, 287 (1990).
45. K. Uehara, A. Maekawa, A. Sugimoto, and H. Inoue, *Thin Solid Films* **215**, 123 (1992).
46. Y. Yamashita, W. Takashima, and K. Kaneto, *Jpn. J. Appl. Phys.* **32**, L1017 (1993).
47. A. Aviram and M. A. Ratner, *Chem. Phys. Lett.* **29**, 277 (1974).
48. N. S. Sariciftci and A. J. Heeger, U. S. Patent, No: 5,331, 183 (University of California, U.S.A.) (1994).
49. H. Yonehara and C. Pac, *Appl. Phys. Lett.* **61**, 575 (1992).
50. S. Morita, A. A. Zakhidov, and K. Yoshino, *Jpn. J. Appl. Phys.* **32**, L873 (1993).
51. C. S. Kuo, F. G. Wakim, S. K. Sengupta, and S. K. Tripathy, *Sol. State Commun.* **87**, 115 (1993).
52. I. D. Parker, *J. Appl. Phys.* **75**, 1656 (1994).
53. G. Yu, C. Zhang, and A. J. Heeger, *Appl. Phys. Lett.* **64**, 1540 (1994).
54. G. Yu, K. Pakbaz, and A. J. Heeger, *Appl. Phys. Lett.* **64**, 3422 (1994).
55. H. Antoniadis, B. R. Hsieh, M. A. Abkowitz, M. Stolka, and S. A. Jehneke, *Polymer Preprints* **34**, 490 (1993).
56. J. J. M. Halls, C. A. Walsh, N. C. Greenham, E. A. Marseglia, R. H. Friend, S. C. Moratti, and A. B. Holmes, *Nature* **376**, 498 (1995).
57. A. A. Zakhidov, H. Araki, K. Tada, K. Yakushi, E. Saiki, and K. Yoshino, *Phys. Lett.* **A205**, 317 (1995).
58. H. Araki, A. A. Zakhidov, E. Saiki, N. Yamasaki, K. Yakushi, and K. Yoshino, *Jpn. J. Appl. Phys.*, Part 2, Letters, 34, L1041 (1995).

Index